MEASURING
SHAPE

MEASURING
SHAPE

F. Brent Neal • John C. Russ

CRC Press
Taylor & Francis Group
Boca Raton   London   New York

CRC Press is an imprint of the
Taylor & Francis Group, an **informa** business

CRC Press
Taylor & Francis Group
6000 Broken Sound Parkway NW, Suite 300
Boca Raton, FL 33487-2742

First issued in paperback 2017

© 2012 by Taylor & Francis Group, LLC
CRC Press is an imprint of Taylor & Francis Group, an Informa business

No claim to original U.S. Government works

Version Date: 20120328

ISBN 13: 978-1-138-07219-0 (pbk)
ISBN 13: 978-1-4398-5598-0 (hbk)

Visit the Taylor & Francis Web site at
http://www.taylorandfrancis.com

and the CRC Press Web site at
http://www.crcpress.com

Contents

Introduction

Some Background

Shape is an elusive concept. Humans depend on the recognition of shapes for many purposes, but find shapes hard to describe in words. Like Supreme Court Justice Potter Stewart's description of pornography, "I know it when I see it," We can't always communicate what we see to someone else, "It's shaped kind of like a thing-a-ma-bob but with a smaller whatchamacallit and no dingus." Dryden (1998) defines "shape" as "all the geometrical information that remains when location, scale and rotational effects are filtered out from an object." But that negative philosophical declaration does not tell how to describe the "information that remains."

Many artists have explored the relationships between objects in the world and our internal representations of them (Figure I.1). René Magritte (1927, 1935) insisted that the names (and presumably other descriptions) of things were inadequate mental representations of our experience of the world; that there are objects that do not require a name; and that any object is not so attached to a particular name that another more suitable name cannot be found. We will see in this book that numerical descriptions of shape are like that too: a multitude of possibilities exist, and no particular one is necessarily the best choice, depending on the purpose for which it is to be used.

Interestingly, the notions of size, location, and orientation are well understood and easily measured. There are several different ways to describe them, but each is straightforward to measure and to communicate. For example, the "size" of an object may be represented by its volume, its maximum dimension, or some other geometric quantity. Position may be reported as absolute coordinates such as latitude, longitude, and compass heading, or relative to some local point or object. But all of these are familiar concepts.

Shape is usually more important than size or position for recognition but is not so easy to describe or to report. Because of the inadequacy of words, many scientific papers include pictures. These are almost universally captioned as "typical" representations of some object of study. But does a two-dimensional image faithfully communicate the shape of a three-dimensional object and show those characteristics that the authors found most important or diagnostic? Can a single instance show the range of variations that exist in nature? And, most important, is the object shown really a meaningful average, or was it selected based on some unspoken, perhaps unrecognized, aesthetic

FIGURE I.1
An example based on Magritte's (1927) message "This is not a pipe," demonstrating that a picture and a word are distinct from the object to which they refer. One works by resemblance, whereas the other is merely an arbitrary intellectual association.

judgment? As editor of a scientific journal for seven years, one of the authors of this book came to understand that the "typical object" illustration was often one that was selected as the prettiest or as the best quality image available.

The lyrics in Arlo Guthrie's "Alice's Restaurant" song include a pretty good description of some scientific reports: "Twenty-seven eight-by-ten color glossy photographs with circles and arrows and a paragraph on the back of each one explaining what each one was." And that approach still doesn't fully describe or communicate information about shape.

So, if words and pictures by themselves can't suffice, what can be done? The physicist William Thomson (Lord Kelvin) said: "When you can measure what you are speaking about, and express it in numbers, you know something about it; when you cannot express it in numbers, your knowledge is of a meagre and unsatisfactory kind; it may be the beginning of knowledge, but you have scarcely in your thoughts advanced to the stage of science" (Thomson, 1889).

We need to describe shape with some numbers. But which ones? That is the subject of this book.

The Plan of the Book, and Why It Was Written

There have been many different approaches to shape. Some methods are widely used (and sometimes misused); the "circularity" of features, based on one of the dimensionless ratios discussed in Chapter 3, is measured in Adobe Photoshop®! Some methods have deep mathematical roots and fiercely loyal adherents, with a body of supporting literature (shelves full of books, as for the landmark and Fourier methods discussed in Chapter 3 and Chapter 4, respectively) that makes it difficult for the beginner to know where to start, or to compare the advantages and drawbacks of the various approaches.

Some methods are more amenable to human visual understanding and some less so. Some have found a comfortable home in a particular field of study (e.g., Fourier methods in sedimentology, landmark methods in the analysis of human form). But this is not proof that they are necessarily the best techniques for any other purpose or even optimum for the ones to which they are applied.

The goal of the book is to provide an overview and comparison. The widest possible variety of approaches to the measurement and quantitative description of shape is presented, with an intentionally diverse set of examples that illustrate and compare the methods. We personally have used all of these methods at one time or another and do not feel any ideological attachment to any of them. Each may be appropriate in a specific case; each may be a poor choice in another. What we hope to do is to provide enough information to showcase each method, its advantages and limitations, and enough examples to illustrate the ways that it can be implemented.

Because of the desire to compare various methods and to apply them to some of the same diverse sets of images and objects, there is inevitably some back-and-forth between ideas and chapters. Some methods are shown in several chapters, including the final one on interpretation and analysis, and some of the examples and measurements are repeated for comparison. Some of the analysis results are shown in the sections on the measurement methods, while the analysis procedures are not formally described until the final chapter.

The emphasis is on methods that are quantitative and produce numbers (thank you, Lord Kelvin) that can be used for statistical analysis, comparison, correlation, classification, and identification. Ernst von Siemens summed this up succinctly as *Messen ist Wissen* ("measuring is knowing") (von Siemens & Coupland, 1893). Qualitative methods such as syntactical analysis are described briefly. For some applications these provide very powerful tools, for example, for matching or recognizing individual objects, but they can be difficult to apply to natural objects or to implement using computers.

The book begins with the assumption that images representing the shapes of interest are digitized for computer analysis as an array of pixels (or, for three-dimensional objects, of voxels) that can be manipulated and measured with software. Virtually all the quantitative methods described are available in existing standard programs. Some are available free (but with the usual drawbacks of free software: limited documentation, variable quality, and unknown support), and some come in widely used professional image analysis software (but with little information about why, as opposed to how, to select a particular option). We do not know of any single program that offers more than a few of the methods and measurements shown here, and offer no endorsement of any particular program(s).

Also, for any analysis of shape measurements, statistical tests are imperative. This is not a textbook on statistics; many excellent ones already exist. We present many statistical comparisons and summaries of various data sets, but use off-the-shelf programs (primarily SAS JMP 9®, but also others, particularly for the nonparametric procedures). There is no attempt or intent to delve into the derivation or mathematics of the statistical tests themselves, beyond a brief description with examples. But it should be noted that because of the non-Gaussian distributions typically encountered with shape measurements, the use of the more familiar parametric statistical tests based on mean and standard deviation is often not appropriate, or at least does not produce the expected confidence values.

A Choice of Methods

As an organizing framework to relate the various approaches to shape measurement to one another, we propose the following.

Figure I.2 shows a classification scheme for different shape descriptors, based on their ability to reconstruct the shape and whether the measurement is based on the entire area or the just the periphery of the object. One way to

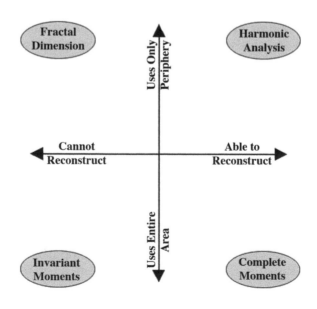

FIGURE I.2
Characteristics of shape descriptors, with examples. The methods discussed in Chapter 3 and Chapter 4 occupy various positions on the graph.

make a choice of descriptor(s) for a particular application is to consider the implications of this figure.

The horizontal axis of the graph distinguishes between methods that completely describe the shape and are capable of reconstructing it from the numbers versus ones that select just a few characteristics (Loncaric, 1998).

Descriptors that cannot reconstruct the shape are reductive. They discard all but a limited set of information about the shape. These descriptors have the advantages of highlighting those aspects of the overall shape for which each descriptor is designed, so that appropriate statistics and decision logic can be applied. Most are also relatively fast to compute, making them suitable for high-throughput applications such as quality control or industrial sorting. These include invariant moments, and classical shape descriptors such as formfactor and aspect ratio based on dimensional measurements (all discussed in Chapter 3). These descriptors will often be the initial choice because of their simplicity, ubiquity of implementation in commercial software packages, and their ability to be understood intuitively, at least if suitable graphical examples are provided.

With reductive techniques, it is often not possible to choose a small (or even a large) set of descriptors that separate different classes of objects or monitor changes in shape, correctly or completely. Human judgment becomes difficult when the desired classifications are visually similar to each other or are visually confusing because of wide variations in shape or in other characteristics such as color. Examples include distinguishing individual grains in different types of sediments or histological classification such as diagnosing cancerous cells in a Pap smear. In these cases, it is often necessary to retain more information about the objects. This will almost always require the use of statistical techniques in order to gain any insight or obtain a result from the measurements. Unfortunately, in some instances the statistical analysis may tend to obscure the visual "meaning" of the descriptors.

The methods that are capable of reconstructing a shape with all of its original details (discussed in Chapter 4) include the medial axis transform, harmonic or Fourier analysis of the boundary, and the complete set of moments. Although these are complete, they also produce a lot of numbers. Humans generally don't do well with large sets of numbers, unless they can be presented as a graph (whose "shape" may communicate information and aid interpretation better than the values themselves). Statistical analysis may be able to select a small subset of the numbers that are satisfactory for recognition or classification in any given instance, but that subset is no longer a complete representation. Furthermore, any relationship between the selected values, or the reason for their selection, and the visual characteristics of the shapes they describe, may not be recognizable or understandable to a human.

The second (vertical) axis of the graph distinguishes between methods that deal with the interior or the boundary of the object (Pavlidis, 1978; Zhang & Lu, 2004).

Lestrel (1997) offers this definition: "Shape is a boundary phenomenon. It refers to the boundary outline of a form in 2D or 3D. Its focus is on curvature." But his book is concerned principally with harmonic analysis, which relies solely on the boundary and its curvature. We prefer to take a somewhat broader view of the possibilities. Moments, for example, are calculated using all of the pixels within a feature.

The choice between descriptors that use only the periphery of an object and ones that use the entire area (or volume) of an object may not be obvious. In most cases it is possible, at a given level of precision and resolution, to convert one way of representing a feature to the other. But this does not alter the fact that some shape measurements use the boundary, while some are based on the entire pixel array. And not all measures lie at either extreme. The skeleton and the medial axis transform depend on both the interior and the periphery. The former is reductive and the latter complete.

Many of the periphery-based measurements, such as fractal dimension, are scale-invariant within the limits of precision. This allows them to be used effectively for classification across multiple images. Typically, the outline or boundary of an object will have a higher density of data about an object's shape than the interior pixels. Every pixel on the periphery may be important, but many of those in the interior are individually of less significance. There are exceptions to this generalization: removing a single interior pixel from a feature can dramatically alter the skeleton, for example.

One consequence of that information density is the need for adequate resolution in the imaging system (optics, camera, or whatever other imaging device is employed). The resolution and noise characteristics of the digitized image must be sufficient to accurately depict the periphery and capture the fine details of irregularity that are significant. It is not always easy to determine the required resolution. Many of the measurements of shape that depend on the periphery—whether reductive like formfactor or complete like harmonic analysis—have values that change as image resolution or noise is varied. That creates a problem.

To illustrate this behavior, Figure I.3 shows several identical circular features (80 pixels in diameter) superimposed on a noisy background with random variations in pixel values. Thresholding the circles based on pixel brightness and measuring the area and perimeter (as described in Chapter 2) gives the results summarized in Table I.1. The area values, which depend on all of the interior pixels, have a mean value close to the ideal and a standard deviation of 0.5%. In other words, there are some pixels that are erroneously added to the feature area and others that are omitted from it due to the random noise, but the additions and deletions tend to cancel out.

The perimeter values, which depend only on the periphery, have a mean value that is 40% too high, with a standard deviation 10 times greater than that for the area. Figure I.4 shows that repeating the same test 100 times produces a Gaussian or normal distribution for the area measurements but not for the perimeter values. The noise pixels never produce a smaller

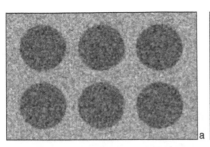

FIGURE I.3
Effect of noise on measurements: (a) six identical circles with additive grayscale random noise; (b) circles after thresholding and filling internal holes. Table I.1 lists the measurement data for area and perimeter, and the calculated formfactor.

TABLE I.1

Measurements of the Circles in Figure I.3

Circle	Area	Perimeter	Formfactor
1	5027	365.36	0.4732
2	5042	344.53	0.5338
3	4988	350.08	0.5114
4	5065	355.92	0.5024
5	5020	375.76	0.4468
6	5030	321.60	0.6111
Average	5028.67	352.21	0.5131
Std. Dev.	25.42	18.66	0.0568
Percent	0.51	5.30	11.07
Ideal	5026.55	251.33	1.00

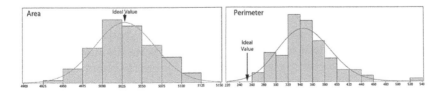

FIGURE I.4
Histograms of measurements for area and perimeter of 100 circles measured as in Figure I.3. The area data cannot be distinguished statistically from a normal distribution (shown superimposed). The curve shown on the perimeter data is approximately lognormal, and the data are skewed to the right.

circumference, and the amount by which they increase it varies widely. A shape descriptor such as formfactor (defined in Chapter 3 as $4\pi \cdot Area/Perimeter^2$), which depends on both the area and perimeter, is biased and is very sensitive to these variations. Harmonic analysis (described in Chapter 4) depends only on the boundary and produces a set of Fourier coefficients that are acutely affected by the noise.

The methods that depend on the periphery of the object are also, of course, very sensitive to any image processing that may be applied in the delineation of the feature to be measured. Chapter 2 describes the principal operations that are used on images to deal with various imperfections in the original image, threshold the feature, and perform the measurements. These all have the potential to affect the periphery more than the interior.

Another Axis

The various approaches to quantitative shape description differ in another important way. This third axis is qualitatively different from the ones in Figure I.2. It is a measure of the correspondence between what the numbers describe and what has meaning to a human observer. Some shape descriptors are more easily understood than others. For example, the ratio of length to breadth, usually called the aspect ratio, is something that most people can "see" in the image, although human vision (as discussed in Chapter 1) is not a very accurate measurement tool and can be easily distracted or fooled.

Aspect ratio is an example of a reductive measure, and because the dimensional measurements of length and breadth on which it is based can themselves be defined and measured in more than one way, it is cited here only as a general indication of a shape descriptor. But the first harmonic coefficient in the Fourier analysis series (which is a complete descriptor based on the periphery) also measures the elongation in much the same manner as the aspect ratio. Once the meaning of that term as the ellipticity of the shape is pointed out, it becomes more accessible to human understanding.

Similarly, human vision is very sensitive to detecting symmetry in shapes. The moments, which depend on the interior as well as boundary pixels, provide one tool to capture this property.

At the other extreme, some methods produce numbers that do not correspond well to human perception and can only be dealt with statistically. Landmark methods that measure all of the pairwise distances and angles between points selected by an expert, and assemble them into a matrix for statistical comparison to other matrices, are a good example. A human encountering the matrix of numbers without the accompanying marked image (which is rarely provided) will not be able to develop a mental perception of the shape nor to compare one shape to another. Even with a diagram, the locations selected for the landmarks often vary with different observers.

Other criteria may also be considered in selecting a method. The complexity of performing the measurement and carrying out the subsequent analysis for recognition or classification was historically a driving force toward using the more simple computer techniques. And some procedures, such as landmark methods, required human judgment such as tracing outlines or marking points, because computer processing was inadequate to accomplish the tasks. Now, with the possible exception of some online sorting or quality control functions that must be performed in milliseconds using dedicated high-speed cameras and array processors, most of the methods shown can be executed in a satisfactorily short time on the typical laptop computer, with minimal human intervention.

Some can even be carried out on a smart cell phone, although it may require the support of a larger computer somewhere in the "cloud." Google Goggles® can recognize landmarks, logos, paintings in a museum, and so forth recorded with a built-in camera, enabling the phone to display immediate relevant information. Smart phones have apps to read UPC codes of products in stores and immediately provide specifications on the product and prices from competitive dealers. Most pocket digital cameras have built-in face-detection algorithms to improve snapshot photos. And so on.

Our Motivation

Shape is the central link in many man-made and natural processes. As shown in Figure I.5, the shapes of objects and structures are primarily controlled by factors such as material characteristics, processing and history, and environmental effects. In industrial processes, many (but certainly not all) of these may be designed and regulated. In the natural world, variations are much wider and may be slower; for example, evolution modifies shape over many generations, and erosion and plate tectonics act over even longer time spans.

Furthermore, shape is the principal governing factor determining the behavior of objects and structures. This includes many physical properties, such as the stiffness of beams and the ability of a leaf to catch water, as well as appearance. Many insects have evolved shapes that are similar to other, poisonous species (Batesian mimicry) so as to avoid predation. The shape of consumer packaging such as bottles is apparently an important factor in successful marketing.

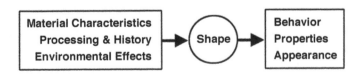

FIGURE I.5
Some of the causal relationships that involve the shape of objects and structures.

For all of these reasons, as well as the unavoidable fact that human recognition of objects is deeply dependent on shape, it is important to understand ways that shapes can be measured.

As researchers and teachers, one in industry and one in academia, the authors have encountered a very wide range of problems that required measuring shape. In some cases the purpose was to recognize something, such as a specific type of defect on the glass used for LCD computer displays or the characteristics of cell nuclei that indicate abnormality and possible cancer. In other cases, it was to correlate variations in shape with either processing history or performance. This correlation lies at the heart of most materials, whether metals, ceramics, or polymers. It also arises in food science (structure is closely related to mouthfeel), agriculture (the effects of soils on fruits and vegetables), genetics (phylogenetic comparisons between species), and even consumer products (the design of automobiles, women's fashions, etc.).

In pursuing these goals, using image analysis including the measurement of shape in our day-to-day research, and in teaching the fundamentals of image processing and analysis to literally thousands of students in the 50-year career of one author, we have discovered (or rediscovered) and used a wide range of methods for measuring shape. We hope to provide others with some guideposts and perhaps shortcuts along the same road. Knowing what the possibilities are, and also the pitfalls, may help others to traverse a satisfactory path from questions to answers.

Acknowledgments

Many of the examples of objects used in this book to illustrate shape measurements and the analysis of the data have been generously shared with us by the original researchers. They are cited with the images, and we offer here an additional thank you for their willingness to provide us with raw images and data. But the measurements and analysis presented here are our own, and we take full responsibility for them. Thanks are also due to Michael Sullivan, Microscopy Lab Manager for Milliken Research, for giving us thoughtful feedback on the manuscript, and to the Milliken i^3 seminar participants for helping one of the authors to clarify the presentation of several concepts.

Grateful thanks are also given to our families, who have supported our efforts throughout and accepted with grace our need to try to explain what we have been doing and thinking about, and the time we have taken from them for this project. Residing 250 miles apart, and both with other responsibilities and travel schedules, we have made extensive use of the Internet (including cloud storage for the chapters as they have evolved) and frequent (and lengthy) telephone conversations, many during the golden time in the evening that we call "happy hour."

Finally, thanks are due to the editors at CRC, who encouraged the project and have had to deal with a text containing an inordinate number of figures. Shape is a topic that just cannot be dealt with using words and equations alone, as most artists, many psychologists, and not a few computer scientists have discovered.

1

The Meaning(s) of Shape

Why Shape Matters

People know a lot about the shapes of objects, yet they have few words in any language to describe shapes. There are a few terms like "smooth versus rough" or "pointy versus rounded" that imprecisely describe differences. But for the most part, instead of adjectives, people use nouns. Object shapes are remembered by storing a mental picture of a (presumably) representative example. If I tell you that something is shaped "like an fish," I am relying on you having a mental picture enough like mine to match the features that I think are important. Presumably fins are important and perhaps a streamlined shape. And an association with water, of course.

Probably for most people, the word "fish" conjures up a mental image of something like a clownfish (Nemo), or perhaps a trout, salmon, or shark. But fish come in many forms (Figure 1.1) that include some that are not always recognized as fish (moray eel, hairy frog fish, sea horse) as well as things that are named "fish" but are not (jellyfish). For some people, the word may recall an image of fish and chips, or, depending on the setting and culture, sushi. Certainly, saying that something is "shaped like a fish" invites misinterpretation.

The artist Georgia O'Keeffe said (1976) "I found I could say things with color and shapes that I couldn't say any other way—things I had no words for."

Some words seem to be adjectives describing shape, but on closer reflection are not. "Round" is usually understood to mean "like a circle," or in the three-dimensional case it might mean "like a sphere." It could also mean "like a cylinder," depending on the circumstances and the point of view. Sometimes "round" or "rounded" is used to distinguish shapes with curved and convex boundaries from shapes with boundaries composed of straight lines, sharp corners, or indentations.

When it is used to mean "like a circle," the word "round" is a reference to a noun. Several of the simple measures that are introduced in Chapter 3 measure the departure of a two-dimensional shape from being circular, and may be given names such as "roundness" or "circularity." Unfortunately, as will be seen, there are many different ways of being unlike a circle. Elongating the

FIGURE 1.1
Fish (from the upper left, left to right): clownfish; flying fish; Atlantic salmon; great white shark; jellyfish; moray eel; fish and chips; sea horse; sushi; stingray; salmon (represented in northwest Indian art and Japanese Kanji); hairy frog fish.

circle into an ellipse, or flattening the sides to form a polygon, or roughing up the periphery to create something like the petals on a flower are just a few examples. The various numeric measures are each sensitive to different ways of being "not like a circle."

The geometric definition of a circle is straightforward: the locus of points equidistant from a center. But an alternative definition that might be considered is a shape whose diameter in all directions is the same. Measuring the diameter of a feature in many directions would then provide an easy procedure by which to recognize a circular shape. Alas, intuition aside, that is not necessarily the same thing. Great Britain has issued several coins in 50p and 20p denominations that are equilaterally curved heptagons (Figure 1.2). The curved sides produce the same diameter in all directions, allowing the coins to roll down slots in vending machines. Any polygon with an odd number of vertices can be used in this way.

FIGURE 1.2
Seven-sided equidiametrical coins.

Measuring versus Comparing

High-precision measuring machines that determine whether manufactured items, such as automobile engine cylinders, are really circular must avoid falling into this sort of trap, by measuring something that seems to correspond to the desired shape, and is relatively simple to measure, but does not in fact provide the required information. Measuring a shape in two or three dimensions to determine whether it is accurately "round" is not so simple and is discussed in Chapter 5. If the actual shape is only approximately circular, describing the deviation from the perfect shape in a concise yet meaningful way presents additional difficulties.

Human vision does not depend upon measurements but is instead comparative. Deciding whether two objects are similar in shape is accomplished visually by mentally dragging one image onto the second, rotating and if necessary stretching it to fit (in other words, discarding size, position, and orientation information), and deciding how closely the two objects match. That is not always a simple task, and as shown in Figure 1.3, it is easy to be confused by mirror images. Differences in color, pattern, or texture also interfere. Generally, the length of time needed to decide whether two objects are the same (or not) is proportional to the amount of translation, rotation, and other manipulation required to make the two images overlap, and women tend to be slower but more accurate than men (Cooper & Podgorny, 1976; Tapley & Bryden, 1977; Blough & Slavin, 1987). We literally "turn things over in the mind" to compare them.

Furthermore, if the comparison is not between two objects both being viewed at the same time, it is necessary to recall an image of the object to be matched. That recollection may, of course, be flawed. Nonetheless, people manage to identify a great many objects based on a remembered shape. The task is much easier and faster if the remembered shape has a name associated with it. Apparently our mental search routine is a lot like computer searching for an image, in which the associated words are used to find examples of a particular class of objects. For example, a Google Images® search for "circle" includes among its top selections things like crop circles and artworks and books whose titles include the word "circle" (Figure 1.4).

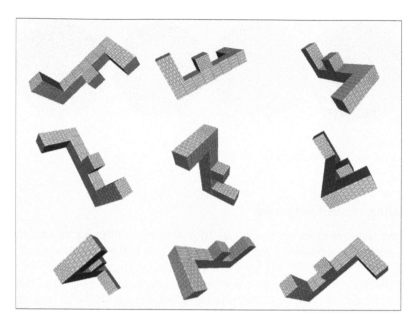

FIGURE 1.3
Some identical and mirror-image shapes.

A further complication in the use of names to describe shapes is pointed out by Magritte (1927): *Les mots qui servent à désigner deux objets différents ne montrent pas ce qui peut séparer ces objets l'un de l'autre* ("The words which serve to indicate two different objects do not show what may divide these objects from one another"). That, of course, is why we turn to measurements and to the various statistical analysis tools described in Chapter 6.

Another problem that arises for many real objects is that we are accustomed to seeing them at an angle that shows several aspects of the shape. As shown in Figure 1.5, the "engineering drawing" views of an object, taken together, define it adequately for someone who has experience in interpreting such views. But many people are not familiar with the conventions of those views, and in any case if only one such view is available, the true three-dimensional shape is not revealed.

Storing a canonical representation of each class of object that may need to be recognized is apparently a successful strategy for humans, but it does not work very well for computers. Matching an image of a new object to such a database, and allowing for variations in size or orientation, is inefficient and time consuming. The time needed to find a match grows as the number of possible reference objects grows. Creating reference books (and more recently online or DVD-based resources) containing collections of images of "representative" identified specimens is a commonly used and accepted method for assisting the recognition process.

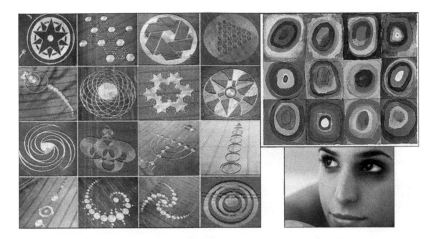

FIGURE 1.4
A few of the responses to an Internet search for "circle": Crop circles (not all of which include actual or even approximate circles); Kandinsky's "Squares with Concentric Circles"; dark circles under the eyes. Other examples include the book cover for *The Tenth Circle*.

Figure 1.6 shows an example, in which the catalog of images is arranged with rules to simplify searching it. At each step of the search, a series of selections with both word description and example illustrations is presented. Selecting one moves on to the next step, until finally a successful identification is made. Similar field guides are available for birds, wildflowers, and so on. Their successful use depends in large measure on whether the illustrated choices at each step are sufficiently close to the target that is to be identified to suggest the proper choice and, of course, on the patience and skill of the user.

As a technical example of this approach, the McCrone Atlas of Microscopic Particles (McCrone Research Institute, Westmont, Illinois, www.mccroneatlas.com), containing thousands of light and electron microscope images of various kinds of particulates, ultimately filled six printed volumes before

FIGURE 1.5
A bucket, shown as idealized top and side views, and as normally seen and recognized.

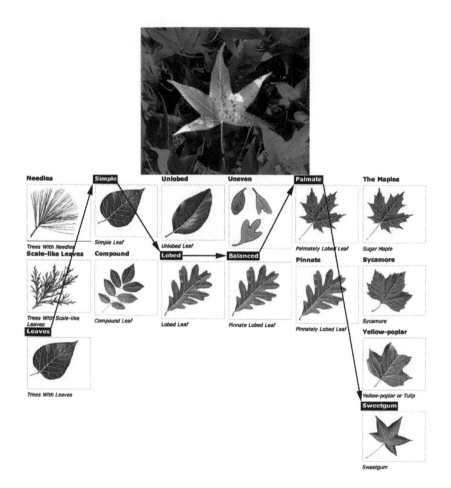

FIGURE 1.6
Leaf identification: an image of a leaf to be identified, and the field guide identification process carried out by selecting a sequence of example images illustrating various characteristics (the sequence Leaves → Simple → Lobed → Balanced → Palmate → Sweetgum is marked).

making the transition to electronic form. Similar atlases of fibers (Figure 1.7) and other items of forensic interest also exist. Like the field guides for real-world natural objects, each of these atlases has an organization intended to help the reader find the image that is most likely a match, based on a combination of word identifiers and pictures, and sometimes measurement data.

These identifying characteristics form the elements of syntactical analysis, discussed in Chapter 3. But they do not lend themselves to mathematical description or statistical analysis to determine, for example, how similar or different two shapes or objects are.

Whether it is called an atlas or a field guide, or is a published scientific paper containing an image described as "typical" or "representative" of a class of objects, this approach is certainly the most widely employed method

FIGURE 1.7
Scanning electron microscope images of various fiber materials (not at the same magnification): (a) wool; (b) cotton; (c) spider silk; (d) extruded man-made polyethylene; (e) cellulose; (f) carbon nanotube.

for transmitting information to other individuals. The problem, of course, is that no single image is ever truly typical or representative of any natural object or structure. In many scientific journals it is understood by the editorial board that a figure has actually been chosen either because (a) it is the best quality image that the authors have been able to obtain; (b) it appeals to them on some aesthetic grounds; or (c) there are some details visible in the image, although often unspecified and possibly not even consciously known to the authors, that they have come to believe are important characteristics for recognition.

The first two reasons are more important than most scientists would like to admit. The third reason raises the concern that there are also other details in the image that may not be important or typical, but which the reader of the article (or user of the atlas) will believe to be important and attempt to rely on in the identification process.

If an image shows something that is not in the database or field guide, some method that reports the types of objects it is most like would be helpful. That is something that people do very well. Seeing a llama or a guanaco for the first time (Figure 1.8), most people recognize similarities to a camel, which is in fact a related species. One characteristic is a relatively long neck in proportion to the body but not so long as (and more curved than) a giraffe. The shape of the ears and muzzle are also distinctive. But the camel's hump is not present.

FIGURE 1.8
(From left) Camel, llama, and guanaco.

The field guide approach has a well-recognized place for recognition of classes of objects, such as animal or plant species, based on representative characteristics. This lies at the heart of Linnean taxonomy. An extension of the approach may be used to recognize individuals within a class. Biometrics based on shape analysis from photographs is an important tool in the study of many wild animal populations. Identification of humpback whales is performed based on the shape of markings on the flukes. The catalog of more than 7000 photographs is primarily used for visual identification, but at least one expert system has been trained for this purpose as well. The initial sorting is done based on a set of categories shown in Figure 1.9 that categorize the shapes of the major features. Additional smaller shape details such as scars and nicks are also important for individual identification.

Notch patterns on dorsal fins are used for dolphin identification and ear shape for wild elephants. Other animal markings are also used to identify

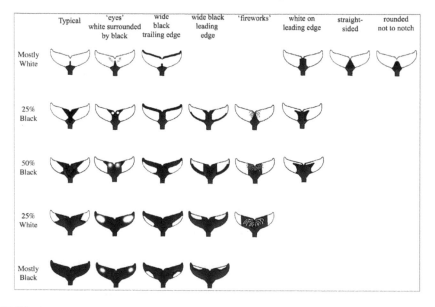

FIGURE 1.9
The principal categories of fluke markings used for identification of humpback whales.

such varied animals as cheetahs, penguins, whale sharks, salamanders, and frogs. Zebra stripe patterns are used much like a bar code to automatically identify individuals in herds in Kenya. This method has not been successful for animals with less distinctive markings or which are more difficult to record photographically. Chapter 4 has an example of tiger identification based not on the stripe pattern but on footprints.

Shape and Human Vision

Human vision is a complex process and a rich subject for study. The visual system involves processing in the eyes, the visual cortex, and other parts of the brain. The literature, which includes physiology, psychology, and many other fields, is vast. The subject is introduced here because some computer image-processing tasks attempt to emulate either the low- or high-level processes that are believed to take place. More comprehensive reviews of the processes involved in vision can be found in books such as Frisby (1980), Marr (1982), Rock (1984), Hubel (1988), Posner and Raichle (1994), Parks (2001), Ings (2008), Gregory (2009), and Frisby and Stone (2010). Culture adds another dimension to a complicated topic: westerners and Asians do not "see" the same things or at least do not extract the same information from images of scenes (Nisbett, 2004; Nisbett & Masuda, 2003).

As several of the referenced books show, illusions are very useful for highlighting particular aspects of the human visual system. Some of the effects arise within the eye itself, chiefly those that cause brightness or colors to be misjudged due to the placement of other colors or brightnesses adjacent to them. These errors derive from the center-surround wiring in the retina that compares the output of one sensor (or group of sensors) with those around them in order to emphasize differences and locate abrupt changes that often correspond to the boundaries of objects. Similarly, in the visual cortex, the interleaving of regions that are sensitive to the orientation of lines and edges causes angles to be misinterpreted when lines with other orientations are nearby. At higher levels of processing, grouping of features biases judgment of size or position, and can create illusions of motion in still images. Figure 1.10 shows a few of the familiar illusions that result.

Psychologists are interested in the extent to which errors of perception arise from physiological causes (e.g., the center-surround wiring in the eye) or from cognitive errors that may be caused by expectation, experience, or knowledge. In one of Tony Hillerman's detective stories, his Navajo detective Joe Leaphorn explains how he looks for tracks. The FBI man asks "What are you looking for?" and Leaphorn replies "Nothing in particular. You're not really looking for anything in particular. If you do that, you don't see things you're not looking for."

On the other hand, Sherlock Holmes often criticizes Watson for "seeing but not observing," which is as good a distinction as any between capturing an

FIGURE 1.10
(See color insert.) Brightness, color, size, and orientation illusions: (a) the gray lines on the left side are usually seen as being lighter than those at the right; (b) the diagonal lines intersected by vertical lines are usually seen as being at a different angle than the ones intersected by horizontal lines; (c) the gray lines are usually seen as outlining wedge-shaped regions rather than being parallel; (d) the gray circle is not usually seen as being in the center of the horizontal line; (e) the red lines in the two spirals are usually seen as orange and magenta rather than identical; (f) the girl's right eye is usually seen as being cyan rather than the same neutral gray as the left eye.

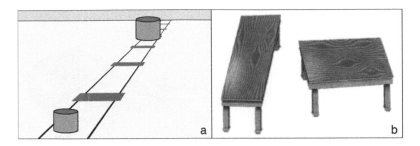

FIGURE 1.11
Perspective and expectation alter judgment: (a) the two cylinders have the same dimensions, but the "railroad track" lines create an impression of distance that makes the upper one appear to be larger; (b) the two tabletops have exactly the same shape and dimensions, but the addition of sides and legs creates an impression of 3D orientation that makes the left hand one appear to be longer and narrower.

image and perhaps making basic measurements on it, and performing the analysis and interpretation that can lead to conclusions or test hypotheses.

As an example of the effect of expectation and the role of prior knowledge, the familiar effect of perspective foreshortening means that objects of the same size appear smaller when they are farther away. If clues in the image suggest that objects are at different distances or the expectation of perspective foreshortening is violated, size and shape judgments are often incorrect as shown in the examples of Figure 1.11.

Many optical illusions involve ambiguity, and often this involves shapes. There are two principal classes of these effects: one in which the same shape(s) can be interpreted in two different ways, and one in which the details are locally consistent but globally inconsistent or impossible. Figure 1.12 shows several common examples of the first group. Some observers lock onto one interpretation and have difficulty seeing the other; those who see both often report that they spontaneously switch back and forth every few seconds. Figure 1.13 shows examples of objects with global inconsistency. Many of these, and similar paradoxical ideas, appear in the art of M. C. Escher.

The description of shape that this text concerns itself with is not necessarily the cognitive or mental description of shape that humans use but rather an objective (and ideally mathematical) description that is fortunately not sensitive to these problems. On the other hand, it means that the various approaches to numeric shape descriptions may not easily translate into human recognition or interpretation. Many of the shape-describing numeric indices require computer processing and measurement, and are best suited for computer recognition or classification of shapes.

A number of studies of human perception, categorization, and recognition of shapes have been published over the past several decades (see, for instance, Fei-Fei et al., 2007; Leonardis et al., 2009). While these are quite interesting, they do not provide very useful guideposts to the development

FIGURE 1.12
Ambiguous shapes that have multiple interpretations: (a) Is the gray dot the near or far point in the cube? (b) Is this a duck or a rabbit? (c) Is this an old woman or a young girl? (d) Does this show facing profiles or bites from an apple (the logo of the Share Our Strength® charity)?

of computer-based methods for shape recognition or description. There are two major classical theories of human vision. One, the Gestalt approach, emphasizes the overall forms of objects, which are resistant to change, and tend toward simple, symmetrical, patterns. The psychological literature supporting Gestalt interpretation is extensive but does not seem to lead to numeric descriptions or computational procedures that can be adapted to computer characterization or recognition.

The second classical approach is a bottom-up methodology often ascribed to Hebb (1968), which argues that form is perceived as the sum of parts, which are individually simple and perhaps universal. Examples are angles, lines, curves, and basic geometric shapes like circles, squares,

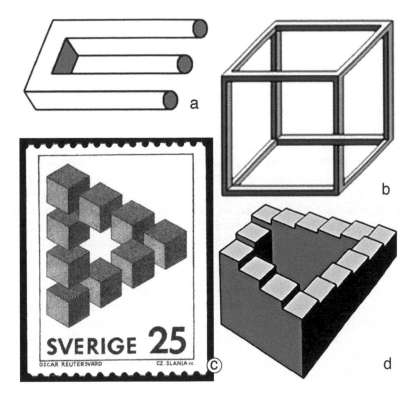

FIGURE 1.13
Shapes that are locally consistent but globally impossible: (a) the fork; (b) the cube; (c) a Swedish stamp from 1982; (d) the endless stairway.

and triangles. Low-level steps in the vision process extract these parts, and higher-level steps examine their groupings and arrangements. This approach does encourage some of the approaches to shape measurement introduced in succeeding chapters, for example, wavelet analysis, but it is not the purpose of this text to explore the fascinating field of the human visual system or to necessarily duplicate either its methods or performance in computer algorithms.

Classification and Identification

A growing demand for robust and fast algorithms for computer measurement of shapes exists in a number of fields. Optical character recognition, to convert printed documents to editable files, is now quite successful and widely used. This is not only a benefit to offices and libraries. A current cell phone app can use the built-in camera to read text that is immediately translated to a different language and the translation displayed or spoken in real time.

Medical imaging wants to diagnose tumors (including microcalcifications revealed by mammography), deformation of anatomical structures (such as morphometric changes in the brain related to Alzheimer's disease), correlate changes in the shape of cells and organelles with various diseases (such as diagnosing cancerous or precancerous cells in Pap smears), and so on. Engineers want to identify the source of various defects that occur in manufactured items (such as the glass sheets used in LCD displays, blisters in paint coatings, and shape variations in extruded fibers) and to automatically sort objects on conveyor belts. In agriculture, a way to identify weed seeds and pollen grains is wanted, as well as using computer vision to judge when to harvest crops and to control the robotic mechanisms that gather fruit. And, of course, there is great interest in using biometrics, such as facial shape, for security purposes.

Finding a few numbers that describe shape is an important goal for computer vision (Loncaric, 1998). Software that can match numbers or apply statistical tests to find similarities between numerical descriptors is efficient and comparatively straightforward. Of course, selecting the best set of numbers that will provide adequate uniqueness, and at the same time offer some correlation with whatever it is that humans use or understand to characterize shape, is a complex subject.

The desirable attributes for a shape descriptor include being robust (stable to minor variations), enabling classification or recognition to be performed using a partial image (e.g., a partially occluded object), being insensitive to translation, rotation and scaling, and to image resolution (within limits, of course). It would also be nice to have a descriptor that is accessible to human understanding, is reasonably compact (i.e., a small set of numbers), is amenable to statistical analysis techniques, and is complete (i.e., contains enough information to reconstruct the original shape). Obviously, many of these desires are mutually incompatible.

In order to provide a set of numeric values that represent shape and use them so that machine recognition of classes of objects can be carried out, the software must be trained. That usually means that humans must be able to select training populations of objects that share a property that defines each class, and that often implies using numeric shape descriptors that are meaningful to humans. This restriction is a major reason that some of the shape measures that are quite attractive from a mathematical viewpoint (such as harmonic Fourier descriptors) have achieved only limited acceptance or use by humans.

For automatic shape recognition to be broadly useful, it must be possible to incorporate the entire process of image capture, processing, measurement, and interpretation into computer algorithms. Some methods for describing shape rely on the location of key points or landmarks. In most cases, this requires some human interaction to recognize and mark those points. While the recognition of the landmarks may require less skill than recognizing or obtaining meaningful measurements on the entire shape, this is still a setback from the goal of full automation. Substituting automatic landmark location, for instance using cross-correlation, provides a solution in some applications.

It is helpful to elaborate further on several of the uses for shape measurements mentioned in the preceding sections. This is not done in any particular order. First, consider the strategies for classification and identification. Generally, classification (grouping of objects into categories) precedes identification (applying a category label to a specific shape) or selection (choosing objects belonging to a specific category). There are two principal approaches to classification: supervised and unsupervised.

Supervised classification requires training sets of objects that have been previously identified (usually by a human, or at least with human oversight), after which the computer combines various measurements of those objects with appropriate statistical calculations to select the measurements or combinations of measurements that best distinguish them. The success of the method is strongly dependent on the quality of the training sets.

Unsupervised classification simply presents a collection of objects to the computer, which then measures them and attempts to find clusters of objects with similar measured values to each other that are dissimilar to others. If the number of clusters or groups is not known beforehand, the procedure must rely on some statistical criteria to determine the limits for separating groups into smaller clusters.

Chapter 6 illustrates and discusses both supervised and unsupervised classification methods.

Hierarchical Classification

Once classes have been established, it is possible to use them for identification of new objects. There is usually a hierarchy of steps in identification. For example, the sequence Animal → Dog → Collie → Lassie proceeds from the general to the specific. How far it is practical (or necessary) to go along this path varies with the application. For example, it may or may not be necessary to identify a particular collie. There have been several different dogs that played the movie role of Lassie; unlike the various actors who have been James Bond, the dogs are not individually recognized by most viewers. Apparently we don't have well developed sets of criteria to identify collies.

One author has a neighbor who owns two collies, named Alice and Kelly (Figure 1.14). When the dogs are side by side, it is possible to see that they are not identical. The owners know them individually. But although the author has lived next door to them for several years, it has never been important enough to learn their individual markings. So identification stops at the

FIGURE 1.14
Alice and Kelly (or perhaps Kelly and Alice).

level of "collie." In other cases, it might be enough to identify an animal as a dog, as opposed to a cat or a coyote, or even as an animal rather than a peculiarly shaped rock.

The Threshold Logic Unit

The same process applies, of course, to identifying an individual human. The hierarchy is Person → Familiar → Individual. The process of recognition has been evocatively (and sometimes derisively) described as involving something that can be called a "grandmother cell." Whether this corresponds to a physical entity in the brain is not as important as understanding how it can be implemented by logic circuits or software calculations. The McCulloch and Pitts "Perceptron" (McCulloch & Pitts, 1943; Rosenblatt, 1962) was an early attempt to describe the operation of a neuron in the brain as a logical unit. As shown in Figure 1.15, the model and its implementation as a threshold logic unit function by collecting values for a great many (unspecified) measurements, applying weighting factors to each (some positive and some negative), summing the result, and comparing the final value to a threshold. If the value exceeds the threshold, it signals recognition of the individual.

Conceptually, this seems like a simple and straightforward description of the recognition process. There are some input values that are more important than others, some that match the stored information about the individual and some that do not and must be given a negative weight. For example,

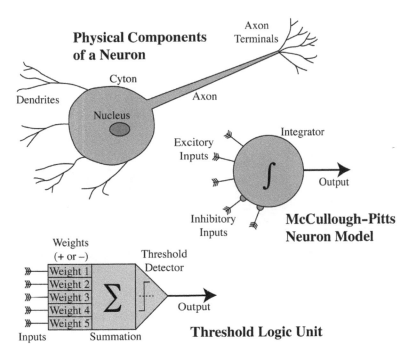

FIGURE 1.15
Diagram of the McCulloch–Pitts neuron model and a basic threshold logic unit.

grandmother is short, has gray hair, wears glasses (which would not be visible if she was seen from behind), often (but not always) wears a checkered apron, and does not have a red mustache (which would have a large negative weight). So from a collection of observed parameters, a quick summation would signal that grandmother had arrived.

In this example, the inputs are binary; a syntactical feature is either present or not. The use of measurements as inputs, such as the person's height in inches rather than the somewhat undefined adjective "short," does not alter the basic approach. An outstanding example of the syntactical approach to identification is the taxonomic classification of plants and animals introduced by Carl Linnaeus in the 18th century. The presence (or absence) of certain structural features can be used to decide on the genus and species identification. No measurement of features is involved.

It is clear from practical experience that recognition of individuals by a human uses a great many different cues and clues, not all of which are available in every situation, and not all of which involve shape. It is also clear that the method is rapid but imperfect—we sometimes fail to recognize a person, and sometimes think that we have recognized someone but are wrong. The recognition process works best for people with whom we are most familiar, because more cues are remembered and their relative importance becomes better defined, which represents a learning process. In terms of the threshold

FIGURE 1.16
Comparison of police artist sketches with photographs. These are considered examples of successful sketch representations. (Images courtesy of Gil Zamora, San Jose, CA, Police Department.)

logic unit, this means that there are more input values, and the weights and threshold value become better established.

When confronted with recognition for persons with whom we are not familiar, there is a tendency for people to select just a few principal features as cues. Figure 1.16 compares the description of criminals to police artists by observers at the scene, with photographs taken after capture. It is evident that the observers noticed and remembered only a few cues in each case. Whether those cues produce a caricature that can help someone else recognize a suspect is another question.

The basic idea behind the threshold logic unit certainly works and has become, with some modifications, the basis for the neural networks that are used in artificial intelligence for a wide variety of purposes, many having nothing to do with images or shapes. For example, a medical diagnostic procedure uses various symptoms and data such as heart rate and blood pressure as inputs, and presents recommended treatment procedures as the result.

Multiple layers of units, with the output from one layer constituting the input for the next, are necessary to implement a complete set of logical functions. The process may be implemented in hardware, but more often the calculations are performed in a general-purpose computer that simulates the hardware functions. In either case, the training consists of selecting the important input values and refining the weights associated with each one. Various strategies are used to accomplish this. Backpropagation of values to correct errors in recognition was one of the first and is still probably the most common method for adjusting the weights and threshold values, but there are many variations in the training algorithms, trading off efficiency, the ability to generalize well to handle marginal cases, and graceful failure modes when a good decision cannot be reached.

One example of the use of the neural net approach is the identification of various types of surface defects that can arise in the manufacture of glass sheets used for LCD televisions and computer displays. Various measurements of the shape of the defect are used to identify the cause of the defect so that corrective action can be taken in the manufacturing process. An entirely different use of the same logic, having nothing to do with images or shapes, is the detection of potential fraud in credit card transactions. The semiautonomous rovers on Mars use onboard logic for local navigation, steering, and obstacle avoidance to reach goals transmitted by radio from Earth.

There is a rich literature on neural nets. Useful introductions and coverage of general methods and applications can be found in Lippman (1987), Weiss and Kulikowski (1991), Anderson (1995), and Gurney (2003). A sampling of more specific papers on image processing and pattern recognition include Weiss and Kapouleas (1990), Zhang et al. (1997), Bischof et al. (1992), Ozkan et al. (1993), Alirezaie et al. (1997), Sinthanayothin et al. (1999), and Egmont-Petersen et al. (2002).

Interest in neural nets is generally divided into two somewhat divergent areas. One is the development of the principles so that they can be applied by computers to practical problems of classification and recognition of objects. This tends to emphasize the mathematical and statistical procedures involved in training the system. The second is the efforts to simulate the actual working of human brains so as to better understand their function. This cognitive modeling must deal with the enormous number of interconnections between neurons in the brain, and the hierarchy from individual neurons through clusters, and so on up to the entire organism. Networks as complex and flexible as a human brain are still far off, but ones that rival those of worms, horseshoe crabs, and fruit flies exist.

In a classical expert system, whether it is implemented by a neural net or sequential logic, a human expert often establishes the criteria and rules. Systems capable of learning by example and creating their own rules are strongly dependent on the quality of the training sets that are used. Examples and strategies for identifying objects based on various shape parameters are illustrated in Chapter 6.

Faces and Fingerprints

The current generation of digital cameras, even quite inexpensive ones, includes logic that locates human faces in images. Since a high percentage of the pictures that people take with these cameras are of people, locating faces so that they can be used to set the camera's focus, and even to take pictures at the moment when people are smiling and do not have their eyes closed, helps casual photographers to get better pictures. The logic by which these tasks are accomplished is highly proprietary to each company, but the computations are capable of being performed by the small processors and limited program memory built into these cameras, and quickly enough to minimize the time delay between pressing the shutter and capturing the picture. Yang et al. (2002) provide a useful survey of the principal methods, which must deal with faces in various sizes and orientations, and different lighting conditions.

For security purposes, there is also a desire to have computer programs not only locate faces in scenes but identify individuals. This can be done to recognize and admit those with clearance for certain locations, or to recognize and detain those considered to be suspicious. In the former case, the number of faces that must be matched is relatively small, but in the latter, the library of faces may be very large, and it must be anticipated that some efforts will be made at disguise to make recognition difficult.

One relatively simple approach that has been somewhat successful is the use of landmarks (Evison & Bruegge, 2010). Finding key reference points such as the inner and outer corners of the eyes allows measuring vertical and horizontal distances as illustrated in Figure 1.17. By using the ratios of these distances, such as the ratio of the distance between the eyes and lips to the distance between the point of the chin and the hairline, the image scale becomes unimportant. Furthermore, the ratios are relatively insensitive to the viewing angle, as the head is tipped or turned. Also, the reference points selected are not easily changed by disguise, and by using many of them, it may be expected that at least some will be visible even if others are obscured because of clothing, glasses, facial hair, or viewing angle. The use of these ratios is not capable of conclusively identifying individuals from a large

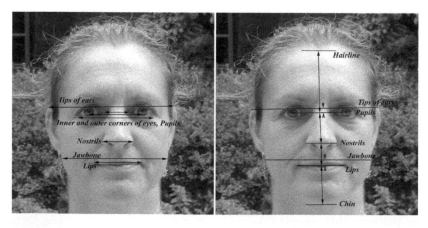

FIGURE 1.17
Examples of some vertical and horizontal distances used in face recognition.

library of faces, but it is able to quickly pull up a small number of candidates for review by a human, who can then make a final decision.

Face identification is now even part of common programs such as Facebook®. The ability of the program to identify people on the street from their Facebook image succeeded in more than 30% of trials (Acquisti et al., 2011). The growing importance of the field is indicated by the fact that as this book is written Google has acquired Neven Vision, Riya, and PittPatt, and deployed face recognition into Picasa. Apple has acquired Polar Rose and deployed face recognition into iPhoto, and Facebook has licensed Face.com to enable automated tagging.

Fingerprint identification by computer is also widely (mis)understood as an example of automated pattern or shape recognition. Even for relatively complete, well-recorded fingerprints, such as those documented on finger-print cards or obtained by computer scanners directly from fingers (but which are certainly not typical of crime scenes) the process is largely manual. Image processing as described in Chapter 2 may be employed to clean up an image, improve the visibility of details, and even extract the skeleton of the ridge markings, as shown in Figure 1.18. However, marking the location and orientation of the significant minutiae is generally performed by a human. Given the data for 10 to 16 minutiae, whose positions are given by the angle and distance from the center or core of the pattern, a computer search is made of records to locate a small number (typically about 10) of the ones that have the greatest number of matches. The images of those prints are then displayed for the human operator to compare visually to the actual print and to determine whether a match exists.

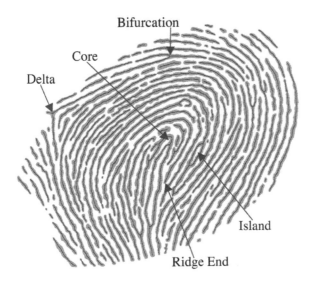

FIGURE 1.18
The principal types of minutiae that are used in a fingerprint image for identification.

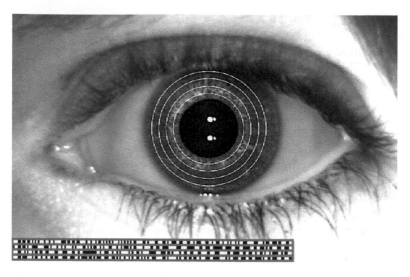

FIGURE 1.19
Extracting the binary code for an iris.

Another form of biometric identification is an iris scan (Daugman, 2004; He et al., 2008; Hosseini et al., 2010). The pattern of tears that form in the pigmentation of the iris during gestation are unique to an individual, do not change over time, and can be photographed through glasses or contact lenses. It is now even possible to capture an image with a smart phone and extract the pattern by applying a wavelet transform to the contrast alterations along concentric circles (Figure 1.19). This code can then be used to identify a person, much as a bar code is used to identify a product in a store. This can be used, for instance, either to control entry to secure locations or to search a database of previously convicted felons for identification (a reference database is gradually being assembled from prison populations).

Biometric identification can also be performed on the shape of the network of blood vessels, for example, those on the retina of the eye (Sims, 1994) or the hand (Hsu et al., 2011). Babies in hospital nurseries are sometimes footprinted for identification purposes.

The General Problem

Identifying individuals with biometric makers depends on very specialized applications, and these are not performed by humans or computers using the same type of information about shapes that is involved in other tasks. The general problem of object classification and recognition involves several steps. One is selecting and measuring meaningful shape parameters, and the second is combining the resulting data with a robust and efficient procedure to arrive at a conclusion. These topics are dealt with in the chapters that follow.

It is worth emphasizing one more time that human methods are certainly not the same as those implemented in computer algorithms (Mumford, 1991). For one thing, human vision is not a quantitative tool for measurement. As already pointed out, vision is comparative. We can tell whether one object is larger or smaller than another, or "rounder," or more red, by placing them side by side in our vision or memory, with plenty of opportunities for flaws and mistakes. And the decision process for identification is not based on any statistical method nor does it yield a reliable estimate of confidence in the result.

One branch of mathematical shape description is applicable to man-made objects possessing simple geometries that can be broken down into a small number of canonical Euclidean shapes (spheres, cylinders, planar surfaces, etc., for 3D objects; or circles, triangles, squares, etc., for 2D features). There is a well-developed algebra based on set theory and mathematical morphology that is applied to those problems (Ghosh & Deguchi, 2008). However, it is not suitable for use with natural objects, which do not in most cases consist of combinations of simple, regular shapes, and which furthermore have a complex range of natural variation.

Frederick Jelinek, who developed at IBM the first successful speech recognition algorithms, pointed out that it was not necessary for computers to attempt to duplicate human response to natural speech. Rather, it was possible to reduce speech to a set of phonemes and to apply statistical analysis to recognize patterns and sequences of patterns to interpret content. His analogy was that airplanes do not fly by flapping their wings, and so computers should do what they do best and not attempt to emulate human behavior or knowledge to reach the same goals.

That principle also applies to image interpretation and object recognition. Of course, it is interesting to study how humans perform these tasks. Programming computers to attempt to perform such functions may give insight into the processes involved (just as attempting to build machines that emulate the flight of a bird or an insect). But that is not essential or even usually helpful in creating a computer algorithm that can produce useful results. In fact, given a robust algorithm, computers are much better than most people at seeing through the clutter of random variations and stochastic noise to extract a few meaningful measurements that enable successful classification and identification.

The idea of the "grandmother cell" mentioned earlier, and the resulting development of threshold logic circuits and neural nets, are based (very) loosely on a model of human processes. The grandmother cell as an actual physical entity certainly does not exist (Tovée, 2008). The number of separate objects that a person encounters over a lifetime is much larger than the (admittedly large) number of neurons that are available to devote to their recognition, and such a method would be too inefficient to function by finding the correct match.

There does not even seem to be a "face cell" that responds to faces, although babies (human and others) seem to instinctively locate and fixate on faces. Presumably there is some distributed network of neurons that finds patterns corresponding to eyes, mouth, and so forth to locate a face. That would also explain why humans seem to find faces in many scenes, such as the well-known "face on Mars" and the Man in the Moon (but in some cultures the pattern of maria and highlands is interpreted as a rabbit, or a dragon, etc.). Connor (2005) has suggested that as few as 100 cells may be able to encode information to recognize a face.

The idea of neural nets (or the equivalent logic implemented in a traditional computer architecture), described in Chapter 6, can produce important results in some applications, even though it is not always possible to decipher the weights assigned to the various links to assign a meaning to the various calculations. But these are certainly not the only way to create a system for computer image analysis.

Correlating Shape with History or Performance

Shape measurement is also used to quantify the effects of variations in manufacture, environment, and so forth on an object, and to relate the variations in shape to the object's behavior. For example, no two plants of the same species have exactly identical root structures (Figure 1.20), but there are similarities in the branching patterns that can be quantified. The type of soil, nutrients, and water available will modify these patterns in significant ways. This can be thought of as a somewhat simpler example of the nature-versus-nurture arguments about the relative importance of genetic heritage versus teaching and upbringing in the development of human children. For

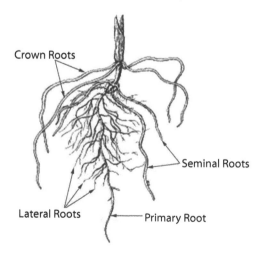

FIGURE 1.20
The roots of a corn plant seedling.

the case of plant roots, finding appropriate shape parameters to describe the patterns that are characteristic of a plant species can provide a convenient tool to study the effect of environmental variables on plant development.

Similarly, the surface of a machined piece of metal is not an ideally flat Euclidean plane, but has roughness characterized by hills and valleys. Measuring the shape of the roughness helps to predict how the surface will perform in its intended application, whether it relates to friction as parts slide across one another, appearance due to the scattering of reflected light, electrical resistance in a switch contact, and so on.

In attempting to find the most useful shape descriptors for making these correlations, there are two basic approaches. One involves measuring everything that can be done, collecting lots of data about the shape as well as the parameters that are thought to control the shape or describe the subsequent behavior from a large number of available but essentially uncontrolled examples. These data are then passed to a statistical analysis program that applies techniques such as stepwise regression, discriminant analysis, or principal components analysis to find which shape descriptors (and perhaps other variables such as size and color) are most closely correlated to history or performance, and to produce equations that allow prediction, along with some measure of the reliability of the prediction.

The other approach is to perform a series of controlled experiments in which one or a few variables are changed at a time, and the changes produced are determined. By designing a well-planned experimental sequence, it is possible to cover a range of variables quite efficiently, and the analysis of the resulting data may be more straightforward than in the uncontrolled case. Both of these methods are used, of course, in many fields that do not involve shape measurement. Drug development by pharmaceutical companies is a typical field in which both approaches are employed.

Once the shape measurement data are obtained, statistical analysis methods are applied as illustrated in Chapter 6. The statistical interpretation is often complicated by nonlinear relationships, and by interactions between variables. Fortunately, there are appropriate statistical tools for treating such cases. The danger inherent in finding relationships between variables is the tendency to mistake correlation for causality.

Shape Matching and Morphing

Human anatomy is pretty standard. Except for cases of injury or trauma, serious disease, or other unusual situations, most of us have the same set of bones and organs, arranged in pretty much the same pattern. Medical students learn to recognize them, particularly when viewed in the standard views used in X-ray, MRI, and other imaging protocols. By learning to recognize the "normal" appearance of the components of the body, they also learn, of course, to recognize deviations from the norm that are of medical concern.

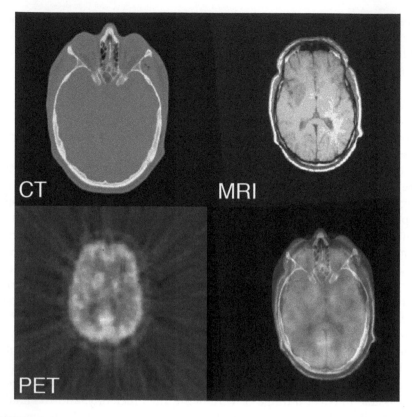

FIGURE 1.21
(See color insert.) Images from computed tomography (CT), magnetic resonance imaging (MRI), and positron emission tomography (PET), and the merged result using color channels.

Although the general shape and placement of the body's parts are consistent, there are slight differences from individual to individual that are not of medical concern but constitute a normal range of variation. Medical imaging software attempts to assist in the visualization process by fitting the major bones and organs in an acquired image of a patient onto a standard reference to make deviations more apparent. This is done by locating key points of reference (such as joints) and using them as control points to align the images, morphing (stretching, rotating, etc.) portions of the image for a best fit.

The special constraints that arise in the registration of medical images obtained from a variety of imaging devices (ultrasound, X-ray, MRI, etc.) and the repertoire of techniques that have been developed to deal with them are beyond the scope of this text. Figure 1.21 shows the case of merging or overlaying images of a single individual obtained from different sources. The original images have different resolution and pixel dimensions, requiring affine transformations (described later) for alignment. A comprehensive review of methods used to align multiple images or to align images with idealized models can be found in Hajnal et al. (2001).

Other examples of shape matching arise in robotics. A simple example is a video-controlled robot arm placing wheels onto a car on a production line. The task is to locate the lug bolts and rotate the wheel so that the holes line up with the bolts. Many robotic tasks utilize video imagery to find parts and places, and these all involve to one degree or another the basic types of shape transformation.

The most basic operations for mapping are translation and rotation, which would be sufficient for the wheel alignment task. This is a common task in robotics used in manufacturing and is typically accomplished using normalized cross-correlation to locate the target (the mathematical basis for this and other operations is shown in Chapter 2). Dedicated hardware, sometimes built into the video camera, can accomplish this operation in a fraction of a second.

If the distance from the camera to the object is not a constant, then sizes will also change. An affine transformation can adjust the scale of the object. It can also take into account some changes in viewing angle as, for example, if the objects are on a plane that is slightly tilted. Physically, an affine transformation preserves the alignment of points (e.g., if three points lie on a line before the transformation, they will lie on a line afterward) and the ratios of distances along a line (e.g., for the same three points, the ratio of the distance between point 1 and point 2 to the distance between point 2 and point 3 will remain constant, although the distances themselves may be altered). In affine transformations, the angles between lines are not preserved, but the change in angles does not vary across the image (a shear transformation). Any combination of translation, rotation, scaling, and shearing can be combined into a single matrix operation, as shown in Chapter 5.

Still more general is a perspective transformation that allows the change in angles to vary with position, corresponding to the foreshortening that occurs when the camera viewpoint changes. Figure 1.22 summarizes these basic mapping operations. These are often used with the landmark methods shown in Chapter 3.

Mappings that do not preserve straight lines introduce a much greater range of possible changes in shape. As shown in Figure 1.23, these morphing operations range from very simple ones that can, for example, correct for lens distortion, to ones that fold portions of the object over itself, or introduce holes inside the object, or disconnect portions of it. Many

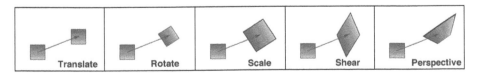

FIGURE 1.22
Basic mapping operations that preserve straight lines.

FIGURE 1.23
General morphing operations that distort the object's shape.

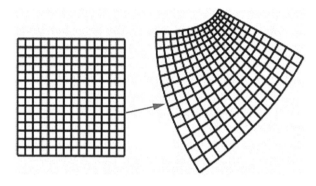

FIGURE 1.24
Conformal mapping preserves local angles.

measurements of shape are insensitive to affine transformations but not to general morphing.

A subset of morphing operations, called conformal mappings, preserves local angles. In Figure 1.24, the lines that cross at 90° angles in the original mesh also cross at 90° after the mapping transformation, although distances and ratios of distances are not maintained, and the overall "shape" has changed. Several of the analysis procedures based on landmarks shown in Chapter 3 use conformal mapping. Some shape measurements are able to survive conformal mapping operations unchanged, but most are not.

Object Recognition versus Scene Understanding

Robotics vision encompasses a broad range of capabilities. "Pick and place" logic that selects parts from a bin and positions them in an assembly, like the wheel alignment task described earlier or the assembly of printed circuit boards, usually requires only recognition of a few well-defined reference points, affine transformations, and minimal computation. A higher degree of flexibility and the ability to deal with some variations in shape are needed for tasks such as robotic butchering of a steer carcass. But all of these operations are required to deal with only one or at most a few objects at a time, and the objects themselves are defined, albeit with varying levels of detail, beforehand.

Autonomous robotic navigation, as demonstrated recently by driverless automobiles able to navigate through a cluttered city environment that is dynamically changing and contains other moving objects, is at a qualitatively different level. These tasks require object recognition using video cameras, laser rangefinders, radar, and other imaging devices, but go beyond that to build an internal geometric model of the relationship between objects, and its evolution over time, in order to understand the scene and to be able to predict object motions. That level of computation is beyond the scope of this text, but obviously starts with a basic ability to classify and recognize objects, based in large part upon their perceived shapes.

One robotic activity still in the "science fiction" category, but not so far out as Asimov's virtually human creations, is autonomous robotic surgery. While obviously of great interest for tasks such as dealing with injuries in dangerous locales such as war zones (or remote ones such as space exploration), this will require a level of visual recognition of objects and structures that surpasses anything yet demonstrated, and demands an extremely high degree of confidence.

2

The Role(s) of Computers

Digital Images

In order for images to be used for measuring shape, they are first stored as a digitized representation. Usually this is an array of pixels, each with the brightness or color of the corresponding point in the original scene. Other, more concise sets of data may also be used, such as the coordinates of a limited number of points to define objects or structures. In a few cases these points, in either two or three dimensions, are input by a human. Mouse-clicking points on the image displayed on the computer screen is the most common method; placing either the object or an image on a tablet and touching landmarks with a stylus, or using a coordinate measuring machine (CMM) to touch points on a three-dimensional object are also examples of manual input. But automatic digitization without human involvement (except perhaps for setting up the initial image acquisition) is usually a preferred method.

As indicated in Figure I.2, measurements of shape rely either on the entire area of an object or primarily on the periphery. All of these require adequate resolution, but the ones that utilize the boundary are particularly sensitive. There are techniques (described later) that represent the boundary in a compact form directly suitable for some forms of analysis, and there are routines that can convert between different representations.

The initial data acquisition rarely begins with a boundary outline. Some measurement devices, such as a metrology instrument that traces the position of a stylus as an object is rotated in order to measure its roundness, do generate boundary data. But in most instances, the beginning point for measurements is a digitized image consisting of an array of pixels. One task of the image processing software is to extract the boundary from that array. For some applications, especially for three-dimensional imaging and measurement, there may be several such images, either from different viewpoints or representing a series of parallel sections through the object.

The pixel array is most commonly obtained from a digital camera, although in some cases analog cameras, or desktop scanners applied to films (or to the objects themselves), or digitization of the analog output from various types of scanning microscopes may be used. For all cases, the important characteristics of the pixel array include the size, the bit depth, and the number of channels.

Pixel Array Size

Digital camera images range in size from a few megapixels upward. The number of pixels required in a given application depends very much on the resolution needed to properly characterize the objects or structures of interest. For example, it makes no sense to mount a 12-megapixel camera (recording a 4000 × 3000 pixel image) onto a light microscope whose optical resolution then corresponds to dozens of pixels. All that accomplishes is to increase the processing time while simultaneously increasing the random noise present in each pixel due to the capture of fewer photons and the generation of fewer electrons. On the other hand, recorded video images (especially from the cameras used in most surveillance installations) have resolution that can barely justify digitizing an image of 400 × 300 pixels, a factor of 100 times smaller, which places severe limitations on attempts to measure shape information sufficient to identify a human face in the recorded image.

Matching the image resolution and number of pixels to the task that is to be performed is a critical first step in any serious effort to perform measurements. In many industrial quality-control operations, the dimensions and objects of interest cover a narrow range of sizes, and consequently it may be possible to work with images having a relatively small number of pixels. This facilitates real-time processing, often using dedicated and highly specific circuitry. In other cases, a need to capture images of many objects at one time, or to include objects covering a wide range of dimensions, means that very large pixel arrays are needed.

High resolution adds cost in terms of instrumentation, processing time, and storage. The resolution needed is application dependent. In the example of Figure 2.1, the delineation of the edge of the key needed to assure that a copy would work in the lock defines the requirement. The original image was obtained by scanning the keys at 1200 pixels per inch (ppi). A lesser resolution of 600 ppi is still adequate, and even 300 ppi is probably sufficient, but 150 and 75 ppi are not.

The basic measurements and algorithms discussed in the following sections do not fundamentally depend on the size of the pixel array, except of course that applying the more complex procedures to large arrays takes more time or processing power, and more memory is needed to hold the entire image (and perhaps several such arrays at once). This is much simpler and faster than having to read parts of the array each time from a disk file. If the objects of interest have a wide range of sizes, the pixel resolution must be adequate to define the small objects and the array must be large enough to encompass the large ones. Large features are more likely to intersect the boundaries of the image, and consequently not be fully defined or measurable.

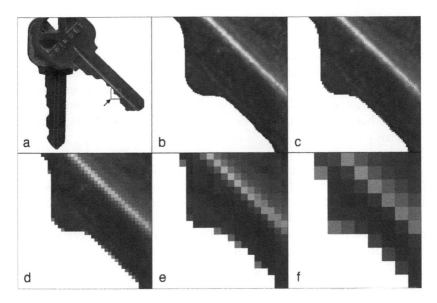

FIGURE 2.1
Effect of image resolution on edge definition: (a) original image with outline showing region enlarged; (b) 1200 ppi image; (c) 600 ppi; (d) 300 ppi; (e) 150 ppi; (f) 75 ppi.

In addition to the size of the pixel array, the number of bytes of storage required for each pixel is important. In the simplest cases, a single byte can represent values from 0 to 255, corresponding to the voltage from the detector and in most cases to the brightness of the point in the original image (although the correspondence is not necessarily a linear one). This range of values corresponds to an 8-bit image ($2^8 = 256$), and such a monochrome image may be sufficient for many applications, particularly when the illumination can be controlled. Examples include the bright-field light microscope and back-illuminated objects whose profiles are to be measured. Some cameras, particularly video cameras, produce signals that do not even justify this much definition. At the other extreme, many metrology devices such as scanning profilometers are capable of recording elevation data with range-to-resolution values that require 4 bytes or even more as illustrated in Chapter 5.

Most good quality digital cameras are capable of measuring light intensity data that require 12 bits (4096 values) to record. This is approximately the same brightness resolution capability as conventional film and has allowed digital cameras to replace most film cameras (also, a 4000 × 3000 pixel array is similar in lateral resolution to that produced by 35 mm film). Since computer memory is typically organized in 8-bit bytes, the usual way to store brightness values when they exceed 8 bits is in 2 bytes (16 bits), although except for the cooled cameras used in astronomy this is not needed. Storing a signal with 12 bits of significant intensity resolution in 16 bits of memory means that the least significant bits contain zeros or random values.

It is sometimes convenient for these cameras to convert the raw data from the sensor to an 8-bit image. Many consumer-grade or pocket digital cameras do this so that the image can be stored efficiently. The reduction in dynamic range can be accomplished by applying a nonlinear (gamma) function to the sensor data (which are linearly proportional to the light intensity) and discarding some of the initial resolution. The resulting 8-bit images are then usually further subjected to lossy compression (JPEG compression) to reduce the file size.

These images are acceptable for snapshots of family and travel, but are poorly suited for scientific purposes requiring measurement because of the lost information, and the fact that the compression alters pixel values, shifts boundaries, and erases some details. It is not possible to recover the lost information, or even to determine what it might have been or where various amounts of detail have been removed. The use of lossy compression such as JPEG is strongly discouraged for scientific and technical applications.

Color

Some scientific applications are naturally suited to monochrome (grayscale) images. Electron microscopes, for example, produce images with no color information, and some light microscope applications use filters in the optical path so that monochrome light is recorded. Medical images acquired from magnetic resonance imaging (MRI) and computed X-ray tomography (CT) record signals proportional to density or water content, and so the images are monochrome (although MRI is capable of recording several different channels of information). But for many applications, images are usually recorded in color. This produces three channels of values, for the red, green, and blue color bands.

Many applications such as remote sensing and astronomy, which rely on infrared wavelengths, present their data as color images, by assigning various wavelengths to the red, green, and blue display channels. Even some instruments that do not record light at all, such as atomic force microscopes, may use colors to combine multiple channels of information (such as surface height, lateral force on the tip, and so forth).

Digital cameras record color images in several different ways, as shown in Figure 2.2. The simplest method uses a single chip with an array of light-sensitive detectors, placed behind a filter wheel that can interpose red, green, and blue filters. Three exposures are recorded, which may have different durations, to balance the intensities in the three channels so that neutral colors (white, black, and shades of gray) are free from color. But these cameras can only be used for static scenes and do not provide the real-time image necessary for making adjustments such as focus or illumination.

It is also possible to use three separate chips with prisms and colored filters so that the three channel images are recorded at the same time and merged electronically. This approach is used in some video cameras but has its own

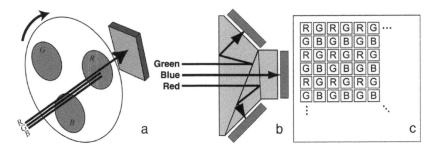

FIGURE 2.2
Digital color camera designs: (a) rotating filter wheel with a single chip; (b) three chips with prisms and filters; (c) single chip with Bayer pattern filters.

limitations including the cost of the additional components and the need to keep them in perfect alignment. Optically, the prism-and-filter arrangement is inefficient, allowing less than 10% of the incident light to be recorded. Also, if wide-angle lenses are used, the different angles of the light passing through the prisms cause color shading across the image (with different shading for the red, green, and blue).

One novel design employs a single chip but at each location stacks three transistors on top of one another. Red light penetrates farther in silicon than blue light, so the topmost transistor primarily responds to blue wavelengths, the one under it to green, and the deepest one to the long wavelength red light. This is very elegant in principle but has suffered from practical difficulties in achieving consistent color balancing.

By far the most common type of design for digital color cameras uses a single chip in which color filters are applied to the individual detectors. The simplest and most widely used filter arrangement, the Bayer pattern, assigns green filters to half of the detectors, and red or blue to the remaining ones (the larger number of green-filtered detectors is intended to correspond to the greater sensitivity of human vision to that portion of the spectrum). The electrical signals are processed with "demosaicing" algorithms that interpolate the values so that the amount of red light (for instance) can be estimated at locations where no red-filtered sensor was present. Each camera manufacturer has its own proprietary algorithms that operate in the camera firmware, but the overall result is to reduce the image resolution by about a factor of two in each direction.

The advertised number of pixels in the camera generally describes the size of the pixel array that is stored, which for most cameras is the same as the number of photosensitive detectors in the chip. The actual resolution of the images is not that good, limited by the interpolation that takes place as well as any other electronic effects (capacitance in the circuitry on the chip or external to it, random noise particularly at low light levels when more amplification of the electrical signals is needed, etc.), and of course by any limitations in the optics.

TABLE 2.1

Some Representative Color Temperatures

Light Source	Typical Equivalent Black Body Temperature
Match flame	1700 K
Candle	1900 K
Sunlight (sunrise or sunset)	2000 K
Incandescent lamp	3000 K
Moonlight	4000 K
Daylight	5500 K
Overcast sky	6500 K
Shade	7500 K
Cathode ray tube	9000 K

The color in the recorded image is dependent on the illumination of the scene. This is usually described as the color temperature of the light source, referring to the spectrum of wavelengths emitted by an ideal black body radiator at that temperature. Most actual sources of illumination do not match this spectrum precisely, but the temperature still serves as a handy approximation. Typical values range from less than 2000 K for a candle, to more than 9000 K for a cathode ray tube, as summarized in Table 2.1. Fluorescent lights (including CFLs) are not in the table for two reasons: (1) their emission spectrum is very different from that of a black body radiator; and (2) many different coatings are used to produce lighting that is considered to be visually similar to color temperatures ranging from less than 3000 K (reddish or "warm") to more than 6000 K (bluish or "cool").

Some cameras allow selecting a color temperature to be used to adjust the measured red, green, and blue intensities, or allow measuring the light from a colorless gray or white reference card to establish the value to be used for performing the correction. Color cameras do not measure color; a spectrophotometer is used to measure spectral intensity. The wavelength ranges of the filters in the cameras in Figure 2.2 are broad and overlap somewhat, so that different combinations of wavelengths and intensities can produce exactly the same output. The goal of color imaging and correction is rather to produce a representation of the scene that visually matches what a human would see.

Camera Specifications

In addition to the "resolution" of the camera, which as noted earlier is generally reported as the number of pixels in the stored image and usually is also the number of individual detectors in the array, there are several specifications that affect the quality of the captured image. Two principal types are in current use: CCD and CMOS. Both function by allowing incident photons to raise electrons to the conduction band in individual transistors. The differences arise from the way that charge is extracted so that it can be converted to a numerical value.

CCDs (charge coupled devices) consist of an array of transistors wired up to their neighbors in columns so that a clock signal can be applied to shift the charge from each horizontal row to the next lower row. When the charge is transferred from the last row, it is shifted horizontally and sent to a separate chip containing an amplifier and digitizer (an analog to digital converter or ADC), producing a string of numbers that are stored in memory or transferred to a computer.

CMOS (complementary metal oxide semiconductor) devices incorporate several additional transistors on the chip for each pixel. Address and control wiring on the chip and separate amplifiers allow each individual detector to be read out directly, much the same as a memory chip in the computer. The additional circuitry takes up some space, which reduces the "fill factor" or the percentage of the area of the chip that is sensitive to incoming photons. Many chips compensate for this limitation by placing small lenses over each detector to collect the light from a larger area.

Variations in these lenses, the separate amplification of each detector output, and the additional capacitance from the wiring on the chip (as well as the different fabrication technologies used for CCD and CMOS wafers) cause the CMOS designs to generally have higher random noise and more "fixed pattern noise" (nonuniformity from one pixel to another) than CCDs. But the possibility to include all of the support circuitry, amplifiers, and ADC on a single chip makes CMOS designs less costly, more compact, and more energy efficient. This is why CMOS dominates the market for cell phone cameras. Some single lens reflex cameras also employ large CMOS chips, using more costly fabrication techniques, and special processing of the image (for instance, to remove the fixed pattern noise in each device). Most cameras intended for scientific or technical imaging use CCDs.

The size of the transistors that collect light photons and accumulate electrical charge ultimately limits the number of electrons that can be held during an exposure, and this controls the signal-to-noise and dynamic range of the image. Technically, this is called the "well size" of the detector, but except for a few specific fields of application (such as the cooled digital cameras used in astronomy) this is rarely specified. Instead, the physical size of the detector may be given. Very small transistors—as small as 1 μm across—in the CMOS chip in a cell phone (further limited in size by the additional control transistors) can barely produce images with 8 bits (256 values) of brightness differentiation. By comparison, the array in a single lens reflex camera may be as large as the 35 mm film such cameras once used, with individual detectors more than 8 μm across (64 times the area, with a correspondingly greater charge capacity). These produce outputs with 12 bits (4096 values) of brightness resolution, comparable to the performance of traditional film.

There are a few other design specifications worth knowing about. Some chips are thinned and allow the light to fall onto the active transistors from the rear (called back illumination). This increases the sensitivity by eliminating absorption in the deposited metal layers that constitute the chip's wiring. Also, some color camera chips use color filter arrays other than the Bayer pattern shown in Figure 2.2. These claim improved performance at distinguishing colors, particularly in the middle wavelengths (the "green" part of the spectrum), but since the final output is still specified in terms of red, green, and blue values, it is not clear that this is an important difference.

Image Processing to Correct Limitations

In order for structures and objects to be measured to determine properties such as size, position, brightness, and of course shape, they must be isolated from the surroundings. Under ideal conditions, with controlled lighting, a flat surface, and good contrast between the background and the object(s), this can be easy. One application that routinely accomplishes this task is scanning pages of printed text and converting the images to words in a file that can be accessed with a word processor or spoken aloud to a person with limited vision.

But most imaging applications, even ones using microscopes or copy stands, and especially "real world" tasks such as crime scene or surveillance photographs, present much greater difficulties. One of the first tasks for computer-based image processing is alleviating some of the problems in as-acquired digital images. The steps in the process include:

1. Adjustments to color values, for instance, to correct for differences in illumination, so that comparisons between images can be made.

2. Reduction of noise, both random noise due to limited signal and imperfect cameras, and periodic noise arising from electronic interference, vibration, and so forth.

3. Correction for nonuniform illumination, so that objects will have the same contrast and appearance wherever they are positioned in the scene.

4. Adjustment of contrast and brightness to make optimum use of the available dynamic range.

5. Correction for distortion, either arising from lenses or a nonperpendicular point of view, so that shapes and dimensions are correctly shown.

6. Removal or reduction of blur due to focus or motion problems.

Of course, not all are needed in every case. A typical workflow performs the operations in the order listed, skipping those steps that are not required. Examples of these processes are shown in the following. Much more complete examples, descriptions, and the algorithms used for implementation can be found in Russ (2011) and many other texts.

Color Adjustment

If the colors in an image are used only to distinguish one object from another or objects from background, it does not matter if the colors are accurate or match the visual appearance of the scene. But if color adjustment is needed, it should be done as the first step in image processing. As noted earlier, digital images describe colors using just red, green, and blue intensity values, and cannot measure colors in the way that a spectrophotometer does. However, in many cases the presence of a few known color standards in the scene can be used to make corrections that allow visual comparison and matching between images. There are several techniques available.

The simplest of these is neutral color correction. Gray objects and regions in the image should have equal red, green, and blue (RGB) values. Locating such regions, which is often done manually, makes it possible to construct curves that modify the red, green, and blue channels of the image as shown in Figure 2.3. In the example, three points marking the darkest neutral region (the shadow under the tire), the brightest such region (the license plate), and a neutral midgray region (the gravel) were selected. By adjusting the RGB values for those points to be equal and interpolating curves for other values, the colors throughout the image are adjusted. Of course, this method depends on being able to locate neutral color regions.

More flexible adjustments are possible using color standards. Because the filters used to determine the RGB intensities in the camera cover broad and somewhat overlapping wavelength ranges, it is not always sufficient to make independent adjustments to the red, green, and blue channels. Instead, a "tristimulus" matrix approach is used in which the corrected red intensity (for example) is calculated from the measured red, green, and blue values, and so on. Determining the correction matrix that modifies all three channels based on the recorded values requires measuring the RGB intensities from known standards.

In a studio, or with a copy stand or other controlled and consistent setup, recording an initial image of a color standard and then proceeding to record a series of photographs allows applying the correction determined from the standard to the entire series. In other situations, placing a color standard in the scene as it is recorded is the preferred strategy. Corrections are especially important for advertising photographs (the catalog must accurately represent the products), in which not only the camera and lighting but also the printing inks and paper must be included in the correction. Accurate color representation is also important in photographing artwork, archaeological artifacts,

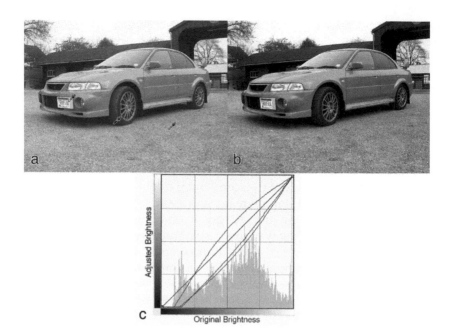

FIGURE 2.3

(See color insert.) Neutral color adjustment: (a) original image, showing locations selected as defining neutral black, gray, and white; (b) corrected image showing that the car is blue, not green; (c) the transfer functions applied to the red, green, and blue channels to make the adjustment.

and other objects of scientific interest. Companies such as GretagMacbeth® provide standards as well as comprehensive equipment and software for these applications.

Correction of colors may also be necessary in order to correctly obtain the shape of an object. In medical imaging, for example, color is an important criterion providing contrast between healthy skin and a healing skin graft. This distinction is used to delineate the boundaries of the graft over time to track healing progress.

Noise Reduction I: Speckle Noise

"Noise" is used to describe the image content that does not represent the actual scene. There are several possible sources, which are best dealt with using different tools. Random "speckle" noise arises partly from the statistics of producing electrons in the individual transistors, partly from thermal noise superimposed on that variability in the process of transferring the electrons out from the chip, partly from the characteristics of the amplifier(s) used to convert the electron charge to a voltage, and partly from the measurement or digitization process. It shows up in the image as a more-or-less random variation in the brightness values of pixels within

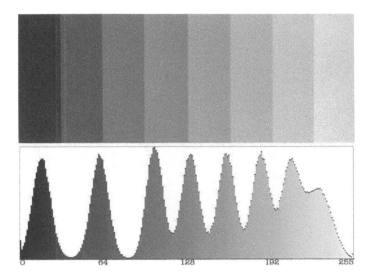

FIGURE 2.4
Image of a step wedge, with its histogram. The spacing and widths of the histogram peaks vary as a consequence of the camera's contrast and gain adjustments. The darkest bands are well separated but the brighter ones are not, and the two brightest are quite overlapped.

a region that should be perfectly uniform, and is evident in the breadth of the corresponding peak in the histogram of the image, as shown in Figure 2.4.

Blurring an image with a Gaussian smoothing filter is sometimes applied in an attempt to reduce speckle noise but is a poor choice because it blurs steps and edges, and can shift or distort boundaries. A median filter, in which the intensity values of pixels in a small neighborhood are ranked into order and the median value in the ordered list replaces the original value of the pixel at the center of the neighborhood, is a preferred tool for reducing random noise. The process is repeated for every pixel in the image, always using the original pixel values. Extreme values are replaced without shifting or blurring steps and edges. Most programs implement the median using an adjustable-size square neighborhood, which is convenient for programming, although an approximately round neighborhood is preferred to avoid directional artifacts.

Figure 2.5 shows the result from applying a median filter with a neighborhood consisting of a 5-pixel-wide circle to the step wedge image and the effect on the histogram. Peaks become narrower and better separated. The boundaries between the bands in the pattern are sharper with the median filter than with the Gaussian blur, as shown in Figure 2.6. A further improvement is possible with a maximum-likelihood Kuwahara filter, which compares each pixel to the possible neighborhoods around it and assigns the pixel a value equal to the mean of the neighborhood that includes it with the

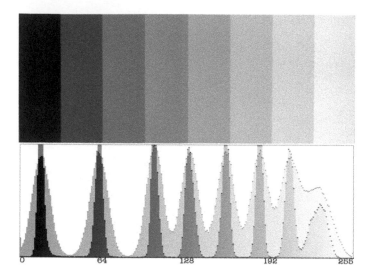

FIGURE 2.5
The step wedge from Figure 2.4 after the application of a median filter with a 5-pixel-wide neighborhood. The histogram of the processed image (shown superimposed on the original) has narrowed peaks, and the speckle noise in the bands is reduced.

smallest variance. This procedure is discussed and illustrated next in the context of edge delineation.

Although the median filter does not shift edges, it does round corners and remove fine lines and detail. More advanced forms of the median, such as the hybrid median (which performs several rankings on different portions of the neighborhood, which are then combined) or the conditional median (which omits from the neighborhood pixels whose intensity difference from the central pixel exceeds some threshold) produce better results.

Figure 2.7 provides a comparison. The original image (the blue channel from an original color image) has both speckle and shot noise (for instance, note the light pixels to the left of the 11 and on the wood frame). The Gaussian smooth reduces the visible noise but also blurs or removes details. The conventional median does not blur or shift the boundaries (for example, the edges of the wooden frame), but removes fine lines and rounds corners. The hybrid median reduces the noise while retaining fine image detail.

For color images, several approaches are used to select the pixel whose RGB values replace the central pixel. The average intensity of the red, green, and blue channels may be used to perform a conventional ranking, or a vector median can be calculated. The latter plots the RGB values for each pixel in the neighborhood in a three-dimensional color space and identifies as the median pixel the one whose coordinates have the smallest sum of Pythagorean distances to the others. This is more computationally intensive.

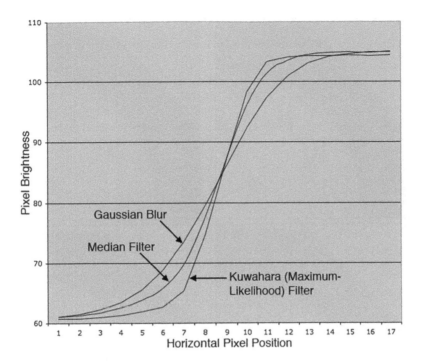

FIGURE 2.6
Brightness profiles across one of the brightness steps in the wedge showing the effect of the Gaussian blur, median filter, and Kuwahara maximum-likelihood filter on boundary sharpness. All of the filters use the same 5-pixel-wide neighborhood.

As shown in Figure 2.7, median filters are also good tools for dealing with shot noise or dropout noise. Most cameras (some very expensive special purpose ones used in astronomy and for space missions are exceptions) have at least a few "dead" pixels out of the millions present. They may either give zero output (black) or be locked to maximum output (white) depending on the nature of the manufacturing flaw. These extreme values are never the neighborhood median, and so are replaced by a nearby value as shown.

Noise Reduction II: Periodic Noise

Periodic noise is best removed in a different way. The power spectrum display of the Fourier transform of an image shows a plot of the amplitude of each of the frequencies and orientations needed to reconstruct the original picture. Periodic information, which is often noise (but which may in some instances be real structure in the scene), appears in the power spectrum as a series of peaks or "spikes" that lie at specific frequencies and orientations. Setting the amplitude of just those frequencies to zero and leaving unchanged all of the others removes the periodic noise without affecting the rest of the image, as shown in Figure 2.8. This approach is particularly

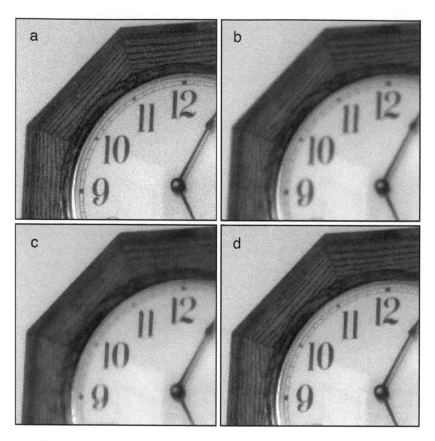

FIGURE 2.7
Reducing speckle and shot noise: (a) original image; (b) Gaussian blur, standard deviation 1 pixel; (c) median filter, radius 2 pixels; (d) hybrid median filter, radius 2 pixels.

effective when electronic noise, flickering lights (as from fluorescent tubes), or vibration are present. It is also able to eliminate the periodic pattern from half-tone printing to recover an image for better visual examination.

Conversely, in the case of a periodic structure such as the fabric in Figure 2.9, the mask removes all of the other frequencies and keeps just those that correspond to the periodicity of the fabric. This removes the random noise and produces an averaged result for all of the repetitions of structure that are present, which can make measurements much easier. Measurements of periodic spacings can be conveniently performed directly on the Fourier power spectrum.

Both automatic and interactive manual tools exist for constructing the mask to isolate the periodic frequencies shown in the Fourier power spectrum, but to be successful in a wide range of situations it is important to understand the steps in the process. The power spectrum represents the amplitude of each frequency as a shade of gray, at a position whose radial

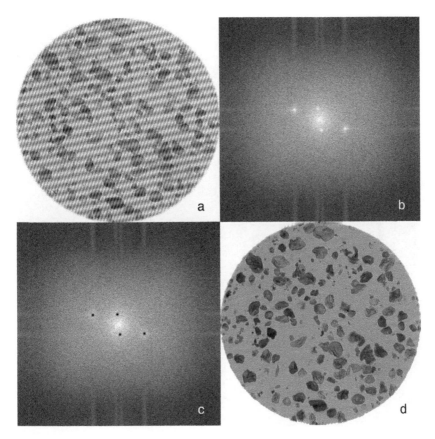

FIGURE 2.8
Removal of periodic noise: (a) original microscope image of particles with electronic interference; (b) Fourier transform power spectrum showing spikes corresponding to the frequencies and orientations of the interference; (c) power spectrum with the spikes removed (amplitude set to zero); (d) inverse Fourier transform showing the image with interference removed.

distance from the center is the frequency and whose direction is perpendicular to the orientation of the sinusoidal lines to which that particular frequency corresponds.

For most images, in addition to whatever periodic spikes may be present, there is a general trend of decreasing amplitude (the brightness value) with frequency (the radius). To isolate the spikes in the power spectrum, it is necessary to compensate for the general variation in the background. Several methods for leveling or removing background from images are shown in the next section, and can often be used with these images. But two of the most straightforward techniques are uniquely well suited to the location of spikes in the power spectrum.

One is to apply a Gaussian blur (a function incorporated in practically all image-processing packages, which is also used in some of the enhancement

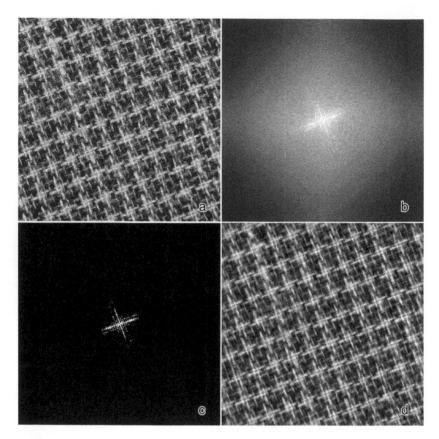

FIGURE 2.9
Filtering to keep periodic structure: (a) original image of woven fabric; (b) Fourier transform power spectrum showing numerous spikes; (c) mask to retain the spikes and eliminate other frequencies and orientations; (d) filtered result.

techniques shown later) to a copy of the power spectrum image so that the details, including the spikes, are smoothed out. The result is the general background shape, which can be removed from the original power spectrum image to obtain the ratio of the original to the background. This levels the background so that the spikes can be thresholded and used to construct a filter or mask (the terms are used interchangeably) to identify the specific frequencies of the periodic structure, which are either kept, as in the example of Figure 2.9, or removed.

The second approach uses a special version of a ranking filter. The median filter, introduced earlier, ranks the pixels in a neighborhood to select the median value from the center of the ranked list. The top-hat filter performs its ranking in two different neighborhoods. One is a round (or approximately round) region, and the second is an annular ring around that central core.

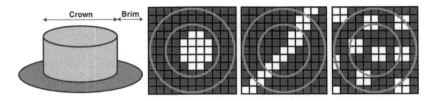

FIGURE 2.10
The top hat filter with a diagram of the crown and brim. The filter detects features that fit inside the crown but not ones that extend beyond it or are separated by a distance smaller than the width of the brim.

These can be visualized, as shown in Figure 2.10, as the crown and brim of a top hat, hence the name of the filter.

By comparing the brightest gray value in the crown to that in the brim, it is possible to find spikes that rise above the local background value. The important variables are the diameter of the crown, the width of the brim, and the height of the crown (the difference between the two values that represents the presence of a spike). For typical Fourier transform power spectra of images containing periodic noise, the spikes are only a few pixels wide, are usually separated by at least a few pixels, and rise significantly above the local background. This method was used in the example of Figure 2.8.

However the periodic frequencies are isolated, the resulting mask is used to reduce the amplitude of the unwanted terms to zero, followed by performing the inverse Fourier transform to reconstruct the original pixel image without those frequencies. It is important to keep the terms near the center of the power spectrum (the low frequencies and the "DC" term at the very center), which define the average brightness of the original image. As shown in the preceding examples, the results are effective at removing periodic noise or isolating periodic structures.

Nonuniform Brightness

The preceding section describes methods for dealing with the variation in background brightness of the Fourier power spectrum. Photographs may also have nonuniformities in brightness that complicate the problem of isolating the objects and structures of interest from their surroundings or background. The variation is usually not as predictable or symmetric as for the power spectrum, and more flexible tools are needed to deal with it. There are three approaches for obtaining a second image that represents the variation and can be used to remove it.

- Record a second photograph of just the background with identical lighting but without the objects of interest.

- Compute a function for the brightness of the background by locating, either manually or automatically, a number of points in the image that should all have the same brightness.
- Use a morphological procedure called an opening to remove the objects of interest from the image, leaving just the background.

When it is possible to obtain a background image, this is the method of choice. In applications such as photographing objects in a microscope, or on a copy stand, or similar situations, it is practical to remove the objects or the microscope slide without altering the lighting or optics, and then to record a second image. The bases of most copy stands are painted a uniform gray for this purpose. Obviously, the method does not often apply for real-world photographs except sometimes in the case of surveillance video, where the camera and lighting do not change as the subjects enter or leave the field of view (automatic gain adjustments in the cameras may prevent this approach from working).

Once the second image is recorded, it can either be subtracted from the original image, pixel by pixel, or the ratio of the original to the background can be calculated. In both cases it is generally necessary to rescale the values and add an offset to keep the results within the range of 0 to 255 that can be displayed by the computer. Figure 2.11 shows an example, in which both the brightness and color are nonuniform in the original image. The question that remains is whether to use the difference or the ratio, and that depends on how the camera recorded the image in the first place.

Camera chips, either CCD or CMOS, are inherently linear, so that their output is a direct measure of the light intensity. If those values are used (e.g., from the "RAW" data that some cameras can export), the ratio of signal to background is the appropriate result. But many cameras, and especially

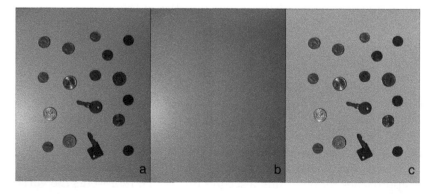

FIGURE 2.11
(See color insert.) Subtracting a measured background: (a) the original image; (b) the background with the same lighting and the objects removed; (c) the result of removing the background.

ones produced for the consumer marketplace as opposed to scientific uses, are programmed to produce images that are similar to the film cameras that they have replaced. Film responds to light intensity logarithmically rather than linearly. If the camera circuitry records the output from the chip using a logarithmic conversion, it is the difference between the two images that is appropriate. In some cases, the most straightforward approach is to try both methods—ratio or subtraction—and select the one that produces a useful result (and in the process identifies how the camera works).

When the nonuniform brightness in the image results from off-center lighting, optical vignetting, or other similar effects, the variation with position is usually gradual and can be well approximated with polynomial functions. If several locations which should have the same brightness, well spaced across the image, can be identified, their brightness values and (x, y) coordinates can be used to calculate a polynomial function of the form

$$Brightness = A + B \cdot x + C \cdot y + D \cdot x^2 + E \cdot xy + F \cdot y^2 + G \cdot x^3 + H \cdot x^2 y + \ldots \quad (2.1)$$

This function can then be used just as a measured background image would be, to remove the variation at each (x, y) position in the image.

Manual selection of the locations to be used in fitting the function may be used, but in some situations automatic selection is possible. This circumstance generally arises when the background is locally brighter or darker than the objects of interest. That makes it possible to subdivide the image into many regions, find the brightest or darkest values in each region, and use those values and their locations to determine the coefficients of the polynomial. Figure 2.12 shows a scanning electron microscope image in which the background near the center is brighter than the objects near the sides, but because the background is always locally darker than the particles, automatic fitting and removal of the background produces a level result in which the objects can be distinguished from the background by thresholding.

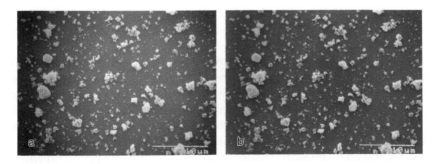

FIGURE 2.12
Automatic leveling of image brightness by fitting a polynomial to the locally darkest points:
(a) original; (b) result.

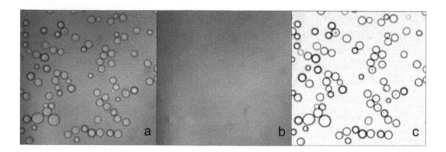

FIGURE 2.13
Background removal by rank filtering: (a) original image of liquid droplets; (b) background produced by a morphological opening; (c) the ratio of (a) to (b) levels the brightness across the image.

If the variation is not gradual, the polynomial method is not appropriate. But a background image may be constructed by removing the objects of interest from the image using image processing. So-called morphological operations compare each pixel to their neighbors in a small region. The example of morphological operations described here is applied to grayscale or color images and involves ranking of the pixel values. (Another class of morphological operations is introduced later in the context of processing thresholded binary images.) Two neighborhood ranking operations were described earlier, the median filter and top-hat filter. Morphological operations use ranking to alter the size or shape of features by adding or removing pixels along the periphery.

The background in the image shown in Figure 2.13 varies in brightness in an irregular pattern not suitable for polynomial fitting, but the borders of the droplets are everywhere darker than the local background. If every pixel in the image is replaced by its brightest neighbor, the dark outlines shrink in size. Repeating the operation or using a neighborhood whose diameter is at least as large as half the width of the lines removes them entirely. This process is called erosion, extending the bright background into the area of the outlines. Because erosion also alters the brightness values and pattern of the background, it is followed by dilation in which each pixel is replaced by its darkest neighbor.

Because the outlines have been removed, they do not grow back in the dilation process, but the background is restored to its original brightness and pattern. The sequence of erosion followed by dilation is called an opening. If the objects are brighter than the background, the procedure is reversed, starting with a dilation (replacing each pixel with its darkest neighbor) followed by erosion (replacing each pixel with its brightest neighbor). The sequence of dilation followed by erosion is called a closing. (Note: Depending on whether objects are considered to be bright or dark compared to the background, the sense of erosion and dilation may be reversed from the example shown here but interchanging the labels does not alter the meaning of the operation.)

Contrast and Brightness Adjustments

Even with automatic exposure adjustments in cameras, an image may not have optimal contrast or brightness for distinguishing the structures or objects of interest. Furthermore, the processing steps described above alter the brightness values. The most important tool for understanding the distribution of brightness values in an image is the histogram, a plot of the number of pixels (i.e., the fraction of the image area) as a function of brightness as shown in Figure 2.4. For a color image, histograms of the individual red, green, and blue channels may be available, but it is generally more useful to examine a histogram of overall pixel brightnesses. This is may be the average of the RGB values or a weighted luminance calculated by combining the RGB intensities in proportion to the response of the human visual system (approximately $0.3 \cdot Red + 0.6 \cdot Green + 0.1 \cdot Blue$, but the exact weights depend on the filters used in the camera, the overall scene brightness, and also vary somewhat from individual to individual).

The first important requirement for a well-exposed image is a distribution of brightness values that covers the full range from 0 to 255 (black to white) without clipping. Clipping occurs when pixels are brighter or darker than the limits of the digitization process, so that information is lost. The pixels are set to the black or white limits, which shows up in the histogram as a large number of pixels with values of exactly 0 or 255. (Note: The 0 to 255 integer range corresponds to the values in an 8-bit image, but by using real numbers such as 195.48 the same range can represent images with a greater dynamic range.)

For an image whose measured brightness values do not cover the full range, the most common adjustment is to reassign the pixel values so that the histogram is linearly stretched to fit, as shown in Figure 2.14. The resulting histogram has gaps showing that there are brightness values with no pixels. The number of missing values is the same as in the original image, but they are distributed uniformly throughout the range rather than lying exclusively at the ends. Since human vision cannot detect small differences, the missing values are not visually evident, and regions with brightness gradients still appear smooth. And since relationships between pixel values are preserved (i.e., if one pixel was originally darker than another, it still is—just by a larger difference), the ability to distinguish one region or object from another is not affected.

The reassignment of pixel brightnesses is not limited to a simple linear stretching of the histogram. Any table of values (called a LUT or lookup table) can be created to provide new values based on the original ones. Displaying this table as a function (as shown in Figure 2.15) is useful. In most cases, preserving the visual sense of the image requires that this be a monotonic rising function, although information can still be lost if initially different brightness values are reassigned to the same value. Shifting the midpoint of a smooth curve up or down (known as gamma adjustment), or creating smooth curves, may improve the visibility of details or objects.

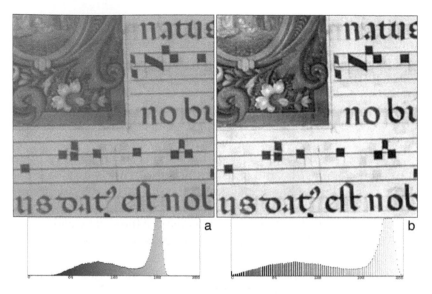

FIGURE 2.14
Linear contrast expansion: (a) original image with its histogram; (b) after linear stretch.

There are several methods for automatically determining the shape of the LUT curve based on the original contents of the histogram. The most widely used is histogram equalization, which shifts the values around so that as nearly as possible there are equal numbers of pixels with each brightness level. This is most easily judged by plotting the histogram as a cumulative graph showing the area fraction of all pixels darker than each brightness level (Figure 2.16); the graph should approximate a straight line, but steps are present due to any gaps in the histogram. Other histogram shapes, such as an approximation to a Gaussian, or matching the histogram of another

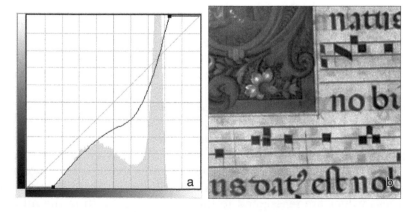

FIGURE 2.15
Nonlinear adjustment: (a) lookup table relating original to resulting pixel values; (b) result.

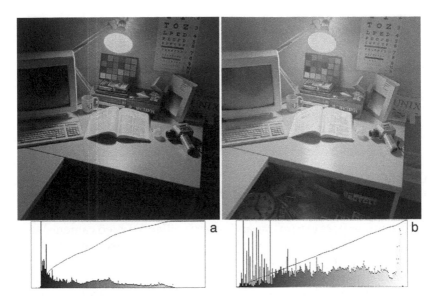

FIGURE 2.16
Histogram equalization: (a) original image with its histogram; (b) result. The monotonically rising line in each histogram is the cumulative plot showing the fraction of the area with darker values.

image (which may facilitate comparison of images taken of the same scene at different times) are also used.

Distortion Correction

Viewing or recording images of surfaces at an angle introduces perspective distortion that alters angles and dimensions. Additional sources of distortion may arise from the optics used. Correction of the distortion is needed in order to make accurate measurements of size, position, or shape.

Len distortions are usually radial and may result in either pincushion or barrel distortion as illustrated in Figure 2.17. Wide-angle lenses (as found in

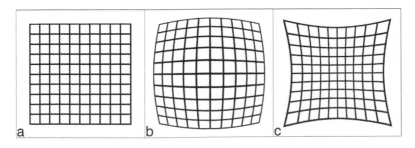

FIGURE 2.17
Lens distortion: (a) original grid; (b) barrel distortion; (c) pincushion distortion.

FIGURE 2.18
Correcting lens distortion: (a) original image showing barrel distortion; (b) corrected result.

many consumer pocket cameras and on surveillance video cameras) usually produce barrel distortion, and many zoom lenses vary from one to the other type with focal length or even exhibit a mixture of the two. It is possible to calibrate a particular lens by photographing a regular grid, and many standard lenses have been measured to produce databases in programs. Figure 2.18 shows one example of such a correction.

Perspective distortion can be more difficult to correct. For aerial or satellite photos, the location and orientation of the camera can be used to calculate the corrections needed (which in the latter case may also take into account the curvature of the earth). But in most instances the correction relies on finding points in the original picture whose actual locations are known. From these coordinates, a transformation matrix is calculated that defines the correction. As shown in Figure 2.19, this can stretch, rotate, and compress portions of the image so that the dimensions and shapes of the features are adjusted to correspond to a perpendicular view. However, shadows, and any structures that extend in front of or behind the plane of adjustment, will still have incorrect angles and dimensions.

Another issue arises when distortion is removed from an image, or when an image is enlarged, rotated, and so forth, in order to line up with another image for comparison. It becomes necessary to interpolate pixel values. The transformation matrix defines the location in the original image

FIGURE 2.19
Correcting perspective distortion: (a) original image; (b) after adjustment. The windows are now all the same size, but note that one side of the building is visible.

that corresponds to each pixel in the new image. These locations rarely fall exactly on one of the original pixels. Instead, they lie "between" pixel addresses. This requires selecting a method to derive the values for the new pixels (Schumacher, 1992).

The simplest method is to select the nearest original pixel to the location specified by the transformation matrix, which is equivalent to rounding off the real number address values that are calculated and just using the resulting integers to select a source pixel. This preserves the brightness values exactly but produces a result in which lines and edges appear stair-stepped. Visually better results are obtained by calculating a new value using several of the original pixel values. The interpolation may use just the four pixels that surround the fractional address point (bilinear interpolation) or may include ones farther away (bicubic interpolation, a common choice, uses 16 neighbor pixels). While this creates a visually smooth result, it also blurs the image slightly and can alter the brightness values and shapes of objects. Figure 2.20 compares these three methods.

Blur Removal

In some cases it is possible to remove or reduce the amount of blur in a captured image due to out-of-focus optics or motion of the camera (or subject). The results of this deconvolution are never as good as capturing a photograph without the blur, but sometimes it is necessary to deal with an imperfect photo. The key to removing the blur is the point spread function (PSF) of the picture. This is an image showing how a single isolated point would be blurred into an out-of-focus disk or a line corresponding to motion. It is assumed that the same PSF applies to every point in the image, which means the technique does not apply to images in which objects at different distances are blurred by different amounts, or to complex motion such as a running man in which the motion vectors of the legs, arms, and body are different.

It is possible to record an image of a PSF in situations such as astronomy (an image of a single star) or microscopy (an image of a single fluorescent bead), and sometimes in surveillance images (an image that includes a small bright light), but it is important that the PSF image have very little noise. It is also practical in some instances to generate an approximate PSF as a mathematical function, such as a line representing motion, a disk corresponding to a camera aperture, or a Gaussian shape approximating general defocus blur.

The process of blur removal uses Fourier transforms. The transform of the blurred image is divided by that of the PSF, point by point (each point corresponding to a specific frequency and orientation). The values of the transforms are complex numbers, requiring complex division, and it is also necessary to avoid dividing by values close to zero, which would cause numeric overflow. When the image contains superimposed noise, the best method for dealing with the noise as well as avoiding overflow problems is a Wiener deconvolution, which adds a constant to the divisor that corresponds

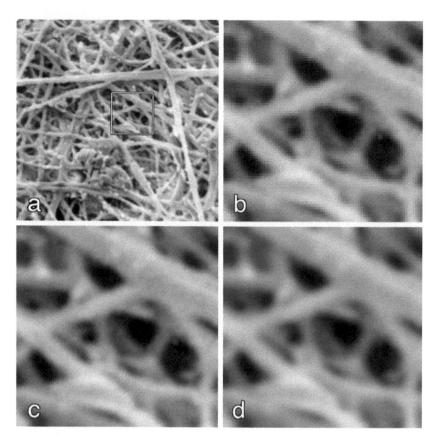

FIGURE 2.20
Image enlargement and pixel interpolation: (a) original SEM image of egg shell membrane with a square marking the area enlarged (courtesy of JoAnna Tharrington, North Carolina State University, Food Science Department); (b) nearest neighbor; (c) bilinear interpolation; (d) bicubic interpolation.

to the noise level in the image. The result, as shown in Figure 2.21, can make it possible to recover useful information from otherwise difficult images.

Image Processing for Enhancement

Images contain a great deal of information, not all of which may be of interest for a particular purpose. In many instances, discarding or suppressing some of the information in the image may make it easier to see, isolate, or measure what remains. It is important to never add anything to an image, but it may be acceptable to suppress some of the information to reveal the

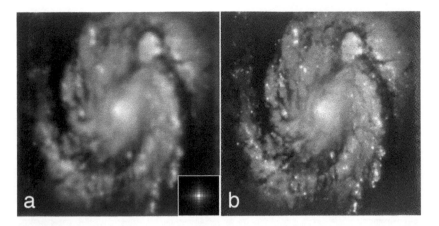

FIGURE 2.21
Deconvolution: (a) original image, with inset showing an enlarged image of a single star, used as the point spread function; (b) deconvolved result, showing increased noise but also improved resolution.

rest. That is the function of enhancement. An extension of this approach finds a particular class of information in an image, such as edges, or texture, and isolates this to facilitate measurement.

The most widely applied enhancement operation, known as an unsharp mask, produces visual sharpening of edges, steps, and fine detail. The method and the name derive from a technique long used in the photographic darkroom. The computer implementation subtracts a copy of the image, which has been blurred to remove the details from the original. As shown in Figure 2.22, this difference consists of just those details, which are then added back to the original image. The result maintains an average contrast between regions but increases the local contrast selectively wherever there is an edge, step, or fine line.

The unsharp mask is an example of a "high-pass" filter. The name derives from the Fourier transform methods described earlier. Edges are fine detail in an image and are represented in the Fourier transform by high frequencies; keeping these frequencies with a filter that "passes" them (while reducing the amplitude of low frequencies that correspond to more gradual changes in brightness) emphasizes the edges and detail in the image. In contrast, a "low-pass" filter keeps the low frequencies, reduces the amplitude of the high frequencies, and results in a blurred image (the Gaussian blur is an example).

Frequency filters can be applied in either the frequency domain, by operating on the values in the Fourier transform, or directly in the spatial domain, by arithmetic operations on the pixel brightness values. For color images, it is important to operate only on the brightness and not on the individual RGB channels. This can be facilitated by transforming the color representation from RGB to another color space such as HSI (hue, saturation, intensity) or

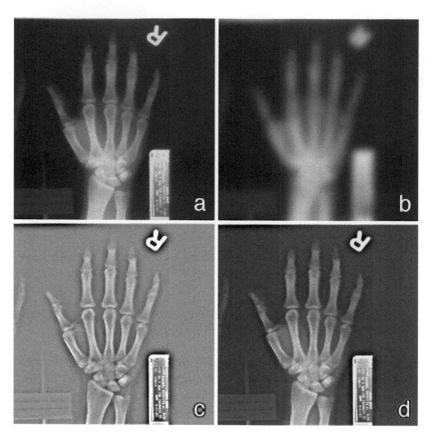

FIGURE 2.22
The unsharp mask: (a) original X-ray image; (b) Gaussian blurred copy; (c) subtracting (b) from (a); (d) adding the difference to the original.

Lab (luminance, and two channels representing variations in red-to-green and yellow-to-blue values). The computer handles the lossless transformations between these various color spaces as needed.

Edges

High-pass filters, such as the unsharp mask, increase the visibility of edges and detail by increasing the amount of brightness change. A different approach increases the sharpness of steps and edges by testing each pixel against its neighbors and reassigning a value that is the average of the neighbors the pixel is most similar to. Figure 2.6 shows an example of the application of this method for sharpening a step in brightness. The result of the Kuwahara maximum likelihood filter is less pleasing visually, because the edges of features are unnaturally abrupt and the pixelation of the image is

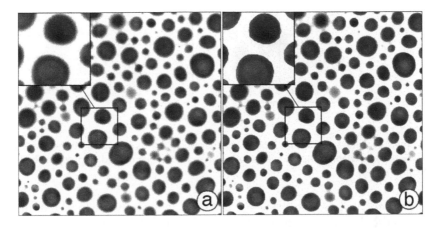

FIGURE 2.23
Maximum likelihood sharpening: (a) original image of bubbles in a foam, showing poorly defined boundaries due to light penetration into the material; (b) result of the Kuwahara maximum likelihood filter.

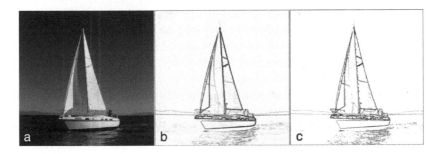

FIGURE 2.24
Edge delineation: (a) original image; (b) Sobel filter; (c) Canny filter.

more obvious, but it can improve subsequent measurement procedures since boundaries are better defined, as shown in Figure 2.23.

Highlighting just the edges and lines in an image makes it appear much like a cartoon or sketch, and can also be useful as a step toward measurement. The most widely used edge delineating function is the Sobel filter, which calculates the brightness gradient at each point in the image and assigns the strength of that gradient to the pixel. As shown in Figure 2.24, the resulting lines show the locations of edges, steps, and lines, whereas interior regions within structures have much smaller gradients of brightness and so are suppressed.

The lines from the Sobel filter are typically several pixels wide, and for measurement purposes it is desirable to have a line that is just one pixel wide, at the most probable location for the edge. This can be accomplished with the Canny filter, a more complex routine that begins by determining the

−1	−2	−1
0	0	0
+1	+2	+1

−1	0	+1
−2	0	+2
−1	0	+1

FIGURE 2.25
Two kernels of values that are multiplied by each pixel and its local neighbors to calculate the vertical and horizontal derivatives of brightness that are combined to produce the Sobel gradient.

local gradient, just as with the Sobel, but then keeps only the local maximum values in the direction of the gradient, as shown in the figure.

The Sobel filter functions by applying two convolutions to the image, and combining the result. Convolutions are most easily understood by considering a small kernel of numbers, as shown in Figure 2.25. Some convolution kernels are larger than the 3 × 3 arrays shown, and combine values from a larger neighborhood. However, the method remains the same. The numbers are multiplied by the values of each pixel and its local neighbors, the results added together, and the result (generally scaled by a predetermined multiplier and additive value to remain within the 0 to 255 range) becomes the new pixel value.

The same procedure is applied to every pixel in the original image, using the original pixel values to generate a new image. In the case of the Sobel, the two kernels produce vertical and horizontal derivatives, and the results are then combined as vectors (the square root of the sum of squares of the values, as shown in Equation 2.2) to determine the magnitude of the local gradient. They can also be used to determine the direction of the gradient vector, which is illustrated next.

$$Magnitude = \sqrt{\left(\frac{\partial B}{\partial x}\right)^2 + \left(\frac{\partial B}{\partial y}\right)^2}$$

$$Direction = \tan^{-1}\left(\frac{\frac{\partial B}{\partial y}}{\frac{\partial B}{\partial x}}\right)$$

(2.2)

Using just one of the kernels in Figure 2.25 calculates the difference between neighboring pixels on one side of each position and the other, giving the brightness gradient in a selected direction. For images in which the structures of interest have a known alignment this makes the detail more visually evident, as shown in Figure 2.26. Notice that the derivative has also eliminated the overall

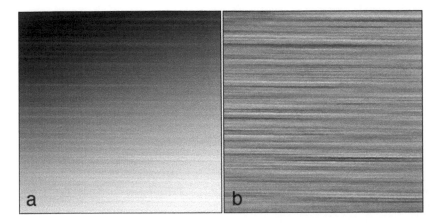

FIGURE 2.26
Derivative to enhance visibility: (a) original image of scratches on a metal surface; (b) vertical derivative filter applied perpendicular to the scratches.

shading from top to bottom in the image. Because the derivative has a low value in a direction parallel to any lines, edges, or structures present, it can be used to suppress unwanted or confusing background detail as shown in Figure 2.27.

Texture

In some images, the objects or structures of interest are not bounded by edge lines or defined by a different brightness (or color). Instead, the visually detectable difference is due to a different texture, as illustrated by the fabric patches in Figure 2.28 and Figure 2.29. In these examples, one of the textiles is characterized by a "coarser" appearance, meaning that adjacent and nearby pixels vary more in brightness. Another patch is identical in spatial characteristics to the surroundings, but has a different orientation in the weave pattern. These differences can be used to isolate the patches by applying various filters. Some of these use kernels like those illustrated, whereas others calculate statistical values within a moving neighborhood.

Spatial texture differences occur in many applications, ranging from aerial photography of agricultural scenes (different planting patterns distinguish different crops) to microscopy of food products (the curds in cheese are smoother than the surrounding matrix). The most common tools for converting the texture to a brightness difference that can be used to threshold the regions for measurement calculate statistical properties of the pixels in a small neighborhood and assign the result to the location of the central pixel. Figure 2.28 shows examples using the variance and the local entropy. Since all of these operations are performed on a local neighborhood several pixels in radius, the exact position of the boundary is correspondingly imprecise.

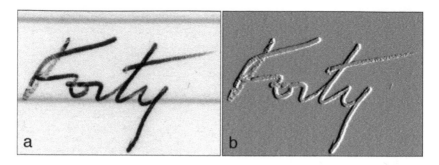

FIGURE 2.27
Derivative to suppress background: (a) original image of writing on lined paper; (b) horizontal derivative filter applied parallel to the lines.

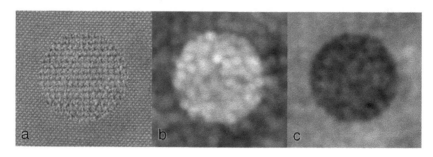

FIGURE 2.28
Texture filters: (a) original image of textile fabrics; (b) variance; (c) entropy.

Texture orientation is present in the examination of muscle tissue, cellulose fibers used in papermaking, and wear patterns on surfaces. It can be revealed by the orientation angle of the brightness gradient vector (whose magnitude is used earlier in the Sobel edge finding filter). Assigning the angle (calculated using Equation 2.2 and scaled to the 0 to 255 range) to pixel brightness values results in different brightness levels as shown in Figure 2.29. Since the vector points perpendicular to the fiber orientation, but may point in either of two directions, the histogram of this image shows values from 180 to 360 degrees that repeat those from 0 to 180 degrees. Reassigning the values to cover just the 0- to 180-degree range produces an image that clearly distinguishes the two regions. The histogram of the resulting image can be used to obtain a quantitative measure of the area occupied by fibers at each angle.

In many instances it is convenient to assign the angles calculated for the gradient to colors. The variation of hue from red through yellow, green, cyan, blue, and magenta then back to red provides a visual indicator of orientation as illustrated in Figure 2.30. In this example, the image of the fibers is used twice: once to obtain the direction of the Sobel gradient and once to suppress to black all pixels that are not part of a fiber. The latter step is needed

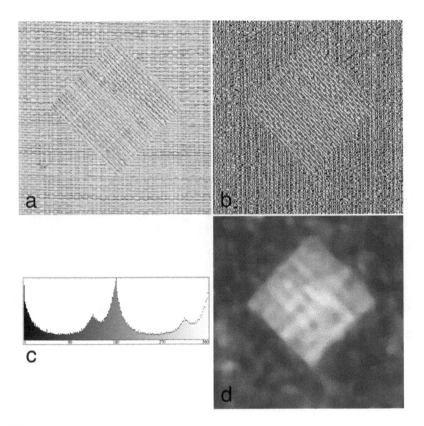

FIGURE 2.29
Texture directionality filter: (a) original image; (b) angle of brightness gradient vector assigned to grayscale values; (c) histogram of (b); (d) reassigning values from 0 to 180 degrees.

since there is a direction angle determined at every pixel location, but in the background area between fibers the values are random due to the presence of finite noise in the original image. Analysis of the directions indicates that the fiber orientations are not uniformly distributed.

Cross-Correlation

Human vision is usually good at picking out specific objects in complex scenes, although camouflage techniques can be used to confuse vision and hide things. Computer-based image analysis often employs cross-correlation to achieve a similar result. The technique can be performed either in the spatial domain of the pixels or in the frequency domain using Fourier transforms. The choice generally is based the size of the target and may use specialized hardware for speed. Small targets are often handled in the spatial domain, whereas large ones are more efficiently implemented with Fourier transforms.

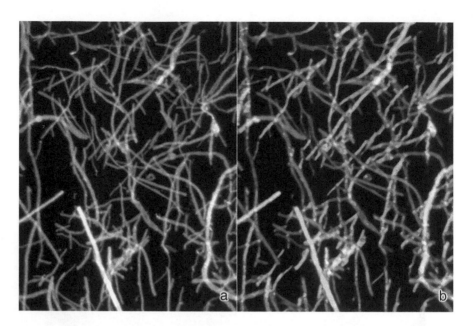

FIGURE 2.30
(See color insert.) Image of cellulose fibers used in papermaking, with a colored representation in which the hue corresponds to the gradient direction (perpendicular to the fiber orientation) at each point along a fiber.

The method can be understood by considering an image of the target on a transparent film being slid across the scene to locate matching patterns of brightness or color. Multiplying the values at each pixel in the target and the scene and summing up the totals provides a measure of the similarity. Locating points with a high similarity score identifies places where the target may be found. In practice, the target is rotated by 180° and the scores are normalized based on the brightness values in the image and target as shown in Equation 2.3. In the summations (j, k) are the coordinates within the target T and (x, y) are the coordinates of pixels in the image I. The method can be extremely fast, and is used in machine vision and quality control applications to locate objects, such as components on circuit boards, holes in machined parts, and so on, in milliseconds.

$$\frac{\sum_{j,k} I_{x+j,y+k} \cdot T_{j,k}}{\sqrt{\sum_{j,k} I_{x+j,y+k}^2} \cdot \sqrt{\sum_{j,k} T_{j,k}^2}} \tag{2.3}$$

Figure 2.31 shows an example. The text consists of letters in a Times Roman font, rotated to emphasize that these are shapes. Using the letter m as

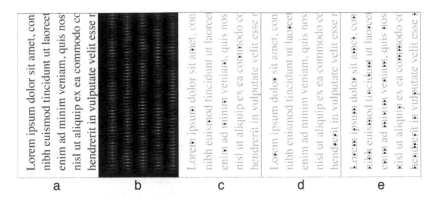

FIGURE 2.31
Cross-correlation to locate specific shapes: (a) original test image; (b) cross-correlation result using the letter *m* as a target; (c) result of thresholding to locate maximum points, which are shown dilated and superimposed on the original for reference; (d) cross-correlation result using the letter *r* as a target, after thresholding just the brightest points; (e) cross-correlation result for the letter *r* with a lower threshold, showing matches to parts of shapes.

a target produces a cross-correlation image with values for every letter position (Figure 2.31b), but thresholding this for just the brightest (maximum) values locates all of the occurrences of the target letter (Figure 2.31c). The letter *r* presents a more complicated situation. Again, thresholding the cross-correlation image for just the very brightest points locates the occurrences of the target letter (Figure 2.31d). But there are parts of many other letters that also match the shape of the *r*, and thresholding the image at a slightly lower level locates them as well (Figure 2.31e).

Cross-correlation depends on the specific shape and brightness pattern in the target. Searching the image with a different size letter, or one from a different font, or one that is rotated at a different angle does not succeed. For tasks such as aerial surveillance to locate military targets, a series of images of each possible object of interest, rotated at several different orientations, may be used. The speed of the operation still makes this practical, and by selecting a threshold that does not require a perfect match it is often possible to detect objects in spite of some attempts at camouflage.

Thresholding and Binary Images

In order to measure objects to determine size, shape, position, or brightness, or even simply to count them, it is necessary to isolate them from their surroundings. The most common way to accomplish this is by thresholding—selecting a range of brightness or color values that represent the objects and clearing all of the other pixels to some background color, usually white but sometimes black or transparent. The pixels corresponding to the objects may either be set to a contrasting color or left unchanged. When the thresholded image consists of just two possible pixel values, such as black for objects and white for background, it is called a binary image. The thresholding operation may be performed manually or using automatic algorithms based on the image histogram. Some methods also take into account the values of neighboring pixels or use independent knowledge such as the permissible size range of objects.

Of course, any of the image processing steps to level the illumination, reduce noise, or perform enhancements as described in the preceding section may be required to modify the pixel values so that the brightness or color values of the objects are different from those of their surroundings. If this is successful, the histogram will show an isolated peak or range of values for the objects that can be selected to perform the thresholding. In most cases, the separation is not perfect and some pixels will be misclassified. Methods for dealing with those cases are detailed next.

Figure 2.32 illustrates a straightforward situation in which the thresholding is successful. The rice grains are brighter than the background (they were dispersed on a flatbed scanner and the lid was left open). The resulting thresholded image shows the grains as black and the background as white.

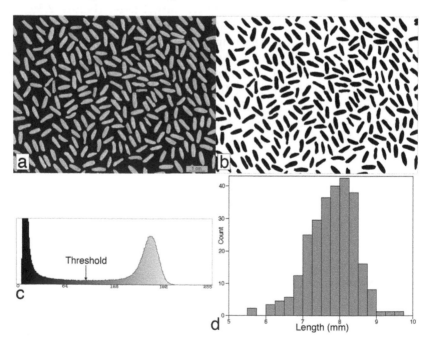

FIGURE 2.32
Brightness thresholding: (a) original image of rice grains; (b) binary image delineating grains; (c) histogram of (a) showing threshold value used to generate (b); (d) measurement results for the length of grains.

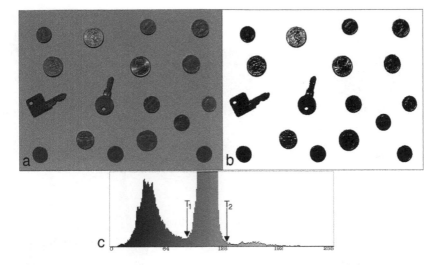

FIGURE 2.33
Thresholding the background: (a) original image; (b) after thresholding and reversing the contrast; (c) histogram of (a) showing the threshold settings on either side of the peak corresponding to the gray background.

Measurement of the grains can then be performed, as described in the next section, to determine their lengths.

In some cases the background is more uniform in brightness than the objects. Figure 2.33 shows a case in which the coins and keys contain both bright and dark pixels, but the background is a nearly uniform gray. Selecting the background and then inverting the image contrast to exchange black and white produces a binary image of the objects. There are isolated pixels within the coins and along the edges that have gray values indistinguishable from the background, which results in small irregularities and voids. Methods for filling these gaps are shown next.

For color images, selecting a range of color values that identify the object(s) of interest can be accomplished in any of the various color coordinate systems. Usually hue–saturation–intensity or Lab coordinates are more useful than red–green–blue. Setting limits on a graphical interface for establishing the ranges interactively can be challenging. Fortunately, it is not always necessary if just one or two coordinate axes are sufficient.

Figure 2.34 shows a photo of a sunflower with sky as the background. The brightness of the flower petals is very similar to that of the sky, so thresholding based on the pixel brightness values is not successful. The histograms show the pixel values in the L (luminance) channel and the b channel of the Lab color space representation. The L channel does not distinguish the flower from the sky, but the b channel, which is the yellow-to-blue axis, does separate the two and allows thresholding to produce a binary image, as shown.

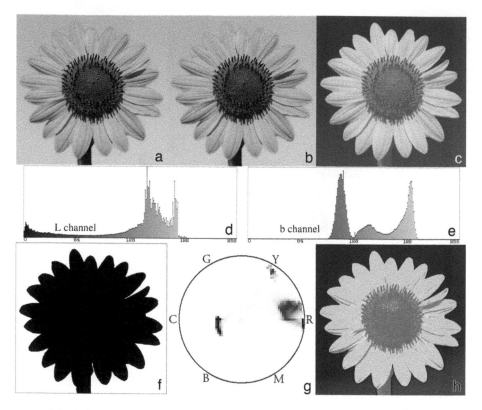

FIGURE 2.34
(See color insert.) Thresholding a color image: (a) original; (b) L channel; (c) b channel; (d) histogram of the L-channel image; (e) histogram of the b-channel image; (f) binary image obtained by thresholding the b channel; (g) histogram of the original showing clusters of pixels with values plotted as a function of hue (angle) and saturation (radius); (h) principal components analysis applied to the original color image, with values along each axis displayed in the red, green, and blue channels.

The color values can also be plotted as a two-dimensional hue-saturation histogram, as shown in Figure 2.34g. The clusters of pixels are shown with values plotted as a function of hue and saturation. The hue values are shown as an angle on the color wheel (red–yellow–green–cyan–blue–magenta–red), and the saturation as the radius. In this color space the sky pixels are partially saturated blue or cyan, while the flower pixels are highly saturated reds and yellows. Thresholding in this color space also produces a successful binary image.

In addition to the standard color space representations, it is possible to compute a set of principal components axes that are rotated in three dimensions to fit the actual pixel values of the image. Unique to each image, these axes provide the optimum contrast and hence the maximum separation for the various structures that are present. Figure 2.34h shows the result for the sunflower image, with the values along each of the three orthogonal principal axes displayed as red, green, and blue intensities.

Automatic Threshold Setting

In order to avoid human bias in delineating objects and to facilitate automation of image analysis, it is desirable to automatically set threshold values. Many of the algorithms were developed for the purpose of discriminating printed characters from paper, as an initial step in the process of converting a scanned image of a printed page to a text file suitable for word processing.

Most of these algorithms assume that there are just two kinds of pixels present—paper or ink—and that any threshold value will divide the pixels into populations that can be compared using standard statistical tests. By varying the threshold value and performing the comparison, the setting that gives the highest probability that the two populations are different is then selected as the threshold, as illustrated in Figure 2.35. Some of the statistical tests make further assumptions about the populations, for instance, that they have a Gaussian distribution of brightness values.

More elaborate algorithms use conditional or fuzzy logic to classify the probability that each pixel represents objects or background. Some include assumptions about the shape of object boundaries (e.g., smooth or irregular). It is possible by using the histogram in a subregion that is moved across the image to apply a threshold that varies according to local conditions. In general, the more that is known about the nature of the image and the objects, the more constraints can be placed on the thresholding process and the better the results become.

Figure 2.35 shows one example of an automatic thresholding algorithm. It uses a Student's *t*-test to calculate the probability that two groups of pixels are statistically different and chooses the setting that gives the greatest likelihood. Furthermore, it excludes from the calculation those pixels that have a high value of the brightness gradient (calculated using the Sobel filter), which eliminates from the calculation pixels that lie on or near edges and may have brightness values that fall between the two principal values.

A second method, illustrated in Figure 2.36, finds the threshold setting that gives the smoothest boundaries around the objects. This assumption is justified by the independent knowledge that the objects in this image are droplets of oil in mayonnaise, and that surface tension should produce smooth boundaries. Such an assumption would not be applicable in other circumstances, such as images of crushed powders, which typically have rough and irregular boundaries.

Morphological Processing I: Erosion and Dilation

Figure 2.37 shows an enlarged portion of the thresholded binary image from Figure 2.33. Due to random speckle noise in the original image and

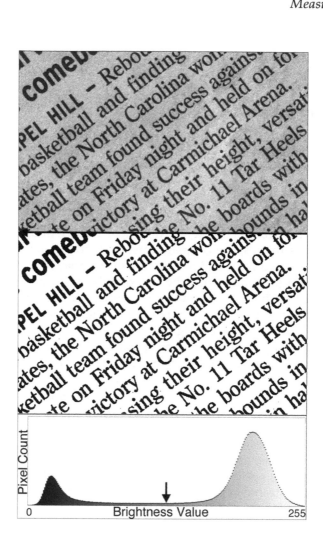

FIGURE 2.35
Thresholding printing. The original (top) is scanned from newsprint. The automatic threshold setting that produces the result shown is marked on the brightness histogram at the bottom.

reflections from surfaces of the objects, there are several defects present that affect the accuracy of the representation, and would alter measurements of size and shape. The usual approach to these defects applies the morphological operations of erosion and dilation. Note that the name "morphological" for this class of techniques implies that shapes, and particularly the shape of boundaries, are altered. There is a risk that the incorrect use of these methods can alter the shapes in ways that degrade subsequent measurements, especially those of shape.

Morphological operations applied to binary images either remove (erosion) or add (dilation) pixels at the boundaries between black and white pixels. In the simplest or classical version of these operations, erosion changes

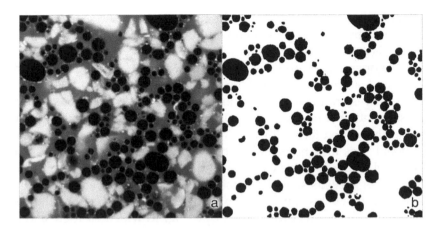

FIGURE 2.36
Automatic thresholding based on prior knowledge: (a) original image of oil droplets in mayonnaise; (b) thresholded binary.

to white any black pixel that touches a white pixel (where touching includes either sharing a common edge or sharing a common corner). Dilation does the opposite, changing to black any white pixel that touches a black one. These are shown in the figure. (As for the similar operations described earlier on continuous tone grayscale images, which replace each pixel with the brightest or darkest neighbor, the terms erosion and dilation may be interchanged depending on whether features and background are considered to be black on white or the reverse.)

Both erosion and dilation change the size or objects or structures, as well as the shape of the boundary. Consequently, they are usually used in

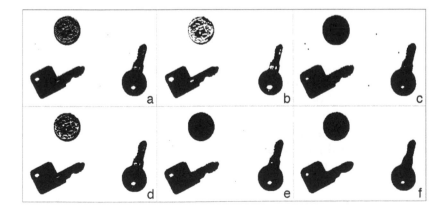

FIGURE 2.37
Morphological erosion and dilation: (a) original binary image; (b) erosion of the original; (c) dilation of the original; (d) opening (erosion + dilation); (e) closing (dilation + erosion); (f) multiple iterations as described in the text.

combination so that sizes are approximately restored. The sequence of erosion followed by dilation is called an opening and that of dilation followed by erosion is called a closing. As shown in the figure, these produce different results. An opening removes isolated pixels, whereas a closing fills in small gaps within features. Both smooth boundary irregularities.

In some cases, multiple iterations are used (i.e., several erosions followed by the same number of dilations or vice versa). This is equivalent to using a larger neighborhood to test for the presence of pixels of the opposite color. Conditional tests may not simply look for the presence of a single neighboring pixel of the opposite color but rather count them and proceed only if the number exceeds some threshold. Both closing and opening may be required to remove stray pixels and to fill holes. Figure 2.37f shows the result of applying a closing followed by an opening, in which each erosion and dilation applied two iterations with a conditional test requiring more than two of the eight neighboring pixels to be of the opposite color. The result has a very clean appearance, but the outline of the keys has been smoothed so that important details of the tooth positions are lost.

Morphological Processing II: Outlines, Holes, and Skeletons

The outlines of objects or structures in binary images can be delineated using the same logic as in erosion, by keeping only those black pixels that touch a white neighbor. Outlines are particularly useful for some measurement purposes (e.g., perimeter length and boundary shape) and also for determining adjacency (when different objects touch).

Holes within features that are larger than can easily be filled by dilation or closing (or whose filling by that method would be accompanied by other undesirable changes to the object, such as removing details from the boundary) can be filled by locating all groups of background pixels that do not connect to the boundaries of the image and including them in the features. In some cases additional tests on the size and shape of the hole features may be used to control which ones are filled and which are left unchanged.

Figure 2.38 illustrates the application of filling holes. The original image of droplets has well defined borders for the features of interest, but after leveling the background contrast and thresholding as shown in Figure 2.13, the centers are not filled in. Filling the holes and then applying a watershed operation (described later) produces an image with separated features for counting or measurement.

A special form of conditional erosion can be applied to objects or structures in binary images to reduce them to skeletons. The process iteratively removes feature pixels with any background neighbor, except that pixels cannot be removed if doing so would separate the object into multiple parts. The final result is the midlines of features, pixels that lie on the centers of inscribed circles, and are equidistant from the edges, as illustrated in Figure 2.39.

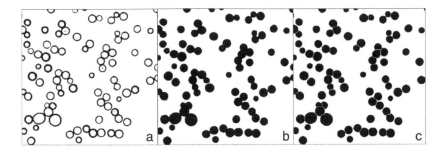

FIGURE 2.38
Filling holes: (a) the binary image obtained by thresholding the leveled image from Figure 2.13, showing just the bubble borders; (b) after filling holes; (c) the individual bubbles separated with a watershed operation.

The skeleton is used for measurements such as the length of extended objects such as fibers and especially for characterizing some aspects of shape. In particular, the number of end points (skeleton pixels with exactly one neighbor) and branch points or nodes (skeleton pixels with more than two neighbors) are readily identified and counted to describe the topology of an object.

The order of connections between nodes and branches, supplemented in some cases either by the length of the branches or radius of the inscribed circle at each node, or by adding the locations of corners—locations with large local values of curvature along a branch—can be combined into a graphical format for use in syntactical analysis (Chapter 3) for the comparison and classification of some types of shapes.

FIGURE 2.39
Several shapes (shown in gray) with their superimposed skeletons.

The Euclidean Distance Map and Watershed Segmentation

The erosion and dilation illustrated earlier are directionally biased. On a grid of square pixels, the neighbors that share a side are closer together than ones that touch at their corners, so that adding or removing pixels in the 45° directions covers a 41% greater distance than in the 90° directions. When erosion and dilation are applied with neighborhood sizes of more than a few pixels in radius (or iterated multiple times), this bias can cause undesirable distortion in object shapes.

There is a different approach to morphological operations that instead constructs a map (called the Euclidean distance map or EDM) of the Pythagorean distance from each pixel within an object or structure to the nearest background (white) pixel. Several algorithms exist that can construct the EDM very rapidly, with floating point rather than integer precision. Erosion can then be performed to any desired distance simply by selecting all pixels with EDM values above a specified level. Constructing the EDM for the background pixels makes dilation possible in the same way.

Performing erosion and dilation using the EDM is preferable to the classical methods because it is faster and more isotropic, and produces results based on actual distances from the original boundaries. Figure 2.40 compares the results of the EDM method to those from classical morphological erosion and dilation (which changes the pixel from white to black, or vice versa, if any neighboring pixel is of the opposite color), and using conditional thresholds requiring more than one, two, or three neighboring pixels to be of the opposite color.

When applying erosion and dilation (or their combinations of opening and closing) to an image using the EDM, it is possible to set the threshold at distance values that are not restricted to integers. In the example of Figure 2.41, the gaps within features are filled but important details are not removed from the boundaries by using a closing with a distance of 1.5 pixels.

Another use of the EDM is for separating touching features. In the image of Figure 2.36, the thresholded binary image of the approximately spherical droplets of oil shows some of them touching or slightly overlapped (due to the thickness of the section represented by the transmission image). In

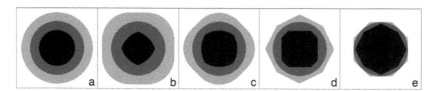

FIGURE 2.40
Comparison of erosion (black) and dilation (light gray) applied to a circle (medium gray). From left to right: (a) by thresholding the EDM; (b) classical morphology testing against any neighboring pixel; (c) conditional threshold of 1 (requiring more than one neighboring pixel of the opposite color); (d) conditional threshold of 2; (e) conditional threshold of 3.

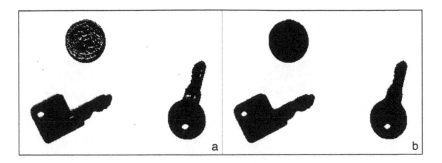

FIGURE 2.41
Morphology using the Euclidean distance map: (a) original (from Figure 2.37); (b) result of a closing with a distance of 1.5 pixels.

order to properly count or measure them, lines of separation are needed. If the values in the EDM are visualized as the elevation of a surface, each convex feature becomes a mountain peak. Where two features touch, there is a watershed line where the two mountains touch or intersect. This is the line where rainfall, running downhill on the imaginary surface, meets from two different mountain peaks. Removing the pixels along the watershed lines separates the features as shown in Figure 2.42. (Figure 2.38c also shows the application of a watershed after thresholding.) The technique is not restricted to ideally round features, but does require them to be sufficiently convex and regular that there is only a single peak in the EDM.

Boolean Combinations

In addition to the various morphological operations described earlier, another class of procedures applied to binary images creates Boolean combinations. If multiple binary images representing objects or structures in the same region exist, then they may be combined to isolate the objects of interest using the logical rules AND, OR, and ExOR. Using the convention that black pixels represent objects and white ones are background, the results of these basic Boolean operations are shown in Table 2.2. The operation is performed pixel by pixel and the order of the images does not matter. The functions may be combined with the logical NOT, which simply reverses all pixels from black to white, and vice versa.

These operations are often applied to aerial or satellite imagery recorded in multiple wavelengths, usually covering both the visible and the infrared. Thresholding these and performing Boolean combinations makes it possible, for example, to select features that are bright in either one (OR) or to select features that are bright in both (AND), as illustrated in Figure 2.43. Another common operation is to compare two images to detect changes or motion (ExOR).

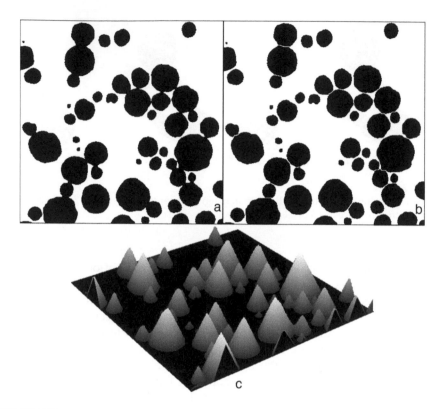

FIGURE 2.42
Watershed segmentation: (a) detail of binary image from Figure 2.36; (b) after watershed segmentation; (c) the EDM of (a) represented as an elevation map, showing mountains for features.

Figure 2.44 shows an example in which the same original image is processed to derive two different binary results, one highlighting the textured areas and the second highlighting the dark region. In both thresholded binary images, a morphological closing was applied to fill in small gaps. By combining the two binary images using the logical relationship "Textured and Not Dark," just the region of interest can be selected.

TABLE 2.2

Basic Boolean Logical Operators

Image 1	Image 2	AND	OR	ExOR
■	■	■	■	□
■	□	□	■	■
□	□	□	□	□

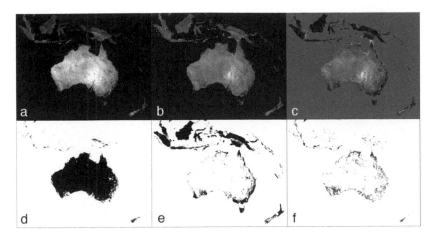

FIGURE 2.43
Satellite images of Australia: (a) red, (b) green, and (c) blue visible light images; (d) thresholded bright pixels in the red channel; (e) thresholded dark pixels in the blue channel; (f) AND combination of (d) and (e).

Encoding Boundary Information

As described in the Introduction, one category of shape measurements relies on the boundary of a feature for analysis. There are techniques that extract the feature's boundary and store it in a more compact form for analysis (Figure 2.45). The full pixel map allows each pixel to have a range of gray scale or color values associated with it, but the various boundary representations are concerned only with the periphery.

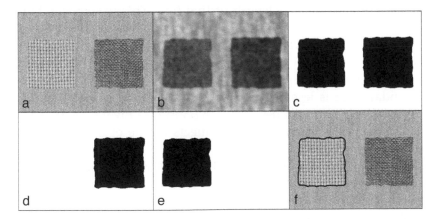

FIGURE 2.44
Boolean combination: (a) original image; (b) processed to convert texture to brightness; (c) thresholded binary from (b); (d) original image thresholded to select dark region; (e) Boolean combination "(c) AND NOT (d)"; (f) outline of region in (e) superimposed on the original image showing the "textured and not dark" patch.

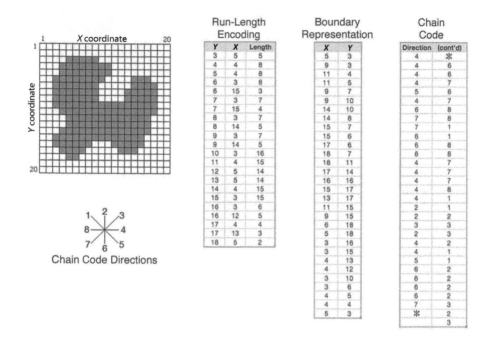

FIGURE 2.45
Alternative representations of a feature in a pixel array: the full pixel map, run-length encoding, boundary representation, and chain code.

Run-length encoding stores the leftmost pixel address of each line in the feature and then the length of that line. This is compact, but has little application in the analysis of object boundaries (it is the basis for fax transmission, however, and the sum of the lengths gives the area). Boundary representation is a list of the (x, y) coordinates of the boundary pixels in order along the periphery; this contains full information about the boundary but is less useful than chain code representation. Chain code is a series of links connecting each boundary pixel to its next neighbor along the periphery; each link connects in one of the eight possible directions (45-degree steps).

In principle, chain code can be started at any location on the periphery and the sequence of links can be rotated to shift to any other starting position. The most common convention is to start at the upper-left corner (the pixel encountered first in going through the image in "reading order"). Another choice is to rotate the digits in the code to the obtain the numeric representation with the smallest value. Also note that the numerical codes associated with each direction are arbitrary, and may use the values 0...7 rather than 1...8, depending on the implementation.

Measurement

Measurements of objects or structures in a thresholded image (collectively called features in the following sections) must begin by considering the nature of pixels. Enlarging a digitized image on the computer screen shows the pixels as little squares, each containing a uniform brightness or color. In a digital camera, the transistors that detect the incident light do so by collecting photons over an area. So for some purposes it is convenient to treat pixels as little squares. Measuring the area of a feature is most often done, for example, by counting the pixels that comprise it and multiplying by the pixel area.

But this approach becomes a bit troublesome when using the pixels to define the boundary of the feature. For example, if the sides of the little squares are counted to determine the perimeter of a feature, the result is the same for each of the features shown in Figure 2.46a, which is quite wrong. Another way of looking at pixels is to treat them as points and to define the boundary as a series of links in the chain connecting their centers, as indicated in Figure 2.46b. That produces a more sensible description of the boundary, and by counting the 90-degree and 45-degree links separately (the latter are 1.414 times as long as the former) provides a simple (and often used) measure of the perimeter length. But the chain encompasses a different, smaller area than the value obtained by counting pixels.

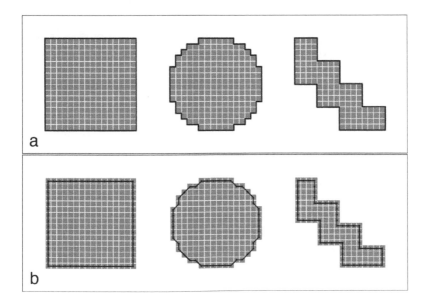

FIGURE 2.46
Different pixel interpretations: (a) as squares; (b) as points. Both the perimeter and the area values are affected.

Arriving at a consistent set of measurements first requires deciding whether pixels are points or little squares. There are other possible interpretations as well, which subdivide each pixel into a series of smaller squares and interpolate a smoothed "super-resolution" boundary that treats pixels along straight edges differently from those at corners or ends. These can produce more consistent results, for example, giving the same dimensions for a feature as for the hole measured by connecting the pixels that neighbor the feature.

Most image measurement programs are inconsistent in their treatment, using different interpretations for different purposes, and this has consequences. For example, one common shape measurement shown in Chapter 3 combines the area with the square of the perimeter. If the area is measured by counting squares, the result is larger than the interior of the chain code. On the other hand, the length of the perimeter obtained by summing the individual 90-degree or 45-degree links overestimates the length of lines depending on orientation. This causes a bias in the resulting values. For some purposes it may not matter if the same bias is present in a series of measurements, but the amount of bias depends on the feature size, shape, and orientation, and ideally should be avoided. Most of the procedures used in the sections that follow use the point and chain interpretation for perimeter. The area, however, is calculated as the sum of the pixels; a few other exceptions are noted.

Counting

One of the first things that is often done with images of objects is simply to count them. This requires as a first step labeling the pixels with a feature identifier, so that all connected pixels have the same label regardless of the intricacy of the shape. Pixels may be considered to be connected and part of the same feature if they touch either along a common edge or at a common corner. This is called 8-connectedness because each pixel has 8 touching neighbor positions. An alternative interpretation, called 4-connectedness, considers pixels to touch only if they share a common edge.

Most image measurements use 8-connectedness for features. By this convention, the left-hand collection of pixels in Figure 2.47 is a single feature, topologically in the form of an *O*, rather than four separate ones, and the right-hand pixels form a continuous line. But the same convention cannot be used for the background, because in that case the interior of the *O* is connected to the outside, and the "line" does not separate the regions on either side. So, if 8-connectedness is used to define features, then 4-connectedness rules must be used for the background, and vice versa.

Once labeling has been accomplished, counting is simple—unless the image is a sample of a larger field of view and some of the objects to be counted extend beyond the boundaries of the image. The correct method for dealing with that case is to count features that intersect two edges, for example, the right side

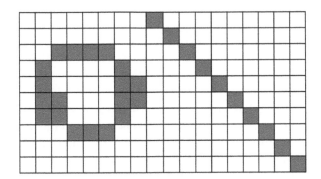

FIGURE 2.47
Pixels demonstrating the consequences of 8- and 4-connectedness as discussed in the text.

and bottom, and not to count ones that intersect the other two (left edge and top). This works because the features that are not counted in one image would be counted if additional images were taken to cover the entire scene.

When measurement of features is performed, features that intersect any edge of the image cannot be measured since there is no information about what lies outside the image area. Consequently, either a smaller region within the image may be used for measurement, leaving a guard frame that is wide enough to prevent features that lie partially within the counting frame from intersecting the edge of the image, or a statistical correction may be made. This is done by measuring every feature that does not intersect the edge of the image, and assigning an effective or adjusted count to compensate for the probability that other, similar features would have been rejected because they intersect an edge. The calculation is based on the feature size, F, and the image size, W, as shown in Figure 2.48 and Equation 2.4.

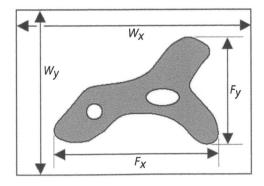

FIGURE 2.48
The adjusted count is calculated for each feature based on its projected dimensions, F, and the size of the image, W.

$$Adjusted\ Count = \frac{W_x \cdot W_y}{\left(W_x - F_x\right) \cdot \left(W_y - F_y\right)} \tag{2.4}$$

Measuring Size

If pixels are interpreted as little squares, the area of a feature is determined by counting the connected thresholded pixels and applying a calibration factor based on the size (or spacing) of the pixels. If instead the pixels are treated as points, the area inside a polygon with n vertices (the number of pixels that constitute the chain code) is calculated from the coordinates as

$$Area = \tfrac{1}{2} \sum_{k=0}^{n-1} \left(x_k \cdot y_{k+1} - x_{k+1} \cdot y_k\right) \tag{2.5}$$

As described earlier, the perimeter may be calculated directly from the chain code by counting the links between edge-sharing pixels and those between corner-sharing pixels, multiplying the latter by the square root of 2 and adding. If a bounding polygon with a reduced number of vertices is used (as discussed later), the perimeter is

$$Perimeter = \sum_{k=0}^{n-1} \sqrt{\left(x_{k+1} - x_k\right)^2 + \left(y_{k+1} - y_k\right)^2} \tag{2.6}$$

Perimeter is often a problematic measurement because the defining pixels along the boundary of the feature are most sensitive to any changes in thresholding or to subsequent morphological processing. Additionally, for many natural objects it is observed that the length of the perimeter depends on the image magnification. (The variation of the perimeter length with scale is one of the shape descriptors described in Chapter 3.)

The maximum caliper dimension (also called the maximum Feret's diameter or maximum projected length) can be estimated with good precision by using rotated axes and finding the maximum and minimum coordinates as a function of angle. The calculation need only be performed for the chain-code pixels along the periphery of the feature, and the rotated coordinate values are

$$x' = x \cdot \cos\theta + y \cdot \sin\theta$$
$$y' = y \cdot \cos\theta - x \cdot \sin\theta \tag{2.7}$$

This calculation can be performed for a series of rotation angles to determine the greatest difference between maximum and minimum. If steps of 10 degrees are used, the greatest potential error is the cosine of 5°, or 0.996. In that case, an object with an actual maximum dimension of 250 pixels would be

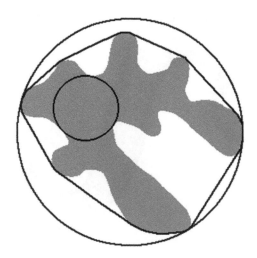

FIGURE 2.49
An arbitrarily shaped feature (gray) with the superimposed outlines of the minimum circum-scribed circle, bounding convex polygon, and maximum inscribed circle.

estimated to have a maximum dimension of 249 pixels. Considering the inher-ent limitations in accuracy of a pixel image, that is often an acceptable error.

If the coordinates of the extreme pixels at each rotation angle are saved, they become the vertices of a bounding polygon, also called the convex hull or taut-string perimeter of the feature (Figure 2.49). The area and perimeter of this region can also be determined using Equation 2.5 and Equation 2.6, and combined with the actual feature area to calculate several of the shape descriptors in Chapter 3. Some other size measures that are sometimes used are the diameter of the smallest bounding circle or the diameter of the larg-est inscribed circle. Both may be used to calculate shape descriptors and are illustrated in Figure 2.49.

The minimum caliper dimension ("width") is not so easy to determine. The smallest projected dimension from the series of rotations is not useful, because the error from even a small misalignment can be large. Instead, the skeleton (which selects pixels along the feature midline) can be combined with the Euclidean distance map (which measures the distance of those pix-els from the nearest point in the background). Figure 2.50 illustrates this, with a few of the inscribed circles superimposed. Either the maximum or the mean of the EDM values along the skeleton may be used as a measure of the width of a feature, and the standard deviation can be used as a measure of the variation. This is also the basis of the medial axis transform (MAT) discussed in Chapter 4.

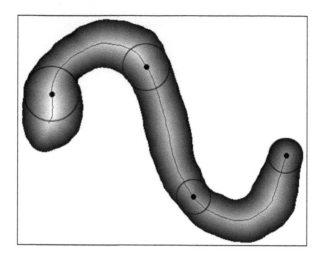

FIGURE 2.50
An irregularly shaped feature with its EDM and the skeleton superimposed, as discussed in the text.

Measuring Location

The positions of objects in a scene can be defined in several different ways. The most commonly used is the centroid, the point at which the binary feature would balance if cut from a uniform sheet of cardboard. For features in which the original brightness information has been retained in the thresholding process and converted to a density value, the density-weighted centroid may be calculated instead. This is used, for example, in determining the location of spots in two-dimensional gel electrophoresis, in which each spot's position corresponds to the characteristics of the protein molecules and the integrated density measures the amount present.

The centroid (or weighted centroid) is determined by summing the (x, y) coordinates of the pixels. Two other possible position markers are the centers of the circumscribed and inscribed circles shown in Figure 2.49. These require somewhat more calculation. The circumscribed circle is fit using the vertices of the bounding polygon, and for irregularly shaped features is insensitive to the orientation. The inscribed circle is centered at the maximum value of the Euclidean distance map for the pixels in the feature. It is the only one of these location points that is guaranteed to lie within the feature. For a nonconvex shape like that in Figure 2.49 the centroid and the center of the circumscribed circle may lie outside the feature's boundaries.

For some applications, such as the gel example, or for the analysis of aerial or satellite images used to map locations for a geographical information system, the absolute coordinates of objects are important. For many other applications, it is instead the relative location of features within the image.

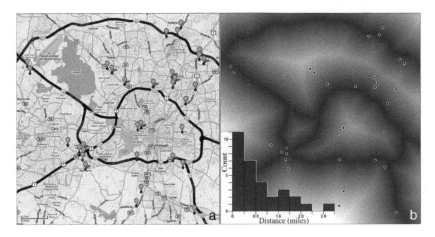

FIGURE 2.51
Measuring distances from a line: (a) Google® map of Raleigh, North Carolina, with limited-access highways and shopping centers marked; (b) shopping center locations overlaid on the EDM of the area around the highways, with a graph showing the number of shopping centers as a function of distance from a highway.

Examples include the distance of organelles within a cell from the cell membrane or the proximity of shopping centers to major highways. These distances can be determined by using the location of the object (based on any of the criteria) to sample the value of the EDM of the background space around the membrane or highway, as shown in Figure 2.51.

It is also interesting in many cases to look at the distances between like objects in an image. If they are randomly distributed, the mean value of the distance from each object to its nearest neighbor is given by the statistics of a Poisson distribution as

$$Mean = \frac{1}{2}\sqrt{\frac{Area}{Number}} \tag{2.8}$$

A measured mean value greater than this indicates a self-avoiding distribution, like cacti growing in the desert or students seated in a classroom. A measured value less than the Poisson mean indicates clustering, like stars in galaxies or students at a party. Figure 2.52 shows that the ratio of the actual value to the Poisson mean can be used as a measurement of the distribution of lipid droplets in a custard.

Measuring Density

As noted earlier, digital cameras are not spectrophotometers and cannot measure color in the sense of a plot of intensity versus wavelength. With good standards and corrections, the images can be adjusted to visually

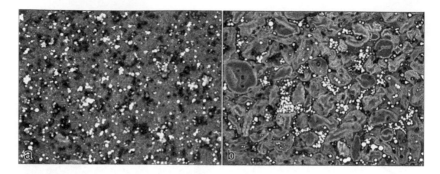

FIGURE 2.52
Clustering of lipid droplets (small bright circles) in custards: (a) well distributed–slightly self avoiding (ratio = 1.048); (b) poorly distributed–clustering (ratio = 0.576).

match colors in different scenes or under somewhat different lighting conditions. In most cases of image analysis, variations in color are most useful for distinguishing the various structures or objects present, as described in the section on thresholding.

Grayscale brightness can be utilized for measurements. The gel electrophoresis example mentioned earlier is one example. Optical density is calculated from the fraction of incident intensity that passes through an object or film. Grayscale brightness is similarly related to density in X-ray images and tomographic reconstructions, and brightness is related to elemental composition in some scanning electron microscope images.

$$OD = -\log_{10}\left(\frac{Transmitted}{Incident}\right) \tag{2.9}$$

A digital camera with 12 bits of precision (one part in 4000) can be used to measure optical density values of 3 or more, but a scanner with 14 bits of precision (one part in 16,000) can read values greater than 4, as is needed for the X-ray films used in medical applications. When using the optical density to determine the mass density of an object or region, it is important to convert the brightness of each pixel to density, and then sum or average those values over the area of the feature, because of the nonlinear relationship between brightness and density.

Measuring Shape

Shape, of course, is the topic of this book. The introduction points out that shape is a complex and sometimes subtle concept, and because there are multiple definitions, there are many ways to perform the measurements. Both the definitions and measurement procedures are deferred to the following chapters. Some of the measurements depend upon the measures of size, position, and density shown earlier. Many use other procedures, but still begin with the pixel-based delineation of the structure or object.

Two-dimensional shapes (for example, characters or symbols printed on a page, or the silhouettes of objects dispersed on a surface) present challenges for measurement, characterization, and recognition. The challenges are much greater for three-dimensional objects. Even when the full 3D object is available, for instance when represented as a voxel array or a set of surfaces, measurement is apt to be limited by the precision of the representation (for example, the voxels are typically much larger in size than the resolution achieved in 2D pixel images). But in many cases the difficulty is further compounded by the availability of just one or a few 2D images that provide limited information about the 3D object.

Sections and Projections

The most common types of 2D images of 3D objects are either projections (views of the exterior of the objects from one direction) or sections (intersections of the object with a plane). Either of these may be randomized or not. For example, both engineering drawings of manufactured objects and the section images of the human body obtained with computer-assisted tomography are generally restricted to views along three principal, orthogonal axes. But in microscopy of virus particles in the electron microscope, many identical particles are oriented randomly on a support film to produce multiple views that can be combined to produce a 3D model. Microscope sections cut through biological tissue produce random 2D cross-sections that do not show the full 3D size or shape of cells or organelles, but in many cases information from multiple random cross-sections can be interpreted to gain useful shape and size information.

The key point is that while both individual projection and section images provide only limited information about 3D objects, the combination of information from multiple images may be able to generate the desired result. If the orientations of the projection views, or the placement of the section planes can be controlled, the results may be obtained more efficiently than with random images. However, enough random images may be used to arrive at the same correct conclusions. In both cases, the key to success is the proper interpretation of the data, and learning to ask the right questions.

For example, projected images can be used to measure size in a number of ways, the most common of which is a maximum or mean caliper dimension. This is sometimes called a diameter, but as pointed out in the Introduction the terms *radius* and *diameter* imply a center point from which the radial distance is measured or through which the diameter passes. That point is not determined from the projected images of shapes other than spheres.

The maximum and mean caliper dimensions of an object can be robustly determined by making a series of measurements in different orientations and may be correlated with some properties of objects (such as the size of sieve openings through which particles can pass). But many measures of size, and most of shape, are not revealed by these images. The surface area of the object, and any shape descriptors that are related to that surface, are not necessarily visible in the projections. Indentations on the surface will generally be hidden in the projected outline, and the roughness or irregularity of the surface will be underestimated.

If the image of the 3D object is not simply a projected silhouette but shows the illuminated surface facing the camera, and if enough is known about the lighting, it may be possible to recover the local slope and/or elevation point by point. There are several techniques, discussed in Chapter 5, that are typically applied to surfaces that are approximately flat, to produce "two-and-a-half D" data sets in which the elevation is stored as a function of (x, y) location. These are used to measure the roughness of man-made surfaces as well as the topography of the earth. The techniques are basically the same regardless of scale.

For fully 3D objects, recovering the complete 3D coordinates and finding efficient ways to represent the data are usually accomplished by tomographic reconstruction, or by imaging a series of closely spaced parallel planes and combining the 2D images into a 3D voxel array.

Stereology and Geometric Probability

Examination of thin slices through complex structures is routinely used in microscopy to study the internal structure of materials (ceramics, metals, polymers) and tissue samples of plants and animals. Either the slices are very thin so that light (or electrons) can penetrate through and produce a transmission image, or the surfaces produced by the cutting and polishing are flat and are examined by reflection of incident light (or secondary electrons or ions). The resulting images show the intersection of the structure with the plane of sectioning.

Relating measurements made on the 2D image to the 3D structures that must have been present to produce the visible intersections is based on geometric probability, and the field that applies the specific rules that have been derived is called stereology (from the Greek for studying or describing a solid, i.e., three dimensions, not to be confused with stereoscopy, which is observing three dimensions using two eyes or points of view). Although most commonly applied in microscopy, the geometric principles hold at any scale and can be applied to geological sections revealed in quarries and road cuts, or to astronomical views of galaxies, and so forth.

Generally speaking, stereological relationships apply to global properties of structures or to statistical descriptions of a population of objects. As shown in Figure 2.53, passing a section plane through a randomly distributed set of

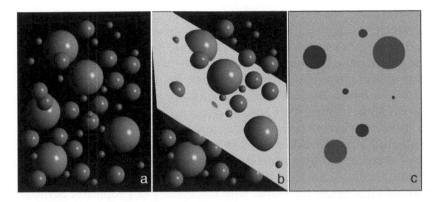

FIGURE 2.53
Intersecting spheres with a plane: (a) a three-dimensional volume containing three sizes of spheres; (b) a planar section through the volume; (c) the image of the section, in which neither the sizes nor the relative numbers of the spheres are directly shown.

spheres of various sizes does not intersect most of them at their maximum diameter. The resulting image shows circles, but the diameters of the circles are not those of the spheres. Furthermore, because the probability that the plane will intersect a sphere is proportional to its diameter, the number of intersections with the smaller spheres is proportionately less than the number of intersections with the larger ones, so even the observed number of each size class of spheres is biased.

Methods for "unfolding" the distribution of intersections as a function of diameter to recover the size distribution of the spheres that must have been present in order to produce the observed result have been known and used for more than 80 years. The advent of desktop calculators in the 1960s made the calculations practical and the method gained in popularity. The method was extended to cover shapes other than spheres, and tables of coefficients were derived and published (Wasen & Warren, 1990; Wasen et al., 1996a, b, c).

But two serious problems arise:

1. The unfolding procedure is mathematically ill-conditioned. Because the calculation involves many subtractions to correct for the number of small circles that are produced by intersections with larger spheres, the precision of the derived size distribution is much poorer than the precision of the original distribution of circle sizes. This effect becomes even more pronounced for shapes that are less regular and symmetrical. In practice, this means that a very large number of individual intersections must be measured and that the sampling must be uniformly random.

2. The coefficients used for the unfolding make the important assumptions that the 3D shape is known and that all of the objects have the same shape.

Unfortunately, the second problem turns out to be an extremely limiting requirement. Except in a few cases, such as spherical bubbles that are controlled by surface tension, the shapes of natural objects are not perfectly uniform or consistent. In many cases, the shapes vary with size, and at best can be only approximated by a regular geometrical form that can be looked up in a book of tables. Because of the ill-conditioned mathematics, even slight shape variations can produce large variations in the unfolded results.

As these restrictions became better appreciated, a different approach to stereological interpretation gained favor. Emphasis shifted toward describing global structural properties such as the volume, surface area, or length of structures. These are robustly determined by straightforward techniques, and are often strongly correlated with the history or properties of the structures. Also, much attention has been devoted in the last 30 years to efficient and unbiased sampling methods (Baddeley & Vedel Jensen, 2005; Howard & Reed, 2005).

Volume

Volume is usually measured as a fraction, that is, the percentage of an organism, cell, rock, or loaf of bread occupied by some recognizable structure, such as bone, nucleus, mineral, or pore, respectively. For images that represent random samples of the structure, the volume fraction estimate can be obtained in several ways, as illustrated in Figure 2.54. The oldest method sums the areas of the dark regions and divides by the area of the image. In the pre-digital-imaging era (indeed, in the pre-photograph era), this was done by tracing the image onto paper, cutting, and weighing. Now it is conveniently determined from the histogram, which counts the pixels with values that correspond to the structure of interest. In the example, this is 36,931/160,000 = 0.231.

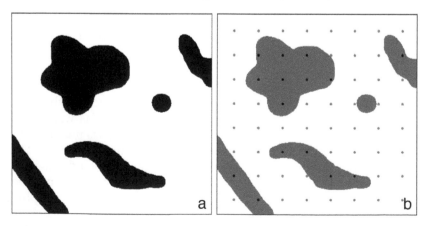

FIGURE 2.54
Measuring volume fraction: (a) the dark structure(s) cover 23.1% of the image area; (b) 15 of the 64 grid points (which have been dilated for visibility) lie on the dark regions.

For many years, still before digital imaging, a useful device for performing this measurement consisted of a motorized microscope stage with a turns counter and a second counter that allowed a human to depress a key to engage it when the structure of interest (usually called a "phase" although not necessarily a phase in the chemical sense) was passing under crosshairs in the eyepiece. The ratio of the two numbers gave the result. About 60 years ago, a simpler method, still implemented manually, came into use. A grid of points in the eyepiece enabled very rapid counting of the number of points that happened to lie on the structure. In the example (Figure 2.54b), a regularly spaced grid of 64 points has been superimposed on the image (eyepiece reticles typically had no more than a 5×5 grid of points for rapid visual counting). Since 15 of the points lie on the dark structure, the volume fraction is estimated as $\frac{15}{64} = 0.234$.

These two results are deceptively close. The precision of a counting experiment is given by the square root of the number of "events" or "hits," which in this case is 15 ± 3.87, meaning that the volume fraction estimate is 0.2343 ± 0.060%. This assumes that the events are independent measures, which for the image shown would require a much sparser grid so that two points rarely fall into the same structural region, but the intent of the illustration is to show the method. In order to achieve a relative measurement precision of 5%, for example, multiple fields of view would be examined and counted until the total number of events reached 400 (the square root of 400 is 20, or 5%). With the grid shown, this should be accomplished with approximately 30 fields of view.

With a grid having fewer points, obviously more fields of view are required, but this is considered a benefit of the point count method, since it forces examination of multiple locations in the sample. With most real structures, there is no single "average" location and examining multiple images makes the results more representative. Knowing how many fields of view must be examined allows designing an experiment to perform uniform sampling, whether it is a matter of looking at all parts of an organ, selecting multiple animals from a population, or polishing locations at the top, center, and bottom of a steel ingot.

The advantage of point counting is that the measurement precision is known. For the area fraction measurement it is difficult to assign a value to the precision, because it depends very much on the shape of the regions, the number of pixels that lie along the periphery, and the confidence with which the regions have been delineated by the processing steps described earlier. With the manual counting method, the usual convention for points that appear to lie on the boundary and cannot be confidently determined to lie inside or outside the region is to count them as one-half.

Surface Area and Length

Surface area can also be measured in more than one way. Unlike the volume fraction, which is a dimensionless ratio, surface area requires knowing the image magnification or scale. The illustration in Figure 2.55 includes a scale

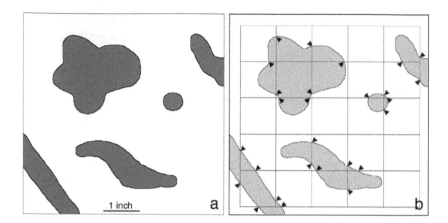

FIGURE 2.55
Measuring surface area: (a) the total length of the boundary line is 23.48 inches, in an image area of 30.864 square inches; (b) the grid lines have a total length of 60.0 inches and intersect the boundaries 27 times.

bar. Measuring the total length of the boundary lines and the total area of the image allows calculating the surface area per unit volume as

$$S_V = \frac{4}{\pi} \cdot B_A \tag{2.10}$$

where the notation S_V represents surface area per unit volume, and B_A represents boundary length per unit area. The factor $(4/\pi)$ adjusts for the fact that the surfaces intersect the section plane at a variety of angles. Notice that the units are consistent: length^{-1}. The result for the image shown is 0.969 square inches per cubic inch.

It is also possible to use a grid to perform a counting experiment. In Figure 2.55b a grid of lines has been superimposed, and the intersections with the boundaries of the structure are counted. In this case the calculation is

$$S_V = 2 \cdot P_L \tag{2.11}$$

where P_L is the number of points of intersection per length of the grid lines. The factor 2 adjusts for the various incidence angles between the grid lines and the boundaries, and again the units are length^{-1}. For the example, the result is 0.90 square inches per cubic inch. Since this is based on a count, the precision is determined by the square root of the number of counts, giving 0.90 ± 0.17 square inches per cubic inch. Once again, the grid is too fine for the structure shown and oversamples the image but is used to illustrate the method.

One of the important considerations in measuring surface area is the need for isotropic sampling. In some cases this can be achieved by random orientation of the specimen and random sampling. Another approach, called "vertical sectioning," uses a controlled set of section planes that are rotated uniformly about a known direction, coupled with a grid consisting of cycloidal arcs that exactly compensate for the directional bias introduced in the sectioning. The calculation using the number of intersections and the length of the grid lines is unchanged.

The length of linear structures is determined by counting the number of intersections with the section plane. The calculation of length per unit volume is

$$L_V = 2 \cdot P_A \qquad (2.12)$$

where P_A is the number of points of intersection per area of image, and the factor 2 adjusts for the variation in angle between the structure and the section plane. As for surface area, it is necessary to perform isotropic sampling, which means that the section planes must be cut at uniformly random angles.

There are a variety of specialized techniques that are applied to carry out the procedures described here, most of which improve the efficiency of the measurements by ordered sampling procedures. Much of modern stereology is directed toward appropriate sampling methods. The generation of various point or line grids, and counting of the events by using a Boolean AND with the structure or its outline, is straightforward and supported in many computer programs.

There are also so-called second-order stereological routines that combine more than one grid and sampling strategy to gain more information about the structure. For example, the combination of a point grid to sample objects with a line grid that measures the distance from the points to the boundary of the object can be used to determine the standard deviation of the size distribution of objects without making any assumptions about their shape.

Topology

The measurement procedures described give the total volume, surface area, or length of a structure occupying a three-dimensional volume, based on the two-dimensional images captured from section planes. Notice, however, that there is no information generated about whether the structure consists of multiple separate islands or one continuous region, whether it is branched, and so forth. These are topological rather than metric properties, and cannot be determined from individual section planes.

The most efficient method for estimating the number of discrete objects and the connectivity of a network requires comparing two section planes that are parallel and closely enough spaced that no elements of the structure can "hide" between them, in other words that the characteristic dimension of the structure is greater than the plane spacing. This "disector" method

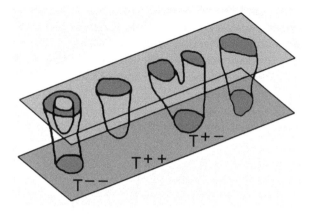

FIGURE 2.56
Counting events with the disector. Features like the one at the right, which pass through both section planes, are ignored. Differences between the two planes are counted.

can be performed either manually (visually) or using software with digital images. It requires counting specific types of events, as shown in Figure 2.56.

The disector can be implemented either by examining two aligned section planes cut through a specimen, or nondestructively using optical sectioning in the confocal microscope, or parallel sections imaged by tomographic methods. In any case, the counting identifies three kinds of events that may be detected between the sections, ignoring any feature that appears in both sections, even if it has changed shape or size.

One event of interest is the end of a feature (it appears in one section but not in the other). This is called a positive tangent event after the surface curvature that occurs at the end, and is marked in the figure as T⁺⁺. Usually rare, the end of a void within a feature is a negative tangent event and is marked as T⁻⁻. For branching networks or nonconvex objects, there are mixed tangent events where saddle curvature occurs, marked T⁺⁻. The number of these events and the volume of the sample examined (the image area times the distance between the section planes) are combined to calculate a topological property:

$$N_V - C_V = \frac{T^{++} + T^{--} - T^{+-}}{2 \cdot Volume} \tag{2.13}$$

where N_V is the number of discrete objects per unit volume, and C_V is the connectivity, the number of redundant connection paths between locations. In many structures of interest, either C_V is zero (the structure consists of separate, unconnected objects) and N_V can be calculated regardless of the objects' shapes or N_V is 1 and the structure consists of an extended, branched structure (such as the pore network in oil-bearing sandstone) whose connectivity can be calculated.

Voxel Arrays

The calculations shown provide a statistical characterization of global structure, but do little to describe the shape of individual objects. That most often requires constructing an entire voxel array, either with parallel sections or tomographic reconstruction, so that the connected voxels comprising an object can be used to carry out measurements.

Most serial section techniques—whether performed nondestructively by optical sectioning or slice tomography, or by sequential polishing or cutting to produce a series of images—generate a set of parallel 2D images that are separated in the third dimension by a distance that is not equal to the pixel size within the plane. In some cases, such as the optical confocal microscope, the resolution in the z direction is only slightly worse than in each (x, y) plane.

In many situations the difference is much greater. The ion microscope has (x, y) resolution of the order of micrometers (μm) and depth resolution of nanometers (nm) as individual atomic layers are removed. Conversely, serial sectioning of opaque materials by ion beam milling coupled with scanning electron microscopy of the revealed surfaces produces 2D images with a resolution of nanometers, separated by distances 10 to 100 times larger. In these instances, the voxels are not cubic but either elongated or shortened in the z direction. This greatly complicates measurements, particularly measurements of shape.

It is possible to use tomographic reconstruction to produce cubic voxel arrays with uniform resolution in all directions. A variety of signals may be used, including magnetic resonance imaging (MRI), X-rays, neutrons, electrons, and even acoustic waves. For example, seismic waves from earthquakes are used to measure the internal structure of the earth. The reliance on natural events restricts the directions in which signal propagation can be achieved, but nonetheless has produced significant mapping of faults in the mantle and the shape of the core.

Three-dimensional tomographic reconstruction requires information generated by the passage of signals in many directions through each point in the specimen. One approach to this is to randomly disperse a set of presumably identical objects so that many images can be acquired in different orientations. This is done, for example, to image virus particles in the transmission electron microscope (TEM), as shown in Figure 2.57. Specifying the orientation of each image requires human interpretation of key landmarks, but the result enables the creation of a voxel array that can be used to render the surface geometry of the virus.

For single objects, the limited ability to tilt specimens in the TEM restricts the orientations that can be imaged to a cone and produces reconstructions that have poorer resolution in some directions. Most microtomographic instrumentation uses a helical scan in which the object of interest (a few millimeters in size) is rotated in a screwlike motion while a cone-shaped

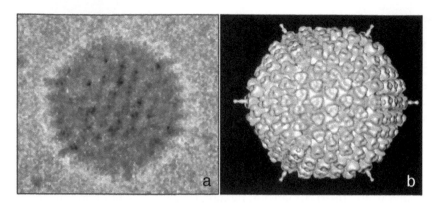

FIGURE 2.57

TEM tomography: (a) one representative transmission image of an adenovirus particle; (b) the reconstructed surface rendering produced by combining many individual projections to produce a voxel array.

beam of X-rays from a point source is used to generate projected images. This gives enough different paths through the specimen to produce good 3D reconstructions with resolution better than 1 μm, as shown in Figure 2.58. Using a synchrotron beam as a source, with specialized devices for orienting the sample in three dimensions, can produce even better results.

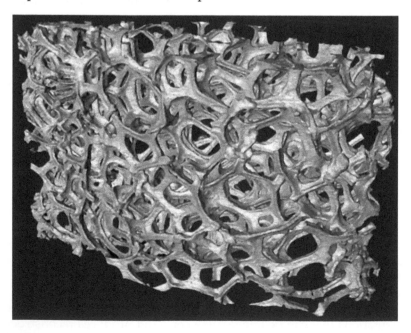

FIGURE 2.58

X-ray tomographic reconstruction of a foamed metal structure produced by helical scanning. (Image courtesy of Skyscan, Kontich, Belgium.)

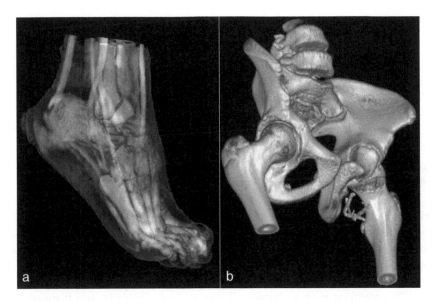

FIGURE 2.59
Viewing modes: (a) volumetric rendering (a human foot); (b) surface rendering (a human pelvis).

Medical imaging may be done using just a single set of parallel slices for reconstruction, but many instruments can generate slices in transaxial, coronal, and sagittal orientations, and these three orthogonal sets of slices may be combined to produce cubic voxel arrays suitable for measurements. However, most use of these data sets is limited to visual examination. The two principal modes of presentation are volumetric, in which ray tracing is used to produce a view through the array in which internal structures are visible, and surface-rendered, in which surfaces defined by an abrupt change in voxel property are shaded to show the appearance with a controlled light source.

Figure 2.59 shows examples of these modes, which are sometimes intermixed or combined. Software that allows interactive rotation, controlled transparency and colorization, and arbitrary sectioning through the array allows users to access details that would be difficult to detect in any other way. Chapter 5 further illustrates these presentation modes.

Three-Dimensional Measurements

Measurement algorithms from 3D voxel arrays are extensions of those used for 2D pixel arrays, but there are some additional complications that must be taken into account. Voxel arrays may be very large, so that they may not be held in active memory and require access from a disk, which slows things down. In addition, many measurements require identifying the connected voxels that belong to an object. In two dimensions, this requires deciding

FIGURE 2.60
Voxels that share a face (6), edge (12), or corner (8).

between 4-neighbor rules in which pixels are only considered to touch if they share a side, and 8-neighbor rules that also treat corner-sharing pixels as connected. In three dimensions, these become 6- (face sharing), 18- (face- or edge-sharing), and 26- (face-, edge-, or corner-sharing) neighbor rules (Figure 2.60). Following connections through a complicated three-dimensional 26-neighbor path to locate all parts of an object is much more difficult than the two-dimensional procedure, which can be accomplished in one scan through the image.

Volume is usually determined by counting voxels, and then multiplying by the voxel size or spacing. This is not sensitive to voxel shape, although the accuracy of the result may be affected if the voxel spacing or resolution varies with direction.

The maximum caliper dimension, determined for two-dimensional shapes by converting the coordinates of the boundary pixels to rotated coordinate systems and finding the maximum difference, can in principle be determined in three dimensions, but the number of rotated coordinate systems is greater, as is the effort to first identify all of the surface voxels. Noncubic voxels complicate the process further. Other measures of size, such as the radius of the minimum enclosing sphere or of the largest inscribed sphere, are also possible but are infrequently used.

Measurement of surface area is particularly troublesome. The chain-code method for determining perimeter cannot be extended to three dimensions for several reasons. First, there is no path from one surface voxel to the next that is guaranteed to reach all parts of the surface and return to the beginning. Second, if the surface is treated as a set of triangular facets whose corners rest at the centers of three neighboring voxels, there are different ways that the facets can be assigned to the points, and they do not in general produce the same result; both the orientation and area of the facets are different. And finally, the comparatively coarser size of voxels relative to the size of the object, as compared to the size of pixels in two-dimensional images, means that the definition of the surface is often poor and the precision of the measurement low. This is all in addition to the problem of determining whether the voxels should be interpreted as points or as solid boxes.

Faced with these difficulties, some programs report surface area as the sum of the voxel faces that constitute the outside of an object. This solves the facet issue but does so at the cost of giving every object a surface area at least as great as that of the bounding box that surrounds it, regardless of the actual extent or shape. The marching-cubes algorithm, described in Chapter 5 and principally intended for rendering surfaces for visual examination, can be used to obtain a better, but still imperfect, measure of surface area. The surface area of the convex hull of a 3D object can also be calculated as the sum of the individual facets.

Measurements of location are usually based on the centroid, which can be calculated simply by summing the voxel locations analogously to the two-dimensional case (or, if the voxel values have meaning, they can be included in the summations to determine a weighted centroid). In many cases, the voxel values can be calibrated. For X-ray tomography, for example, the voxel value represents an absorption coefficient, which may be related to density or to compositional variation. For confocal light microscopy, voxel values may represent fluorescence intensity and consequently can be associated with the concentration of a specific molecule. These simple calibrations ignore second order effects such as absorption of the signal along the path length, beam hardening of X-rays, and so on, but are usually adequate for the purposes intended.

Three-dimensional shape parameters are discussed in more detail in Chapter 5. But many of the approaches to shape measurement become more complicated in 3D. Even simple ratios of size parameters raise questions. In two dimensions, one measure (not the most commonly used one) is the ratio of the diameter of the circle with equivalent area to the longest caliper dimension. Since the volume and maximum caliper dimension can both be determined for 3D voxel objects, a corresponding 3D parameter can be determined with reasonable accuracy.

The most commonly used dimensionless ratio to describe shape in two dimensions is the formfactor, defined as $(4\pi \cdot Area)/Perimeter^2$. This is implemented (often with different names) in most image analysis programs, in spite of the problems associated with 2D perimeter measurement described earlier. Chapter 3 illustrates the application of formfactor and other similar dimensionless ratios, for shape characterization. The 3D analog to this is calculated from volume and surface area, but since the surface area has many more problems of definition, procedural algorithm, and resolution, the results are substantially poorer.

Other approaches to 2D shape, such as Fourier coefficients, moments, and skeletons, can be extended to 3D but with some restrictions. Spherical harmonics, for example, may require the radius from the centroid to be a single valued function. And there are two different types of skeletons that can be formed in 3D, one consisting of lines and one of surfaces. These issues are discussed in Chapter 5.

Short-Range Photogrammetry

A collection of techniques known collectively as photogrammetry can also provide shape information. These methods are generally applied to surfaces, and sometimes to just a few selected points on those surfaces, which must be viewable from the camera locations. Specific methods (considered in Chapter 5) such as stereoscopic viewing, surveying, shape from shading, photometric stereo, and structured lighting all come under this heading.

Stereoscopic viewing locates the same points in images taken from two known locations. Usually the viewing angle between these two views is in the 5- to 10-degree range (matching the vergence of human eyes at typical viewing distances). The disparity or sideways shift in the relative location of the points is proportional to the difference in their distances, and can be used to generate (x, y, z) coordinate values. The accuracy of the z values is poorer than the (x, y) values because of the small viewing angle, but in the more general photogrammetry or surveying application, arbitrary viewing positions or multiple views may be employed to improve the results. The method is used for applications ranging from surface metrology to aerial photography and mapmaking. Stereo cameras are used on the Mars rovers and are found on most depictions of robots (although not on many actual ones).

So-called photometric stereo requires only a single camera location, but acquires multiple images using a variety of known light source positions (usually dozens). The brightness of each point under each illumination condition is used to calculate the local surface orientation, and these are then used to generate a complete set of elevation values. The difficulties with this shape-from-shading approach include sensitivity to variations in surface reflectivity and the need to determine range or elevation by integrating the slope values from point to point, which magnifies uncertainties. A major application of the technique has been to archaeological and art objects in order to avoid any contacting measurements.

Structured lighting operates by shining light in a known pattern (lines, circles, or an array of points) onto a scene, most often a surface, and observing the pattern as it appears from another direction. Measuring the displacement of lines or points from the original pattern allows straightforward calculation of the range of the point on the surface as shown in Figure 2.61. In addition to performing noncontacting measurements of art and archaeological items, this technique is used to determine the shapes of fabricated metal parts and the flatness of milled lumber, as well as measuring the shape of human eyes before laser surgery.

FIGURE 2.61
Structured light: (a) object with projected lines of light, whose lateral displacement measures the elevation; (b) elevation map (with scale) produced by shifting the light pattern to cover all points on the surface; (c) computer model of the surface generated from the elevation values.

Computer Graphics, Modeling, Statistical Analysis, and More

The era is not long past when computer input and output was restricted to text. Although the goals of using images as inputs have not yet been accomplished except in a few specific cases (reading bar codes, recognizing iris scans, interpreting American Sign Language gestures, and so on), most modern computers are capable of displaying elaborate graphics as a way to communicate with users. Some of this is fairly trivial, such as the use of instantly recognizable icons to identify file types, rather than having to read a three- or four-character type descriptor in the file name. But some graphics achieve a richness and flexibility that are much greater. Examples certainly include the characters, objects, and scenes generated in computer games.

Figure 2.61c illustrates the ability of computers to generate images that model shapes and surfaces. Whether the application is engineering (e.g., translating mechanical drawings into finished parts) or entertainment (e.g., the films *Star Wars* and *Toy Story*), understanding of the rules of light propagation and reflection and the creation of a database that specifies the material properties has allowed computers to create images that can hardly be distinguished from actual photographs.

There are two somewhat different types of specifications that are used. One consists of a fully defined geometric model consisting of points and curves. Often, surfaces are created as tessellations of facets, each of which has properties and orientation that are used to calculate the appearance. Blending one facet with its neighbors to prevent abrupt changes may be done with a mathematical interpolation such as Phong rendering or a more complex model such as radiosity. Foley et al. (1996) provide deeper insight into this field. Some of the processing to create the displayed image may be offloaded from the computer to the display circuitry, which may contain

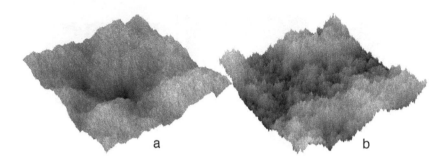

FIGURE 2.62
Generated random fractal surfaces, with the same amplitudes but different fractal dimensions:
(a) dimension = 2.18; (b) dimension = 2.36.

significant processing power for this purpose. Computer-aided design and interactive computer games both rely heavily on this type of rendering.

The second type of surface specification, which may be superimposed on the first, is a statistical description such as a fractal dimension. As explained in Chapter 5, many natural surfaces (and at a fine enough scale, many man-made ones) have roughness that conforms to this geometry. When multiple processes combine to create surfaces, and when large amounts of energy are dissipated locally in the generation, a hierarchy of scales results in which more detail becomes visible as magnification is increased, and the general appearance of the surface remains the same. This is the hallmark of fractal geometry.

Using statistical procedures to generate a surface or to model the scattering of light from a surface, it is possible to simulate a rough surface simply by assigning a fractal dimension, as shown in Figure 2.62. This type of modeling has been used to create scenery for many movies. Other statistical procedures allow a few numbers to be used with procedural algorithms to generate rich and complex graphics, such as fur or hair, and turbulent flows in liquids.

Computer creation of graphics is also very important as a data analysis tool. Humans find it much easier to interpret graphs of data than they do columns of numbers, and practically all data handling and statistical analysis programs, ranging from simple spreadsheets like Microsoft Excel® to programs like SAS JMP®, include routines to present the data and the analytical results in graphical form. Some of the graphs themselves become so complex that it takes specialized knowledge and familiarity to interpret them, whereas others such as regression plots relating one variable to another or histograms showing a distribution of values are ubiquitous. Subsequent chapters will use some of these graphics to summarize and present measurement and analysis results.

Of course, the computers are also invaluable at performing the statistical analysis of measured data. Humans generally find it difficult to deal with the interaction of multiple variables or to see through stochastic variations superimposed on measurements in order to extract the underlying trends. The mathematical techniques for doing so are well established, but too time consuming for manual implementation to be practical for many applications. Computers excel at this kind of analysis.

In the context of shape measurement and analysis, the following chapters will use computer tools for purposes such as

- calculating descriptive summary statistics for groups of objects;
- correlating measurements with history or property data;
- extracting rules to distinguish groups of objects based on multiple measurements;
- identifying efficient procedures for classification of new objects; and
- recognizing individual objects.

3

Two-Dimensional Measurements (Part 1)

Shapes in two-dimensional digitized images may be interpreted in a wide variety of ways. In some applications, they are projections of three-dimensional objects, whereas in others they may be sections through them. This latter case includes elevation profiles across surfaces, as can be provided by devices such as scanning profilometers. Such profiles (and the variety of devices used to obtain them) are particularly important in industrial quality control, where they are used to measure the flatness and roughness of surfaces, the roundness of objects, and so on. The interpretation of these surface profiles is considered in Chapter 5, since they generally arise in efforts to understand the shape of three-dimensional objects and their surfaces.

In this chapter and the next, silhouettes of objects (or in some instances sections through them) consist of groups of touching pixels, subsequently called features (not to be confused with the use of the word *feature* in some texts to mean a calculated combination of measurement values used in a statistical classification procedure). These groups of pixels are thresholded as a binary or black-on-white image, as described in Chapter 2, and are used to calculate various measures of shape.

There are two main classes of shape measurements: ones that attempt to summarize complex shapes in a few convenient but necessarily incomplete numerical values; and ones that contain more values and are sufficient to reconstruct the entire feature, but are generally more difficult for humans to interpret. This chapter deals with measurements of the first type, and the next chapter considers ones of the second. Several of the examples that compare different methods are presented in each chapter.

Template Matching and Optical Character Recognition (OCR)

Numerical values based on measurements are not the only ways to describe shape. When a name can be applied, for instance, when a feature can be recognized as belonging to a known class, that name carries with it a presumed description of shape. As pointed out in Chapter 1, a very human approach to considering shape is the catalog or field guide method, in which presumably representative images of the various types of objects of interest are arranged according to some ordering principles so that they can be efficiently searched

1 2 3 4 5 6 7 8 9 0

FIGURE 3.1
The OCR-A numbers used on checks.

to match an unknown object. That same approach can also be adapted to computer algorithms applied to features in digitized images.

The simplest version of this method attempts to match a shape against a set of templates that correspond to each of the known shapes. The numbers printed on the bottoms of checks are a familiar example of this technique. The only symbols needed are the numbers 0 through 9 (Figure 3.1) and a few special characters, but the OCR-A font used is complete, with a full uppercase and lowercase alphabet. The characters were designed (in 1968) to be readily distinguished by template matching, and have fixed size and uniform spacing to make locating each character as easy as possible.

Subsequent developments created fonts with a more "normal" appearance, and software has been developed that can read most standard fonts, both with and without serifs, and can adapt to different printing sizes. There are several commercial optical character recognition (OCR) software packages designed for use with scanners (primarily intended for conversion of printed documents to editable electronic files and widely used in business). The U.S. Postal Service uses OCR to sort mail, and another significant use of the technology is in reading machines for the blind (pioneered by Ray Kurzweil in 1976).

In addition to commercial software programs, there are several freely downloadable open-source OCR packages. The most widely used, and probably the most advanced, is Google's, which is based on the Tesseract OCR engine developed at Hewlett Packard in the 1990s. This software is used for document analysis, creation of electronic libraries, and even reading text in images captured by camera phones (for translation, recognition of locations, and so on). Like many of Google's offerings, it can be accessed and used online at http://docs.google.com.

Modern OCR software uses more complex algorithms than template matching, but understanding that is still a good place to begin. A simple approach to template matching is shown in Figure 3.2. Each template is placed over the letter, and a score value is determined as the difference between the number of pixels that match and those that do not. This may be normalized by dividing by the number of pixels in the template. Some systems assign different weights to various pixels in the template (as shown in Figure 3.3), representing the probability of a match, and sum the weights. The template with the highest score identifies the symbol. Templates may be created on the fly to match the size of the text and the pixel size in the image. The typical character template consists of fewer than 100 pixels.

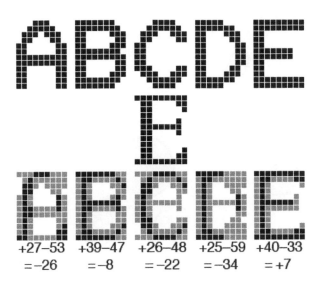

FIGURE 3.2
Matching each of the five-letter templates (top row) with a target letter (center) produces a net score (number of black matched pixels minus number of gray unmatched pixels) that is greatest for the correct match.

FIGURE 3.3
Templates whose pixels have different weights assigned, represented by the grayscale values.

Reading the tightly controlled characters on checks is essentially 100% reliable. Template matching has also been applied to other tasks with partial success. Russell et al. (2009) attempted to distinguish three different rat species based on their footprints. Image processing is used to threshold the prints and to fit ellipses to the marks made by the central pad and toes, as shown in Figure 3.4, which are then matched to templates for each species. They report a success rate of 70% for identifying the species of rat.

Syntactical Analysis

OCR software may also use the topology of letters. This can be derived from the skeleton, or by locating key points such as ends and corners and treating the linkages between those points as elastic, considering only their network

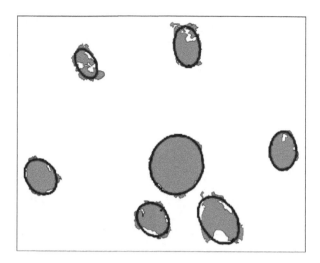

FIGURE 3.4
Footprint of a rat, with ellipses fit to the pad and toe prints that are matched against a template.
(From J. C. Russell et al., 2009, *Ecology* 90(7):2007–2013.)

order. These methods are often called "syntactical." Syntactical models (Fu, 1974, 1982; Pavlidis, 1977; Tou & Gonzalez, 1981; Schalkoff, 1991) break down the shape, often after first deriving the feature outline or skeleton, into the important parts with their relative positions recorded. This recording or graph can be likened to a set of characters that spell a word, and compared to other words on the basis of the number of missing, substituted, or rearranged letters.

For shape recognition, the requirements of syntactical analysis are that the number of primitive elements must be small, but able to be combined to completely or at least adequately represent the important characteristics of the objects. They should be easily segmented and individually recognizable, and preferably correspond to some natural structures present in the objects. As examples throughout this text show, syntactical elements are very useful for recognition of objects, but less so for measurement or for statistical description and comparison.

As an example, Figure 3.5 shows the key elements of the letter *A*. The two branch points at the sides, two end points at the bottom, and the corner or apex at the top identify this letter visually, and no other letter in the alphabet has the same key points nor the same arrangement of them. The "springs" in the figure indicate that the key points can move around quite a bit without changing their usefulness for identification, as long as their relative positions remain in the same topological order. Extracting these key points and their arrangement provides a very robust identifier of the letter, which is quite insensitive to size and font.

Figure 3.6 shows an example of the letters A through E, in different fonts, with the key identifying points of each (ends, branches, and corners). There is actually more information than needed to uniquely identify each character; some

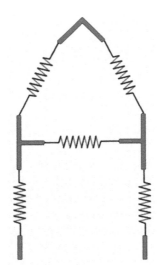

FIGURE 3.5
The important branch and corner points that identify the letter A.

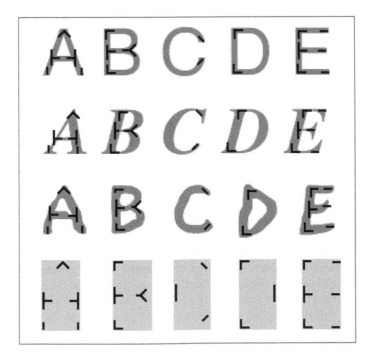

FIGURE 3.6
The letters A through E with their key branch, corner, and end points.

FIGURE 3.7
One CAPTCHA format with distorted and crowded characters.

of the points can be missing (e.g., due to extreme alterations in letter shape in handwritten characters) without compromising readability. Syntactical analysis finds many other uses. For example, characterization of the shape and density patterns in human chromosomes is routinely used in karyotyping.

The ability of human vision to use syntactical clues for recognition is used in many Web pages as a challenge-response test to ensure that a person is entering information rather than an automated "web-bot." Several slightly different "CAPTCHA" methods present distorted characters as shown in Figure 3.7, which can be deciphered by a person but not reliably by an image-processing algorithm.

Modern text recognition systems also improve their results by using context. For example, a shape initially identified as an 8 may be changed to a B when it is found in a string of letters and (since it is a capital letter) at the beginning of a word. It is worth noting that context is also important for human recognition of characters, as illustrated in Figure 3.8. Finally, most OCR software identifies words (sets of characters set off by spaces), which are then compared to a dictionary to find the most likely match. Of course, this requires dictionaries for each language. Typical accuracy rates for printed text exceed 99%; recognition of hand printing and cursive handwriting are still areas of active development. Rather than relying on template matching, much of this research is based on determining the topology of the characters from the skeleton, as discussed in Chapter 2.

Reading License Plates

Although OCR is a well-developed field, the task of reading license plates is much more complicated. Automatic license plate readers are now coming into widespread use, principally by police departments. Some are mounted at fixed locations to record the identity of cars entering or leaving an area. Others are mounted on the trunk lids or roofs of police cars, looking downward to automatically scan the plates of every car that passes by, with the intent of spotting cars reported stolen. This is a demanding application.

The first problem is to locate the plate on the back of the car. Different makes and models position the plate in various locations, and many cars also exhibit other text (such as bumper stickers) that can create confusion. License

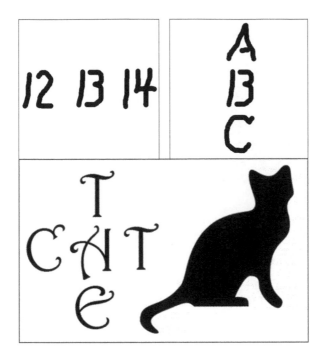

FIGURE 3.8
The same character shape may be recognized as either a number or letter depending on the context in which it appears. Similar duality has been exploited in a clever sign for "The Cat" Coffee Shop.

plate frames (or dirt) may cover part of the plate. Because the cars are in motion, very short exposures are needed to record a sharply focused image, and lighting conditions may be far from optimum. Some camera systems use infrared lighting to illuminate the plates, and some states use paints that provide high contrast in the infrared, but this also requires special optics in the cameras.

Once the image of the plate has been captured, software must deal with different colors for characters and background (some plates have pictorial backgrounds that further increase the difficulty), somewhat variable character shapes, and different spacings or layouts (even for plates from a single state), as indicated by the examples in Figure 3.9. Numbers and letters may be present in different sizes. And even if the characters can be read, it is also necessary to identify the state or country that issued the plate. All of this processing must happen within a few seconds, so that (for example) the number can be compared to records of stolen vehicles and the police officer alerted to the presence of the car. Systems are available commercially that claim better than 96% overall accuracy, meaning that the individual characters must be located and read with better than 99.5% accuracy.

FIGURE 3.9
A few examples of car license plates, with different layouts, characters, artwork, and colors, illustrating the difficulty of automatically reading the plate number.

Universal Product Code (UPC)

Automatic bar code readers are used not only on packaged food at the supermarket checkout counter, but also on many other products (this book, for example), and also for other inventory and tracking tasks. The original bar code format was developed by IBM for grocery stores (introduced in 1974) to speed the checkout process and keep track of inventory, but quickly spread to other applications. As shown in Figure 3.10, it consists of 12 digits (one of which is a checksum digit). Each digit is represented by transitions from black to white, or white to black, in the bar pattern, which is symmetrical (the left half is a mirror image of the right half, with the contrast reversed) to allow scanning in either direction.

UPC bar codes are typically read by a linear scanner, and so are a very limited example of shape analysis. But a wide variety of two-dimensional codes have been developed that are capable of much greater information density. Figure 3.11 shows a few examples. Datamatrix can store up to about 2000 characters in a square and is often used in industry to identify items during manufacture. Maxicode holds 93 characters and is used, for example, by UPS for package tracking on high-speed conveyor lines. QR Code can hold 1520

FIGURE 3.10
Example of a UPC bar code.

FIGURE 3.11
Examples (from left to right) of Datamatrix, Maxicode, and QR Code.

alphanumeric characters and has elaborate error detection and correction. One current use is encoding a Web address printed in advertisements and on real estate "for sale" signs that can be photographed with a cell phone to access additional information.

Other two-dimensional patterns are used for hospital patient records, drivers' licenses, tax return forms, and many other aspects of modern life. In all cases, the information is contained in the pattern of black and white marks, along with some fixed pattern shapes such as the edges, squares, and concentric rings shown in the examples, which are used to align the image of the pattern with the reading template, or vice versa.

Many other forms of data encoding using patterns are utilized. One that hides information in the pattern of yellow dots from high-quality ink jet printers records the serial number and location (e.g., from the IP or MAC address) of the printer, in order to foil counterfeiters who attempt to print copies of legal tender (Chim et al., 2004; Beusekom et al., 2010). All printer manufacturers include the algorithms, which are highly confidential, in their firmware.

Cross-Correlation

Cross-correlation is a particularly powerful form of template matching. Although it is often performed in frequency space using Fourier transforms, it can be conveniently understood (and is sometimes carried out) based on the pixel array. Figure 2.31 illustrates the process of using cross-correlation to locate a particular letter in text. The template may contain only black pixels to define a shape, but may also have grayscale values at each location to represent a probability or significance, as, for example, the pixel weights shown in Figure 3.3. Likewise, the image being scanned for matches to the template may be a thresholded binary image, but in many cases is also continuous tone grayscale.

To compensate for the possible differences in brightness and contrast in the template and the image being scanned, the preferred method of normalized cross-correlation is used. This is done by subtracting the mean value and dividing by the standard deviation of the pixel values both in the template, t, and at the corresponding pixels in the image, f. At each location, the degree of match between the template and the underlying pixels at each location (x, y) is calculated by summation over all the template pixels as shown in Equation 3.1:

$$\frac{1}{n-1}\sum \frac{\left(f(x,y)-\bar{f}\right)\cdot\left(t(x,y)-\bar{t}\right)}{\sigma_f \cdot \sigma_t} \qquad (3.1)$$

The result of the cross-correlation operation can be displayed as a gray-scale image in which each pixel location has a value corresponding to the degree of the match. Peaks in this image indicate locations where matches occur. The example in Figure 3.12 is a scanning electron microscope (SEM) image of a Nuclepore filter containing latex spheres as well as debris, with an irregular background pattern of holes. The template image shown is used to perform cross-correlation, with a top-hat filter (described in Chapter 2) applied to locate the peaks, which are then thresholded. The result locates all of the particles without confusion from the background or debris.

For target features that are less regular in shape than the spheres shown in the image, it is necessary to use several templates, corresponding to the variation in appearance. For example, for aircraft that may be parked at various angles (Figure 3.13), a set of templates rotated in about 15- to 20-degree steps can be used, and the results of each cross-correlation combined with a

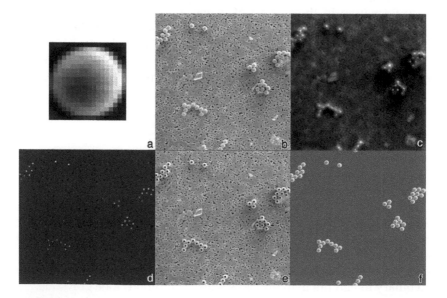

FIGURE 3.12
Locating features by cross-correlation: (a) template image (enlarged to show individual pixels); (b) original SEM image; (c) cross-correlation result; (d) top hat applied; (e) locations superimposed on the original; (f) reconstruction showing just the particles.

FIGURE 3.13
Aerial view of the United States aircraft carrier John F. Kennedy docking in Malta, with various airplane types parked at different angles.

Boolean OR. Alternatively, some implementations first calculate a principal axis of orientation for the pixels at each location and adjust the pattern on the fly (Zhao et al., 2006). The speed with which the cross-correlation can be carried out makes either approach practical.

The location of the peak in the cross-correlation image can be interpolated to a fraction of the pixel spacing, which can be useful for precise determination of object location or for aligning a sequence of pictures of a moving object. However, interpreting the peak value in the cross-correlation result to characterize the degree to which the feature's shape diverges from the template is more difficult. Slight changes in size, brightness values, or orientation, as well as shape, can affect the peak value and the shape of the peak. Consequently, cross-correlating features with a circular target (for example) does not quantitatively measure the extent to which a feature differs from a circle. Examples of using cross-correlation for shape matching are shown in Chapter 4 (Figure 4.51) and Chapter 6 (Figure 6.8).

Describing Noncircularity

One description of shape that is usually accepted as visually recognizable is "roundness" or "circularity." Closer examination of the assumptions behind

FIGURE 3.14
Four shapes that can be described by the word "round" (from left to right): a beach ball, a Celtic knot, the Cretan labyrinth, a sunflower.

this acceptance raises more questions, however. The shapes in Figure 3.14 are all "round" in some way: there are few if any straight lines or corners; the exterior shape (the convex hull described in Chapter 2) is approximately equiaxed; and so on. But only the outline of the ball has a circular image. The use of a circle as a measure of "roundness" is common in image analysis, but what characteristic of the circle should be used?

One of the oldest, and still most widely used shape descriptors, is calculated from the area and perimeter of a feature as

$$Formfactor = \frac{4\pi \cdot Area}{Perimeter^2} \tag{3.2}$$

This has a value of 1.00 for an ideal circle and for perfect measurements. Smaller values indicate a greater departure from this ideal. The relationship has been variously named formfactor, circularity, roundness, and so on, by different authors and software providers. Some computer programs omit the 4π, or use the reciprocal or square root of the value, but the intent is the same: to provide a formally dimensionless number that describes in summary fashion the departure of a feature from a circular shape.

The first problem that this measurement encounters with real images and software is the difficulty of defining and measuring the area, and to a greater degree the perimeter. Chapter 2 explains the basic problems that arise with features represented on a finite pixel grid, and the conflicts between treating pixels as points or squares. In practical terms, the value of the formfactor does not reach 1.0 for a circle composed of real pixels, and furthermore the value varies with size as shown in Figure 3.15. These results are produced by software (Fovea Pro, Reindeer Graphics, Asheville, North Carolina) that counts pixels to determine the area and constructs a smoothed boundary for the features using their centers to measure the perimeter. Many programs perform more poorly.

The second problem, which is more important, is that there are a great many ways to "not be like a circle," and this measurement does not distinguish between them. The shapes in Figure 3.16 have identical areas, and values of formfactor that decrease as the apparent irregularity of the boundary

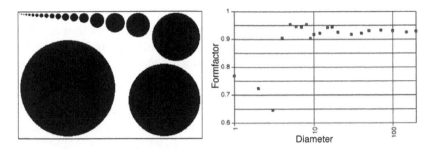

FIGURE 3.15
Circles of varying sizes with their measured formfactor values.

FIGURE 3.16
The shapes have the same area but the increasing perimeter changes the measured formfactor values as shown.

and depth of indentations (and the length of the perimeter) increases. But the features in Figure 3.17 have identical measured values for the formfactor and are not visually the same shape.

As pointed out in Chapter 1, a shape may depart from being a circle in many ways, not limited to being elongated, or having an uneven boundary,

FIGURE 3.17
A set of shapes that are visually dissimilar but have identical values for the formfactor (= 0.44).

or being polygonal, and so on. Formfactor is widely used because it is easy to calculate, based on measurements of area and perimeter that must often be carried out anyway. It is, however, a very limiting candidate for shape characterization. In some instances, formfactor may turn out to have some correlation with the history or properties of objects, but there is no reason to expect it to be an optimum choice.

There are several other easily calculated, formally dimensionless ratios of size measurements that are also used to describe a departure from being circular. These also are called by a variety of different names (some of which are the same ones that other authors or programs use for the calculation in Equation 3.2). Equation 3.3 defines two in the terminology used here, based on the maximum and minimum caliper or projected dimensions:

$$Roundness = \frac{4 \cdot Area}{\pi \cdot Max.Dim.^2}$$

$$Aspect\ Ratio = \frac{Max.Dim.}{Min.Dim.}$$

$$(3.3)$$

Obviously, this list can be extended easily to combine any of the measures of size in a way that allows the units to formally cancel out. It is also helpful to include numerical scaling factors where necessary to produce a value of 1.0 for an ideal circle, but this is not essential. Aspect ratio is sometimes defined differently, as the ratio of the length (the maximum dimension) to the projected dimension in the direction perpendicular to the length; for a square shape, the aspect ratio thus defined is 1.0, whereas the definition in Equation 3.3 yields a value of 1.414.

Like formfactor, these shape factors are not very specific, and each measures something different about the way(s) that a shape can depart from being like a circle. For example, as shown in Figure 3.18, a smooth ellipsoidal feature has a formfactor slightly less than a circle, but the roundness or aspect ratio values vary from 1.0 (the roundness becomes less than 1, the aspect ratio greater than 1). The irregular "gear-shaped" feature has a roundness slightly less than a circle, a formfactor much less, and an aspect ratio near 1.

The roundness and aspect ratio measurements have similar problems of precision as formfactor. Measuring the same circles shown in Figure 3.15 produces values that only gradually approach the ideal 1.0, primarily due to the pixelation of the circles. Figure 3.19 shows the results.

More Dimensionless Ratios to Measure Shape

The definitions in Equation 3.2 and Equation 3.3 attempt only to describe shape departures from circularity. Because of their ease of computation, a

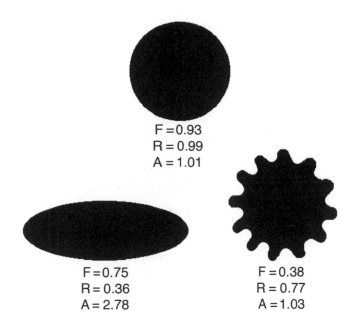

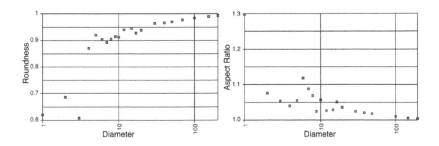

FIGURE 3.18
Three features with the same area, each labeled with the formfactor (F), roundness (R), and aspect ratio (A) as defined in the text.

FIGURE 3.19
Measurements of roundness and aspect ratio for the circles in Figure 3.15.

great many dimensionless ratios of size measurements have been created and often used successfully in specific applications. As for the ones shown earlier, there is no uniformity in the naming of these parameters and no consistency in the definitions. A few examples are listed in Equation 3.4 to illustrate the possibilities and to caution that if any of these are used, the exact definition and measurement procedure must be understood:

$$Curl = \frac{Length}{Fiber\ Length}$$

$$Elongation = \frac{Fiber\ Length}{Fiber\ Width}$$

$$Extent\ 1 = \frac{Area}{Bounding\ Box\ Area}$$

$$Extent\ 2 = \frac{Area}{Circumscribed\ Circle\ Area}$$

$$Solidity = \frac{Area}{Convex\ Hull\ Area} \qquad (3.4)$$

$$Convexity = \frac{Perimeter}{Convex\ Hull\ Perimeter}$$

$$Eccentricity = \frac{\sqrt{a^2 - b^2}}{a}$$

$$Area\ Fraction = \frac{Net\ Area}{Filled\ Area}$$

$$Radius\ Ratio = \frac{Inscribed\ Radius}{Circumscribed\ Radius}$$

Some explanation of these measurements is needed.

Curl compares the length measured either from end to end, or the maximum caliper dimension (which may not be the same) of a shape such as a fiber, worm, or surface scratch, to the length along the irregular path. The latter is usually determined from the skeleton as illustrated in Figure 2.50, preferably with a correction for the fact that the skeleton does not quite reach to the actual ends of the feature. But a spaghetti noodle or a worm can curl up in a great many different shapes that produce the same numerical value for this parameter.

Also for fiberlike or wormlike shapes, the *elongation* is the ratio of the length along the midline to the width of the fiber. The width may be the maximum dimension perpendicular to the midline, or the minimum, or an average along the fiber. As illustrated in Figure 2.50, these widths can be measured from the values of the Euclidean distance map for the pixels along the fiber skeleton.

The *extent* may be defined either as the ratio of the feature area to that of the minimum circumscribed circle, or to that of an enclosing rectangle, and the rectangle may either align with the edges of the image or be rotated to align with the feature. The alignment may be done either based on an axis defined by the feature's moment or longest chord, or rotated to produce the smallest bounding rectangle. These choices often produce quite different measurement values.

The convex hull, also called the taut-string perimeter, bounding polygon, or rubber-band outline, is the basis for several ways to describe a departure from convexity. It is illustrated in Figure 2.49. Both the ratio of perimeter and convex perimeter (*convexity*), and the ratio of area and convex area (*solidity*) are based on this bounding shape. Any feature that is formally convex will have no indentations around the periphery. But these two parameters are very different in their dependence on the interior or boundary of the feature. This is a distinction that applies to many shape factors; as previously noted, those that depend on all the pixels within the feature capture a different characteristic of shape than those that depend only on the pixels along the periphery.

Eccentricity is defined in terms of the major and minor axes (*a* and *b*, respectively) of an ellipse. For irregular shapes, an ellipse may be defined in several different ways, which give rise to different values of eccentricity. Using the maximum caliper dimension as the major axis and calculating a minor axis that gives the same area as the feature is one way, but the minor axis may also be taken as the minimum caliper dimension or as the caliper dimension perpendicular to the major axis (which is not necessarily the same thing). An alternative is to define both the major and minor axes to match the moment of the feature about its major axes while matching the area. Likewise, the first amplitude coefficient in the Fourier series used in harmonic analysis is a measure of the ellipticity of the shape (the zeroth coefficient in the series is the radius of a circle with the same area). These are discussed in Chapter 4, where moments and harmonic analysis are considered.

The *area fraction* is not affected by the overall convex shape of a feature but measures the presence of internal gaps or voids. These may be many separate, small voids or a single large one, or any combination. Figure 3.20 shows two examples, one of which (the flower stem) is not convex. Features that contain internal holes have area fractions less than unity. It should also be noted that in all of the preceding measurements that use feature area, a decision must be made as to whether the area should include or exclude any internal voids.

Finally, the *radius ratio*, also known as the *modification ratio*, has a history that precedes digitized images and computer measurement. It uses the radii of the largest inscribed and smallest circumscribed circles, as illustrated in Figure 2.49. Many man-made fibers, such as those used in carpeting, have extruded cross-sections that are designed with intricate shapes, such as multipointed stars (Figure 3.21). By quickly performing a visual examination of the cross-sections of fibers and using a circle template for comparison, a human observer can estimate this ratio. An increase in the ratio indicates that the die through which the material was extruded is wearing (is "modified"),

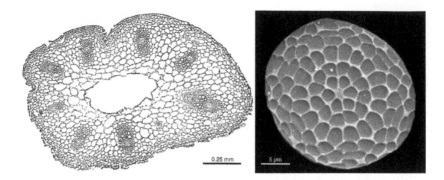

FIGURE 3.20
Area fractions of cross-sections. (Left) Light microscope image of a flower stem showing cell walls (which comprise 31.5% of the area). (Right) Scanning electron microscope image of an "islands-in-the-sea" coextruded polymer fiber, etched to reveal the polymer phase (light gray) that will be chemically removed to leave the fine yarns, which comprise 75.1% of the area. (Image courtesy of Judy Elson, College of Textiles, North Carolina State University.)

increasing the amount of material near the core of the fiber and increasing its stiffness. This quality control measurement can now be performed by computerized image analysis but retains the original name.

The many ways that measures of size can be combined in formally dimensionless ratios can be confusing, and the variety of names that may be associated with the different combinations is even more so. It may be helpful to group them into three rough categories, as shown in Table 3.1. Some of the ratios do not fit neatly into one of these, and it is important to

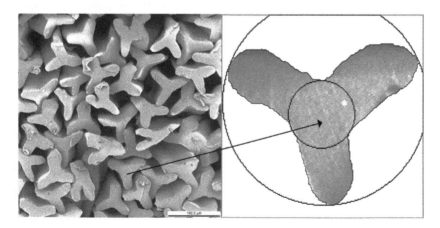

FIGURE 3.21
Scanning electron microscope image of cross sections through trilobate nylon fibers, with an example of the automatically fit inscribed and circumscribed circles. The average ratio for the fibers in the image is 0.326. (Image courtesy of Judy Elson, College of Textiles, North Carolina State University.)

TABLE 3.1

Categories for Dimensionless Ratios

How Much Different from a Circle	How Much Stretched Out	How Fully Filled Out
Formfactor	Aspect ratio	Area fraction
Roundness	Curl	Solidity
Eccentricity	Elongation	Extent 2
Convexity	Extent 1	Radius ratio

point out that many of the ratios can also perform useful service in other categories, such as:

- Aspect ratio can also be a measure of departure from being like a circle.
- Eccentricity can also be a measure of how much a feature is stretched out.
- Convexity can also be a measure of how filled out a feature is.

The point is that these dimensionless ratios appear to be simple (and are certainly simple to calculate) but what they actually measure is not so simple, either in terms of the mathematical description of "shape" or in terms of what people visually extract as significant in a feature's shape.

The diversity of dimensionless ratios (and the medley of names used for them) can be disconcerting. But the various ratios are not entirely redundant, and in any particular situation one or another may be the most useful. For example, Figure 3.22 shows binary images of two common flowers from one author's lawn: a violet and a dandelion. The shapes are visually easily distinguishable, but as shown by the values in Table 3.2, many of the ratios are not significantly different. Of the dimensionless shape parameters listed, only the radius ratio values can separate the two flowers. Many

FIGURE 3.22
Binary images of a violet and a dandelion. The measurement data are compiled in Table 3.2.

TABLE 3.2

Measurements of the Shapes in Figure 3.22

	Violet	Dandelion
Formfactor	0.28	0.28
Roundness	0.75	0.79
Aspect ratio	1.12	1.11
Solidity	0.87	0.88
Convexity	0.58	0.57
Radius ratio	0.35	0.75

software packages provide only a few defined parameters (most commonly the one named "formfactor" in these pages) as a measure of "circularity"; this is unable to distinguish these shapes successfully. The measurement of formfactor was shown to be useful for describing the shape changes during locomotion of amoeba (Ueda & Kobatake, 1983; Soll et al., 1988).

Other suggested shape descriptors may be either more or less complicated to calculate. For example, Carter and Yan (2005) measure the maximum, mean, and minimum radius of all points on the boundary from the geometric center, calculate the deviation in radius (r_{dev}), and from that a dimensionless shape factor (S_F) as shown in Equation 3.5:

$$r_{dev} = \sqrt{\frac{\left(r_{max} - r_{mean}\right)^2 + \left(r_{max} - r_{min}\right)^2}{2}}$$

$$S_F = \frac{r_{dev}}{r_{mean}}$$

(3.5)

Another approach (Poczeck, 1997) measures the deviation of points on the feature boundary from a set of standard images of a circle, triangle, and square (each sized to cover the same area as the feature). Trials with several test powders show that the resulting shape descriptors perform better at discriminating the batches than aspect ratio and circularity (it is not reported exactly what the definition or measurement procedure was for each of these or how the triangle and square were aligned with the feature).

Example: Leaves

In practical terms, since few of these derived dimensionless parameters correspond with any precision or consistency to what people "see" as shape, it is necessary to use statistical methods to determine which, if any, are useful. In Chapter 6, methods such as stepwise regression, discriminant analysis, principal components analysis, and other statistical techniques are used to select those shape descriptors that provide the best classification or identification

TABLE 3.3

Tree Types

Cherry (*Prunus 'Kwanzan'*)
Dogwood (*Cornus florida*)
Hickory (*Carya glabra*)
Mulberry (*Morus rubra*)
Red maple (*Acer rubrum*)
Red oak (*Quercus falcata*)
Silver maple (*Acer saccahrinum*)
Sweet gum (*Liquidambar styraciflua*)
White oak (*Quercus alba*)

in a particular instance. As an example of the use of this technique, leaves from local trees were collected as listed in Table 3.3; Figure 3.23 shows examples of each of the nine types.

Measurement of a broad selection of shape parameters was made. The classical shape parameters described in this chapter were determined, and a software package (JMP® 9, SAS Institute Inc.) was used to select the minimum set of parameters needed. Using just the formfactor, convexity, and radius ratio shape factors shown earlier, the various leaf types are grouped correctly with few individual errors.

Some explanation of the plots in Figure 3.23 is appropriate, because similar plots are used in examples in the following chapters. The axes are not the measured parameters themselves but linear combinations of them, determined by statistical regression, that best distribute the data points and separate the identified classes. In this case, the equations for the canonical axes are

$$Canonical\ 1 = -0.156 \cdot Formfactor + 58.19 \cdot Solidity + 1.956 \cdot Radius\ Ratio$$
$$Canonical\ 2 = -46.44 \cdot Formfactor + 36.22 \cdot Solidity + 20.97 \cdot Radius\ Ratio \quad (3.6)$$
$$Canonical\ 3 = +5.488 \cdot Formfactor - 9.248 \cdot Solidity + 21.86 \cdot Radius\ Ratio$$

Canonical axis 1 is responsible for 60.9% of the classification, while axes 2 and 3 contribute 36.1% and 3.0%, respectively. Because the various measures, as shown earlier, respond strongly to either the boundary or to the full interior area of the feature, it is not surprising that they can combine to sort out these leaf shapes. The differences between cherry, dogwood, and red mulberry are primarily the details along the periphery. The differences between dogwood, oak, and maple are evident in the compactness of the former and the lobed, extended shapes of the latter. Other combinations of dimensionless shape factors, such as roundness, convexity, and solidity, are also able to separate the groups and successfully identify each individual leaf.

The error matrix in Figure 3.23 shows that only 3 of the 146 individual leaves are misclassified. Of course, that does not guarantee that the model derived from this small set of leaves would work for all leaves from these trees, let

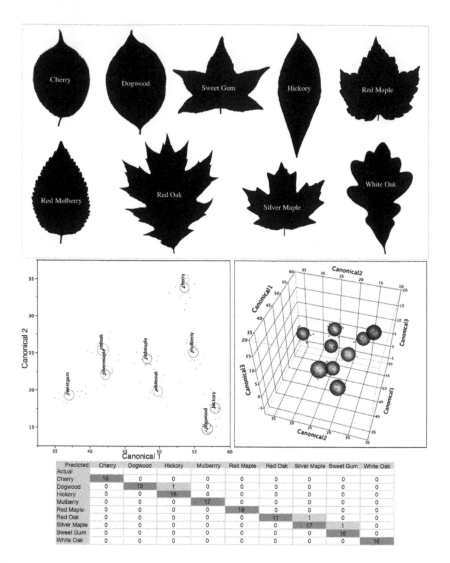

FIGURE 3.23
(See color insert.) Classification of leaves by shape: examples of nine types of leaves (not all at the same scale) with the successful grouping of the leaves based on three dimensionless shape parameters, and the error matrix for the leaf identification.

alone all trees of these species. The issue of selecting an adequate training population and testing it against unknowns is considered in Chapter 6.

Some shape parameters for the leaves include measurements on the skeletons. The skeleton of the shape of each leaf is not quite the same as the vein pattern, although both generally have the same purpose: to provide access to all parts of the area. Figure 3.24 compares the skeleton and vein pattern of a maple and an oak leaf. The vein patterns contain many branches. Although

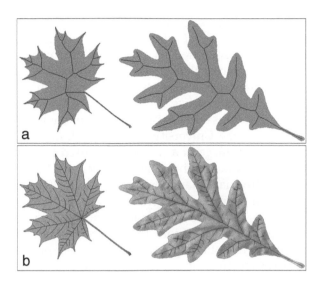

FIGURE 3.24
Silver maple and white oak leaves: (a) the computed skeletons of the shapes; (b) the actual vein pattern of the leaf.

they cannot be computed from the binarized leaf shape, it is possible to directly image them using a light table.

For the oak and maple leaves, the palmate (lobed) shape results in a very branched vein pattern. Measurements of the lengths of the branching veins can be subdivided into the main branch coming from the base of the stem, the second level veins that branch from the main, and the third level veins that branch from the second level (fourth and higher levels are generally not reliably captured by the imaging process and are excluded from the following analysis).

The average number of second-level branches is 7 for both of the two leaf types, but the number of third-level branches is quite different (64 ± 6 for the maple versus 91 ± 9 for the oak). Measuring the length of the branches (normalized to the length of the first-level branch) is also distinctive as shown in Table 3.4. The maple leaf veins are proportionately longer for second- and third-level branches than the oak leaves. Similar comparisons are possible between many other lobed leaves but are more difficult for compact leaves, which tend to have simpler and more common vein patterns.

TABLE 3.4

Branch Lengths for Veins in Leaves

	Oak	Maple
Second-level branch	0.234 ± 0.143	0.369 ± 0.136
Third-level branch	0.038 ± 0.025	0.150 ± 0.048

This is a small illustrative example compared to more ambitious efforts toward automatic recognition of plant leaves. Belhumeur et al. (2008) describe an automatic system based on a tablet PC that can be used to capture an image of a leaf, process that image and perform the necessary measurements, access a database of hundreds of plant types, and show the user images of the 10 most likely matches.

In 2011 this effort matured as the "Leafsnap" cell phone app developed jointly by Columbia University, the University of Maryland, and the Smithsonian Institution that allows an image of a tree leaf captured by the phone's camera to be uploaded. The photograph is matched against a library using shape descriptors computed from landmark points along the leaf's outline (landmark methods are discussed later). The best matches are ranked and returned to the user for verification, along with information and images showing the tree, bark, flowers, and so forth. Furthermore, the GPS coordinates from the phone are used to update a record of tree populations, an example of "crowdsourcing" to collect a broad spectrum of information using human intelligence.

The authors use this app, which successfully identified leaves of each of the nine types in Figure 3.23. But this effort pales in comparison to the potential of online databases of plants. In addition to 90,000 type specimens from the U.S. National Herbarium at the Smithsonian Institution, there are 120,000 images at the New York Botanical Garden, 35,000 at the Royal Botanical Garden Kew, and 35,000 at the Missouri Botanical Garden, as well as other resources worldwide. Expert systems that can negotiate such large data sets will revolutionize the traditional "field guide." Similar databases exist for insects, flowers, birds, and so on.

Example: Graphite in Cast Iron

Sometimes it is possible to create a new dimensionless shape factor that suits a specific purpose. Grum and Stürm (1995) tested several factors combining the feature area, perimeter, and the diameter of the circumscribed circle in an effort to automatically classify the graphite grains in cast iron; examples are shown in Figure 3.25.

The shape ratios used were the *formfactor* and *extent2* as defined earlier and the ratio $(Perimeter/[\pi \cdot D])$, where D is the diameter of the circumscribed circle. They found that none of these individually could distinguish nodules, flakes, and several intermediate stages. Instead, they proposed a new combination by multiplying those individual dimensionless ratios together. The result, as shown in Figure 3.26, is a dimensionless ratio that varies between 0.001 and 1.0, and provides a reliable method for recognizing the shapes of graphite particles that occur in ductile iron and gray iron.

This same problem has been revisited by Prakash et al. (2011b). They measure two dimensionless ratios, the ratio of the inscribed to the circumscribed diameter, and the ratio of the area of the object to the area of the minimum-bounding rectangle. Flakes and nodules are distinguished in a

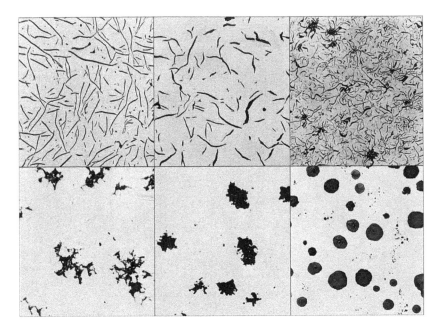

FIGURE 3.25
Cross-sections through graphite grains in cast iron, showing the variation (from top left to bottom right) from flakes to nodules.

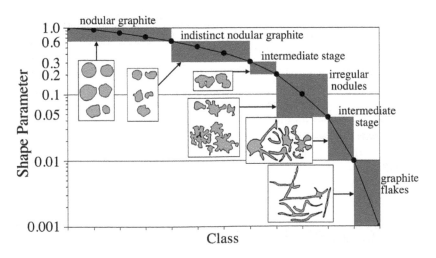

FIGURE 3.26
Classification of graphite grains using a custom dimensionless ratio, as described in the text. (After J. Grum, R. Stürm, 1995, *Acta Stereologica* 14(1):91–96.)

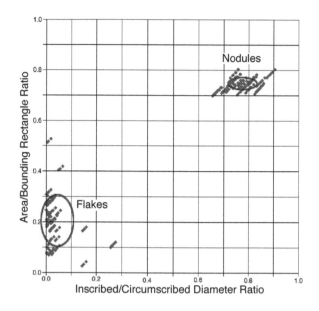

FIGURE 3.27
The data points for the diameter and area ratios for nodular and flake graphite, with ellipses representing two standard deviations. (From P. Prakash et al., 2011b, *International Journal of Advanced Science and Technology* 29:31–40.)

two-dimensional decision space (they report 95% to 98% success rates, similar to results from harmonic Fourier coefficients reported in Prakash et al., 2011a). Their data (Figure 3.27) show that either parameter could accomplish the task independently, and applying principal components analysis to the values yields a linear combination (shown in Figure 3.28) that produces the maximum discrimination between the two groups.

The published data deal only with the two extreme forms (flake and nodule), but using the original series of images (kindly supplied by the authors) a more extensive analysis was undertaken. The variables used by Prakash et al. (diameter ratio and area ratio) distribute the same six discrete classes of shapes as shown in Figure 3.29, with 97.5% of the variability accounted for by the first of the two canonical variables (the horizontal axis); 25% of the individual data points are misclassified by ±1 class and 4% by ±2 classes.

Applying the same statistical analysis to the full set of dimensionless shape descriptors introduced earlier yields the result shown in Figure 3.30. The classes are more widely separated and are not in a linear progression, with 85.6% of the discrimination accounted for by the first canonical variable (the horizontal axis) and 14.4% by the second; 16% of the individual data points are misclassified by ±1 class and none by more than that.

It is interesting in this result to note that the dimensionless ratios selected by the analysis are more independent of each other than those in Figure 3.29.

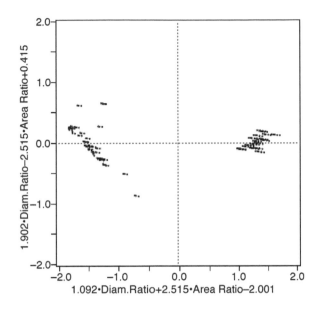

FIGURE 3.28
Data from Figure 3.27 plotted on principal components axes. The combination of measured parameters shown on the horizontal axis accounts for 97.3% of the variation in the data.

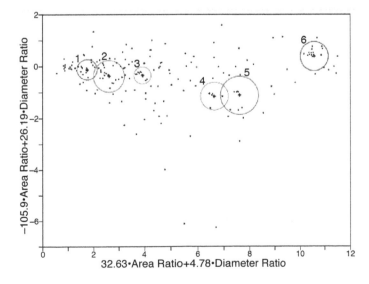

FIGURE 3.29
Measurement data plotted for six shape classes using linear discriminant analysis with the same variables as Figure 3.28.

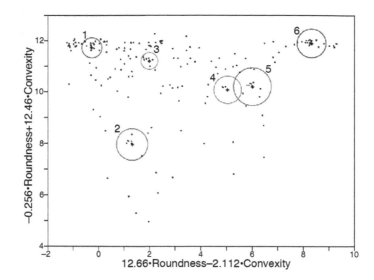

FIGURE 3.30
Similar plot to Figure 3.29 but using the dimensionless ratios that give optimum discrimination.

Roundness is calculated from the area and maximum caliper dimension, while convexity is calculated from the perimeter and the perimeter of the convex hull. All of the variables used in Figure 3.29 are based in one way or another on the full area or interior of the shape, ignoring the perimeter of the features. Using a wide selection of different original measurements to form shape descriptors is usually a good starting point for discrimination.

Dimension as a Shape Measurement

The various ratios shown earlier are used to describe shape, independent of size, because they are formally dimensionless. The units in the numerator and denominator cancel out, leaving a pure number. The precision with which the number can be determined does depend on size, because of the effects of the finite pixel grid. Shape measurement or matching often involves prior filtering to remove or suppress high-frequency terms (e.g., noise, fine-scale detail) so that the measurements deal only with large-scale shape features. But in many cases with natural objects, the fine-scale detail and irregularity are important components of what humans mean by "shape."

A different approach to shape description uses dimension itself as a parameter. In familiar Euclidean geometry, dimensions are integers: zero for points, one for lines, two for surfaces, three for solids. A newer approach to the idea of dimension can be credited to Benoit Mandelbrot, whose classic book *The Fractal Geometry of Nature* (1982) looked back to mathematical questions raised decades before, and outward to the fact, as he wrote it, that "Clouds are not spheres, mountains are not cones, coastlines are not circles, nor does lightning travel in a straight line." Mandelbrot coined the term "fractal," which, among other references, implies that dimensions need not be integers but can be fractional.

To introduce the application of fractal geometry to the shape of features in two dimensions, it is instructive to consider an earlier question studied by Richardson (cited in Mandelbrot, 1967): "How long is the coast of England?" As an armchair scientist, Richardson used maps and dividers to measure the lengths of coastline and the borders of countries. His important discovery was that the result he obtained depended on the scale or resolution of the map, and on the stride length set into the dividers. Shorter stride lengths were able to follow more of the irregularities of the coastline, producing a longer total measurement. When plotted as log (total length) versus log (stride length) the data lay on a straight line. Furthermore, the slope of that line varied for measurements on different coastlines, such as the heavily indented fjords of Norway or the relatively smooth beaches of Florida.

As indicated in Figure 3.31, the slope of Richardson's line can be used to calculate a fractional dimension greater than 1.0 (which corresponds to a Euclidean line) and less than 2.0 (at which point the line has spread out across an entire plane, whose topological dimension is 2). There are "monster" curves discovered by mathematicians such as Peano and Hilbert in the late 19th century, that fill a plane entirely and have a fractal dimension of 2. Perfectly Euclidean boundaries such as the borders of the state of Colorado produce a horizontal line (slope of 0) on a Richardson plot, indicating that the perimeter value does not change with measurement scale.

Note that the dimension is for the boundary, ignoring the internal portion of the feature. When applied to different coastlines, the more "uneven" the

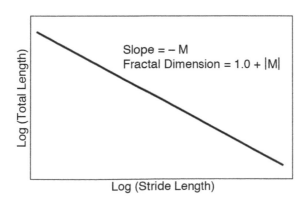

FIGURE 3.31
A Richardson plot and the resulting fractal dimension.

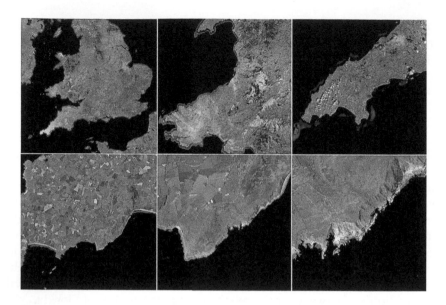

FIGURE 3.32
A series of images of the coastline of Britain, each at a scale four times greater than the last (i.e., one-sixteenth the area).

shape of the coastline the higher the number. Using Richardson's data, the west coast of Britain has a dimension of 1.125, whereas that for the South African coast is 1.005.

Another way to understand this behavior, which is observed for a great many natural phenomena covering an enormous range of dimensions, is that as magnification is increased, more and more detail is resolved, while in a statistical sense the degree of irregularity stays more or less the same. Figure 3.32 shows a series of images of the west coast of Britain, each one a factor of 4 times more magnified than the last. As the scale changes, the irregularity of the coastline stays approximately the same, and would presumably continue to do so until the level of sand grains, or perhaps the atoms of which they are composed, is reached.

Although it offers a useful way to understand the idea of a fractal dimension, the Richardson plot is a poor way to measure it in a digitized image. The "striding" process along a boundary works with maps and dividers, but in a pixel array there will not in general be a location exactly where it is needed, and interpolation imposes a smoothing or fitting of a straight line segment that will bias the data.

A method that is closely related to the Richardson method uses blurring of the image to reduce detail along the boundary. If the image is smoothed using a Gaussian blur of increasing radius and then thresholded, the result is a series of images with progressively reduced perimeter. Plotting the perimeter against the standard deviation of the blur radius on the usual log–log axes produces a linear graph whose slope is related to the fractal dimension. As

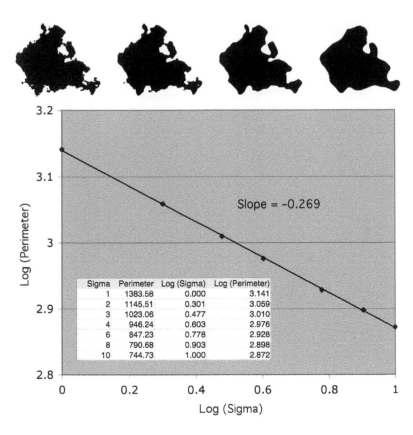

The following table appears within the plot:

Sigma	Perimeter	Log (Sigma)	Log (Perimeter)
1	1383.58	0.000	3.141
2	1145.51	0.301	3.059
3	1023.06	0.477	3.010
4	946.24	0.603	2.976
6	847.23	0.778	2.928
8	790.68	0.903	2.898
10	744.73	1.000	2.872

Slope = -0.269

FIGURE 3.33
Progressive smoothing. A feature is shown after smoothing with a Gaussian blur having standard deviation (sigma) values of 1, 2, 4, and 8 pixels, followed by thresholding. The slope of the log–log plot of the perimeter versus smoothing sigma gives the dimension ($D = 1.269$).

for the Richardson plot, the dimension is equal to 1.0 plus the absolute value of the slope; in the example of Figure 3.33 the dimension value equals 1.269.

There are several other ways that fractal dimensions can be determined (Flook, 1978; Clark, 1986; Russ, 1994). For an image consisting of an array of pixels the method that is most efficient and gives the most precise value uses the Euclidean distance map. Technically, the result that is produced is a Minkowski dimension, which is not identical to the Hausdorf dimension produced by the Richardson technique. For practical purposes these differences are not important, provided that the numerical values that are to be compared are all generated in the same way.

The Minkowski approach can be understood as sweeping a circle along the boundary with the center following the line, and measuring the area the moving circle covers (often called a Minkowski sausage). Plotting the area of the sausage against the diameter of the circle, again on log–log axes, produces a graph whose slope yields the fractal dimension of the coastline. The

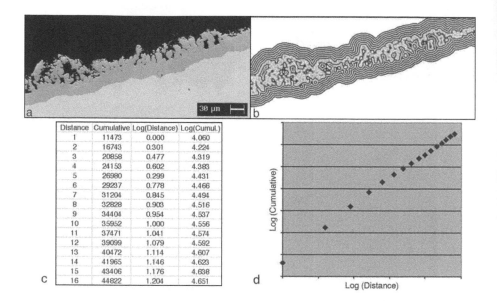

Distance	Cumulative	Log(Distance)	Log(Cumul.)
1	11473	0.000	4.060
2	16743	0.301	4.224
3	20858	0.477	4.319
4	24153	0.602	4.383
5	26980	0.299	4.431
6	29237	0.778	4.466
7	31204	0.845	4.494
8	32828	0.903	4.516
9	34404	0.954	4.537
10	35952	1.000	4.556
11	37471	1.041	4.574
12	39099	1.079	4.592
13	40472	1.114	4.607
14	41965	1.146	4.623
15	43406	1.176	4.638
16	44822	1.204	4.651

FIGURE 3.34
Calculating the Minkowski fractal dimension: (a) original image (section through an oxide coating on metal); b) the Euclidean distance map on both sides of the boundary, contour-shaded to show the shape of the sausage; (c) the data as described in the text; (d) graph of the data.

Euclidean distance map provides an easy way to measure this. Applied to all of the pixels inside and outside of the feature border, the distance map sets each pixel value to the distance from the nearest point on the border.

For the highly irregular boundary such as the one shown in Figure 3.34, counting the pixels with EDM values less than or equal to each distance value constructs a table of the cumulative count versus distance. Plotting the log–log results generates the graph. The fractal dimension, equal to 1.0 plus the slope, is 1.518 for the example in the figure.

As shown in Figure 3.35, the count of the number of pixels as a function of brightness represents how many lie at each distance from the boundary. Plotting the cumulative count as a function of distance gives the area of the sausage. On log–log axes, the trend is a straight line, adding 1 to the slope gives the dimension ($D = 1.242$ for the example feature).

Another method used to determine a fractal dimension employs a technique suitable for implementation on pixel arrays called mosaic amalgamation (also known as box counting). This was introduced by Brian Kaye (1989). The pixels comprising the periphery of an object are progressively coarsened into blocks of 2×2, 3×3, 4×4, and so forth to count the number of coarser grid squares through which the periphery passes, as shown in Figure 3.36. Counting the number of these grid squares and plotting the log of the number versus the log of the block size generates a line. The magnitude of the

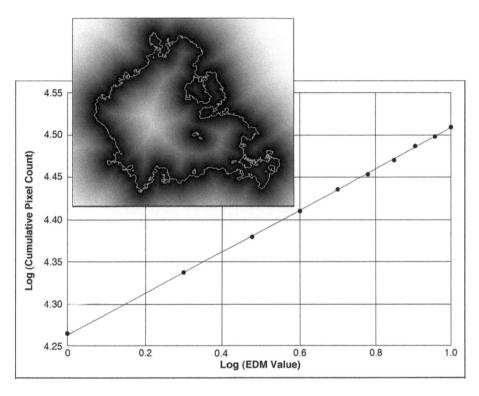

FIGURE 3.35
The Euclidean distance map applied to the pixels inside and outside of the feature, and a log–log plot of the cumulative count of the number of pixels as a function of distance.

slope is the fractal dimension. In this example, the numerical value (1.235) is similar to but not identical to the value produced by the other methods. Technically, the value determined by this method is a Kolmogorov dimension, still another way that mathematicians have described the "monster" curves that are not defined by Euclidean geometry.

Using the Fractal Dimension

Many publications have shown that various natural shapes are fractal, at least over some range of distances (Chermant & Coster, 1978; Mecholsky & Passoja, 1985; Sander, 1986; Kaye, 1989; Mecholsky et al., 1989; Smith et al., 1989; Przerada & Bochinek, 1990; Sapoval, 1991; Wehbi et al., 1992; Orford & Whaley, 2006). Clouds are fractal (Lovejoy, 1982), as are patterns of sea ice (Ivanov et al., 2007). Figure 3.37 shows scanning electron microscope images of various particles, which can be distinguished by their fractal dimensions.

Fewer publications have tried to correlate the dimension with the history or the properties of the objects. Fractal shapes are reported to have diagnostic possibilities for cancer detection (Bauer & Mackenzie, 1995; Baish & Jain,

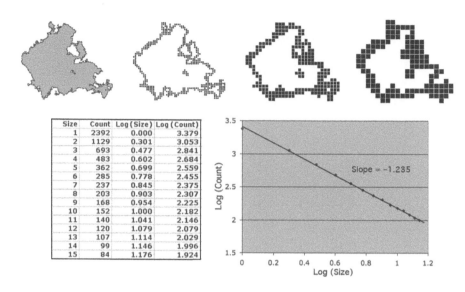

Size	Count	Log (Size)	Log (Count)
1	2392	0.000	3.379
2	1129	0.301	3.053
3	693	0.477	2.841
4	483	0.602	2.684
5	362	0.699	2.559
6	285	0.778	2.455
7	237	0.845	2.375
8	203	0.903	2.307
9	168	0.954	2.225
10	152	1.000	2.182
11	140	1.041	2.146
12	120	1.079	2.079
13	107	1.114	2.029
14	99	1.146	1.996
15	84	1.176	1.924

FIGURE 3.36
Mosaic amalgamation. The number of grid squares of increasing size that are crossed by the outline of the feature is plotted against the size of the square. The examples show the steps with size equal to 1, 4, 8, and 12 pixels.

2000; Kikuchi et al., 2002; Jayalaitha & Uthayakumar, 2007; Martin-Landrove et al., 2007; Abdaheer & Khan, 2009). At larger scales, the shapes of some lakes are fractal and the dimension has been observed to correlate with altitude (Goodchild, 1980). The dimension increases with the energy absorbed in the brittle fracture of metals (Chermant & Coster, 1987; Mecholsky et al., 1989; Fahmy et al., 1991; Ray & Mandala, 1992). Fractal dimension is able

FIGURE 3.37
Scanning electron microscope images of particles, from left to right: volcanic fly ash (fractal dimension = 1.18); sunflower pollen (fractal dimension = 1.27); talcum powder (fractal dimension 1.38).

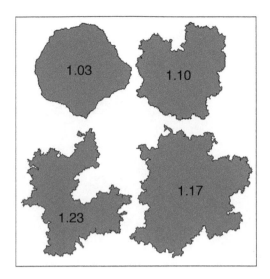

FIGURE 3.38
Several shapes, with their fractal dimensions.

to distinguish the wear patterns on stone tools arising from different uses (Bueller, 1992; Stemp et al., 2010). Whether a correlation observed implies some causality is a more difficult question to answer.

Because so many natural objects have a fractal geometry associated with their shapes, human vision has learned to recognize it. People do not measure the dimension numerically, of course. But given a set of shapes with varying dimension, the increase in dimension is perceived as an increase in "roughness" or "irregularity" of the boundary as shown in Figure 3.38. The fractal dimension is only associated with the irregularities of the boundary and not other larger-scale aspects of shape such as whether the overall shape is more or less equiaxed, branched, and so on.

Fractal geometry is also used to generate shapes. If strict rules are followed, the results have unlimited complexity but are orderly and immediately recognized as artificial. The shape in Figure 3.39 is an example: the von Koch snowflake is created iteratively starting with an equilateral triangle, and then replacing each line segment with one having a superimposed smaller triangular facet. The perimeter increases at each iteration by $\frac{4}{3}$, approaching infinity and giving rise to a fractal dimension of log(4)/ log(3) = 1.2619....

There are many such shapes, but the most interesting are ones in which the rules are statistical rather than fixed. By introducing probabilities for each operation, shapes are generated that mimic a great many natural phenomena, from the branching patterns of trees and rivers to pore structures in sandstone, the path of lightning strikes, and even the fluctuations of the stock market. Figure 3.40 shows an example. In diffusion-limited aggregation (DLA), individual particles are allowed to walk randomly on a surface until

FIGURE 3.39
The first six steps in generating a von Koch snowflake.

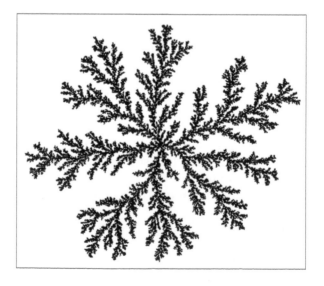

FIGURE 3.40
Randomized fractal with a dimension of 1.70, which models the chemical process of diffusion-limited aggregation, as when a crystal grows on a surface (such as frost on a window).

FIGURE 3.41
Generated randomized fractal surface creating an artificial landscape.

they encounter a growing cluster, to which they then adhere (Vicsek, 1992). If the sticking probability is 100%, the resulting fractal has a dimension of 1.70. Lower sticking probabilities allow the particles to penetrate deeper into the gaps and bays to generate a more compact cluster with a lower dimension.

The measurement and generation of fractal surfaces, as opposed to planar shapes, is discussed in Chapter 5. Fractal geometry is used, for example, to generate realistic artificial landscapes in many movies (Figure 3.41).

Skeletons and Topology

The boundary fractal dimension is concerned only with the periphery of a feature, and the "roughness" or irregularities that are present. It is insensitive to whether the feature is elongated, equiaxed, branched, has interior voids, and so on. The basic topological characteristics are something that human vision is very quick at picking up, and constitute one of the things that people mean by the concept of "shape." The features in the top row of Figure 3.42 are topologically equivalent. Each one can be changed to the other by stretching or bending, without any cutting or joining. The three features in the bottom row are topologically distinct and cannot be bent or stretched to match.

The skeleton, introduced in Chapter 2, removes the outer details of a shape, leaving just a central line of pixels. As shown in the example in Figure 3.43, the result is a set of 8-connected pixels (i.e., pixels which touch neighbors at corners as well as sides to form a continuous line) in which most of them

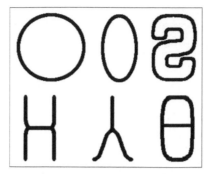

FIGURE 3.42
Two-dimensional features as described in the text.

have exactly two touching neighbors. The pixels in the skeleton with a single neighbor are the end points and correspond to the ends of branches in the original feature. Pixels in the skeleton with either three or four neighbors mark nodes, again corresponding to the shape of the original feature. The pixels with two neighbors form the branches that either connect one node to another or to an end. Identifying skeleton pixels by their number of neighbors locates these topological indicators and allows them to be counted.

Euler's rule for the topology of two-dimensional features is

$$Loops = Branches - Ends - Nodes + 1 \qquad (3.7)$$

Consequently, counting the number of separate branches, end points, and nodes determines the number of loops present and describes the topological

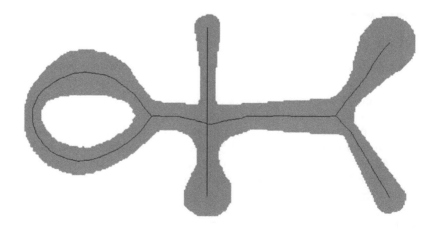

FIGURE 3.43
A feature (gray) with its skeleton superimposed. The skeleton has one loop, seven branches, four ends, and three nodes.

shape of the object. There are some caveats in applying this relationship. For example, the skeleton of the letter O is a circle, which has one loop, but which appears to have no ends, no nodes, and to consist of one branch. But a branch is defined as a line segment with two ends, so there must be a point somewhere on the circle where the ends meet at a (virtual) node. Also, a solid circular disk will skeletonize to a single pixel at the center, which seems to be a single end point. In fact, it must be considered as a (very) short branch with two ends.

The greatest difficulty with using the skeleton to describe a feature's topology is not related to the Euler rule, but to the practicalities of constructing it. Many programs create the skeleton by sequential erosion using classical erosion rules that test only nearest-neighbor pixels. For features on a finite pixel grid, this can produce artifacts in the form of short lines that reach toward every irregularity on the boundary, even if it consists of a single pixel.

This can be dealt with in several ways by

1. Smoothing the feature before skeletonizing, for example, by applying a morphological opening;
2. Constructing the skeleton using the Euclidean distance map (in which case, the result is the medial axis transform and consists of the ridges in the distance map, as illustrated in Chapter 4, Figure 4.1); or
3. Removing short terminal branches (branches that terminate at an end point, rather than connecting two nodes).

Of course, this latter method requires some independent knowledge or arbitrary judgment about the length of lines that should be ignored. In the case of the von Koch snowflake shown in Figure 3.44, the irregularities down to single pixels are significant. It is only the finite resolution of the pixel grid that keeps even smaller irregularities from being present and generating more and shorter branches in the skeleton.

Because the distances to the four corner-sharing neighbors are different from the distances to the four edge-sharing pixels, if integer arithmetic is used and only nearest neighbors are considered, the skeleton lines are often directionally biased as shown in Figure 3.45; this does not affect the topology of the skeleton (the numbers of loops, branches, nodes, and ends), but it does distort the branches, altering their orientation and length. In both cases, the number of end points (22) provides a rapid way to count the number of teeth on the gear.

In addition to extracting the basic topology from a feature, measurements can be performed on the skeleton and its components, and if the skeleton is not biased as shown in Figure 3.45b, ratios of these measurements can also be used to describe shape. For example, the skeleton in Figure 3.43 has a ratio of the total length of terminal branches (the four arms with end points) to internal branches (the three segments with a node at each end) equal to 0.756. The

FIGURE 3.44
The von Koch snowflake from Figure 3.39 with its skeleton superimposed. Branches reach to end points at every irregularity that is resolved in the pixel array.

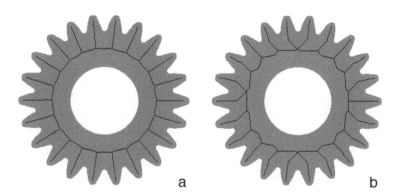

FIGURE 3.45
Binary image of a gear skeletonized using software that employs (a) floating point arithmetic and the Euclidean distance map, and (b) using integer arithmetic and nearest neighbor tests.

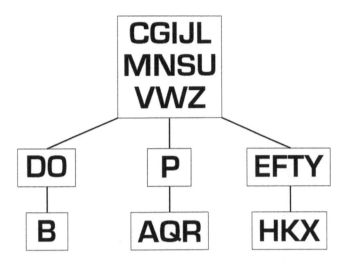

FIGURE 3.46
Letters of the alphabet grouped according to the number of ends and loops.

ratio of the skeleton length of a fiber to the projected length is used for the *curl* dimensionless ratio defined in Equation 3.4. As with the ratio measurements introduced earlier, creating various combinations that utilize skeleton measurements and inventing names for them is practically unlimited.

To emphasize how human recognition of shapes depends on basic topology as captured in the skeleton, Figure 3.46 shows the capital letters of the alphabet grouped according to the number of ends, loops, and branches. Twelve of the letters have two ends, with no loops or branches. There are smaller groups with one or two loops, one loop with either one or two ends, or no loops and three or four ends. The letters with the same topology (able to be stretched or bent to match) are still distinguished by the location of the branches or corners, and especially the ends. These letters are printed in a sans serif font; the removal or elimination of the short segments introduced by serifs can be carried out as described earlier. Figure 3.47 shows an example, in which terminal branches shorter than 25% of the character height are removed.

These are the characteristics used in the syntactical description of shapes shown in Figure 3.5 and Figure 3.6, with the additional step of locating corner points. Corners are generally defined by an abrupt change in direction of the skeleton (or sometimes of a boundary), but locating these by measurement requires dealing with the local pixel array. A line consists of individual links from one pixel to another, which are connected either at 0-, 45-, or 90-degree orientations. As shown in Figure 3.48, this makes it necessary to examine directions over a distance of several or even many pixels (Freeman & Davis, 1977).

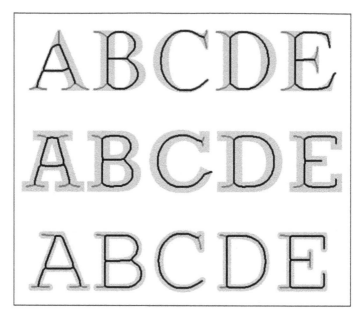

FIGURE 3.47
Letters printed in three different serif fonts (light gray), with the skeletons (medium gray), and the result after pruning short terminal branches (black).

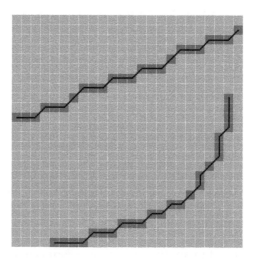

FIGURE 3.48
Two lines represented on a pixel array. The upper one represents a straight line at an angle of 22°, the lower one a 90° curve indicating a corner.

The upper line in the figure is nominally straight, at an angle of about 22 degrees to the horizontal. The changes in direction from one link to the next alternate between left and right deviations, with some straight connections interposed. In the lower line the cumulative deviations are all in one direction, resulting in a curve. By summing these changes over many pixels, the curve, which amounts to a 90-degree change in overall direction, can be detected. The distance over which the averaging must be done depends on the overall scale and size of the object. The use of a biased skeleton such as the one in Figure 3.45b defeats attempts to locate corners.

When applied to feature boundaries or skeletons, irregularities arising from noise also create problems for defining the local curvature. Calculation of bending energy is dependent on the scale chosen as representative of the boundary. Bending energy is defined (Sonka et al., 2008) as the sum of squares of the curvature over an increment of boundary length, L. The sampling length is usually specified as the total line length divided into a number of equal increments so that shapes with different perimeter or skeleton lengths can be compared. For the two lines in Figure 3.48, the bending energies calculated with a sampling length of 1 pixel are the same, $13 \cdot (\pi^2/16)/24$, but with a sampling length of 24 pixels the top line has zero bending energy while the bottom one shows a bend of 90 degrees and a bending energy of $(\pi^2/4)/24$.

When calculated for a set of specific objects, this parameter may give adequate discrimination to enable classification or identification. The smoothed curve may also be used to locate corner points (points of high local curvature) as part of the construction of a convex hull.

Branching Patterns

Many naturally occurring structures consist of branching networks, with skeletons that represent their shapes. Some of these are three dimensional but others, like river patterns and the formation of frost on surfaces, are two dimensional, and some of the three-dimensional ones can be studied using two-dimensional sections or projections.

In biology, the network of neurons (Smith et al., 1989), blood vessels (Sandau & Kurz, 1994; Kurz & Sandau, 1997), air passages in the lung (Bennett et al., 2000) (Figure 3.49a), tree branches (Figure 3.49b), and so on are examples (Bittner et al., 1989). In geology, it is the structure of rivers (Cieplak et al., 1998; Banavar et al., 1997; Rodriguez-Iturbe & Rinaldo, 1997) and their tributaries (Figure 3.49c), as well as the pore structure in sandstone. The fractal branching of lightning is particularly striking when captured in the form of a Lichtenberg pattern (Figure 3.49d).

These structures have a finite width, but analysis of the skeleton often shows them to possess a fractal geometry. The recursive branching reaches close to all parts of the area in which the structure exists. Different trees obviously have different patterns, but all seek to place leaves where they can capture the maximum amount of light. River patterns vary according to the

FIGURE 3.49
Fractal branching patterns in nature: (a) air passages in a cast of a lung; (b) oak tree branches; (c) aerial view of a river pattern; (d) Lichtenberg pattern of electrical discharge captured in a Lucite block.

slope of the land, amount of rainfall, and the types of rock or soil, but all provide drainage of an area.

For a river pattern, as an example, increasing the resolution of the aerial photographs or the map will refine the measurement of the area of the drainage, but the value will tend toward a limiting value, with less and less error or uncertainty. Conversely, the length of the river and its tributaries continues to increase with resolution, as more irregularities and smaller creeks are revealed. Plotting the total length as a function of the minimum resolved distance (on the usual log–log axes) produces a straight-line relationship for a fractal with a slope related to the dimension. It is observed (Feder, 1988) that rivers in the eastern and western regions of the United States have different fractal dimensions, presumably because of the differences in topography, soils, vegetation, river length, and rainfall in the regions.

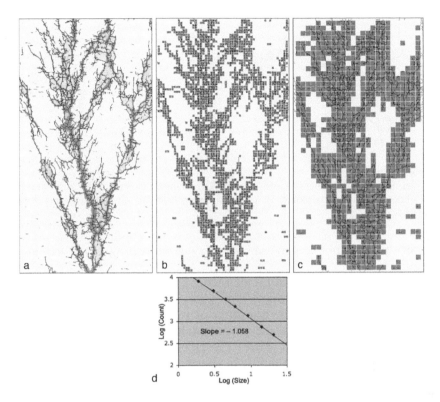

FIGURE 3.50
Box counting applied to a cross-section through a stress-corrosion fracture: (a) the image of
the fracture with the superimposed skeleton; (b, c) two different size sets of grids showing the
boxes through which the skeleton passes; (d) graph of the results.

For digital images of a branching structure, one practical measurement
technique is box counting. Figure 3.50 shows the method, as applied to a
stress corrosion crack in stainless steel. Brittle fracture in materials has been
well studied as an example of a fractal process, and Chapter 5 introduces
measurement techniques that can be applied to images of the surfaces. The
box-counting technique counts the number of grid squares through which
the branching structure passes as a function of the box size. The magnitude
of the slope of the line through the points on a log–log plot gives the dimension of the pattern. Because the image shows a cross-section through the
fracture, this is exactly 1.0 less than the dimension of the fracture surface.

Branching structures that closely mimic those found in nature are easily generated by simple, probabilistic rules. For example, Lindenmayer or
L-systems consist of a grammar with just a few primitives, such as move
forward, turn right or left by some specified angle (or sampled from a probability distribution of angles), and return to a previous position. With a
few sequences of rules and a set of probabilities for selecting each one, a

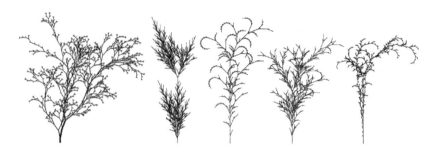

FIGURE 3.51
Examples of branching structures generated by L-systems by varying a few rules and angle probabilities. (From P. Prusinkiewicz, A Lindenmayer, 1990, *The Algorithmic Beauty of Plants*, Springer, New York.)

recursive procedure can generate the various branching structures shown in Figure 3.51.

If nature has encoded the rules for branching in the genetic code, as seems likely, it may not take very many characters in that code to specify the distinctive shapes of plants. The grammar or rules in these constructions can be thought of as the elements for syntactical analysis of the forms. However, analyzing the structures to extract the rules is much more difficult than using the rules to generate a family of shapes.

Not all branching patterns are fractal, of course. Many naturally occurring branching patterns are composed of parts that yield to other simplifying descriptions. Figure 3.52 shows the roots of a plant (the individual roots were cut off and arranged on a flat bed scanner for imaging). The skeleton of the image captures the essential branching pattern. Removing the node points disconnects the links in the skeleton. Measuring the lengths of only the terminal branches (not those that extend from one node to another) produces a distribution that is log-normal (the distribution of the logarithm of the lengths is Gaussian), which allows the data to be summarized by mean and standard deviation.

A log-normal distribution is usually attributed to the combination of many independent random variations that combine in a multiplicative manner (the conventional normal or Gaussian distribution results from the combination of many independent random variables whose effects are added together). Similar log-normal distributions are often found in natural systems. Examples include the distribution of sizes of fruits and flowers, rainfall amounts, lengths of latent periods for infectious diseases, distributions of mineral and petroleum resources in the Earth's crust, and even the length of words in the English language, and perhaps in others (Limpert et al., 2001).

In addition to the skeleton, the outline of a feature can also provide a topological characterization. The grains in steel are approximately polyhedral, and a polished section through the structure shows polygonal shapes. In the example of Figure 3.53, the 832 grains that do not intersect the edge of the

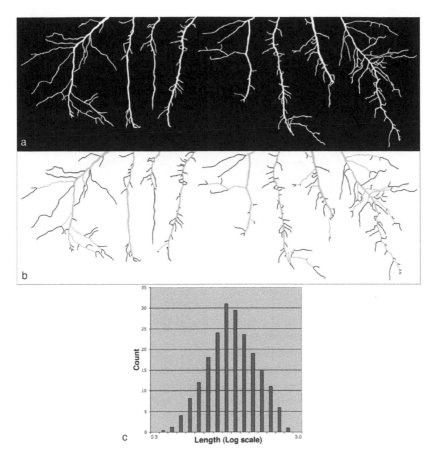

FIGURE 3.52
Root terminal branches: (a) scanned image of plant roots; (b) thresholded image with overlay showing terminal skeleton branches; (c) distribution of the number of terminal branches as a function of the logarithm of length.

image are shaded according to the number of neighbors each one has. This is equivalent to the number of sides and the number of corners. (Another way to count the number of corners or sides would be to use the corner-locating logic from Figure 3.48.)

A distribution of the number of sides (Figure 3.54a) shows that this is another quantity that has a log-normal distribution (which is an indication that the steel is fully annealed and in thermodynamic equilibrium). As shown in Figure 3.54b, the size of each grain section and the number of sides are correlated (in spite of the evident scatter in the data, the R^2 value is significantly high). One reason for this is simply that if the section plane intersects a corner of the polyhedral grain, it is likely to produce a small section with fewer sides than an intersection through the center of the grain. A second reason is that larger grains in the 3D structure have more neighbors, and that

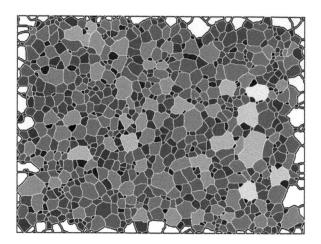

FIGURE 3.53
Grain structure of a low-carbon steel with shading to indicate the number of neighbors around each grain (edge-touching grains are omitted since the number of neighbors cannot be determined).

the small grains that fill in the space around the larger ones are limited to fewer neighbors and hence fewer sides.

This property of the number of sides (or corners) measures something quite different from the dimensionless ratios introduced earlier. Figure 3.55 shows plots of the number of sides versus formfactor, solidity, and radius ratio. In all cases the R^2 value is very small, indicating that there is not a significant correlation.

The number of sides is also important in the analysis of the structure of the diatom in Figure 3.56. The structure consists primarily of six-sided openings, but there are some with five or seven sides. The interesting relationship is that these occur in pairs, so as to maintain the overall structure. The one opening with eight sides has two five-sided neighbors.

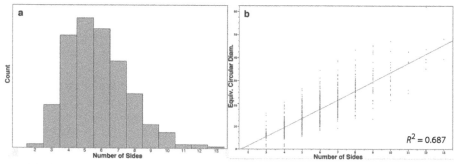

FIGURE 3.54
Histogram showing the distribution based on the number of sides on each grain in Figure 3.53, and a plot showing the correlation between the number of sides and the size (equivalent circular diameter) of each grain.

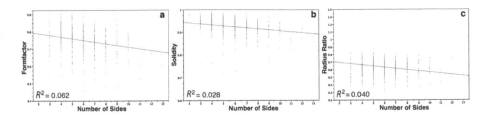

FIGURE 3.55
Regression plots of formfactor, solidity, and radius ratio do not show any significant correlation with the number of sides.

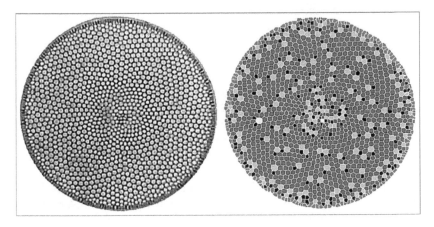

FIGURE 3.56
Image of a diatom (left) and the result of shading each opening according to the number of sides. The lighter seven-sided cells are always adjacent to a darker five-sided one.

Landmarks

A fundamentally different approach to capturing shape information uses landmarks (Kendall, 1989; Bookstein, 1991; Dryden & Mardia, 1998; Lele & Richtsmeier, 2000; Lestrel, 2000). By definition, these are arbitrary locations selected by an expert as representing significant and consistently identifiable points. Bookstein (1991) states that "thoughtful reduction of pictorial data to named landmark locations is the essential first step in any analysis of form." Typically the points have names, such as the "tip of the nose." Often the selected points lie on the boundary of the shape, but internal points may also be used.

Furthermore, in each image the landmarks are recognized by and their locations marked by an expert or at least a human with considerable experience. The result is a set of homologous points whose distances and angular relationships to each other can be analyzed by statistical means. Homologous points are locations selected based on the assumption that they are consistently recognizable and meaningful on different objects, or one object over time, based on ontogeny or phylogeny. The way the points are "pushed around" by ontogeny or phylogeny are then measured.

Of course, this is somewhat circular reasoning, as many critics have pointed out. Also, the definition of homology has evolved and changed over time from "same-function" to "same-appearance." They are sometimes described as "biological eigenvalues." Three types of homologous points are proposed by Bookstein (1991):

1. Boundaries between distinct regions
2. Maxima of boundary curvature
3. Extremal points

Notice that these are all associated with boundaries, and are consequently sensitive to the ability to obtain adequate resolution and definition of the object outlines. The widest application of landmark methods has been to two-dimensional images, but the basic idea and most of the mathematics involved extend straightforwardly to three dimensions. Studies of the 3D skull shape in hominids has been simplified by applying landmark methods to section contours in specified locations and orientations, but there is no fundamental reason not to use full (x, y, z) coordinates.

Landmarks may oversample some areas relative to others, and overlook and ignore much of the fine detail associated with shape. They represent the object by a selectively incomplete mapping of form onto a measurement template. The success of the method depends very much on the adequacy of the selection.

Figure 3.57 shows the landmark points from the mandible of a mouse used in studies of the changes in shape reported in Lele and Richtsmeier (2000). Locations such as numbers 5 and 6 are defined as the "uppermost" or "lowermost" points on the curved rim of the bone, and can shift significantly

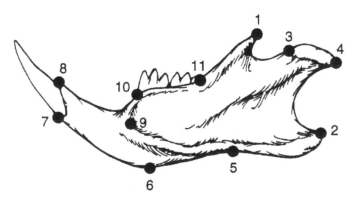

FIGURE 3.57
Landmark points on the mandible of a mouse. (From S. R. Lele, J. T. Richtsmeier, 2000, *An Invariant Approach to Statistical Analysis of Shapes*, Chapman & Hall/CRC, Boca Raton FL.)

with minor orientation adjustments to the bone. The authors also report that differences in landmark locations due to different observers may result from prior knowledge, and they suggest that this "inter-observer error" should be avoided by having a single person collect all of the data.

There have been some attempts to replace the manual marking of point locations by using cross-correlation of the image with fragment images of the idealized landmarks. The greatest use of this technique has been in medical imaging, where the purpose of obtaining the landmark locations is to morph the image to align with a reference image or diagram of the organ or body. This requires distorting the image so that the points are all aligned and is called a generalized Procrustes method. The name refers to the legendary Greek robber (eventually killed by Theseus) who captured travelers and forced them to fit his iron-framed bed by either stretching them or cutting off excess parts.

In addition to the obvious desire to eliminate the need for human marking of locations, the cross-correlation method has the advantage that it can use information from multiple pixels to locate (potentially with subpixel accuracy) landmarks that are not points but extend over a finite area, such as the end of a bone, the tip of a finger, or a branch in a blood vessel. Another approach to locating internal point locations uses the branch points in the skeleton of the shape (O'Higgins, 1997).

The origins of the landmark method are usually credited to D'Arcy Thompson (1917), although it has an earlier connection to the Bertillon system adopted by the French police in 1888. Alphonse Bertillon (1890) advocated using systematic measurements of the head and body of criminals as a means of identification (and also of finding characteristics of head shape he believed were associated with criminality), but the difficulties of reproducing the measurements by different officers, and changes over time, caused the system to be abandoned in favor of fingerprints, which are another example of landmarks (as shown in Figure 1.18). (Sir Francis Galton's book *Finger Prints* was published in 1892, but Mark Twain had described their use to identify a murderer in *Life on the Mississippi* in 1883.)

Thompson applied grids to images such as the heads of different primates (Figure 3.58) or to the shapes of different fish (Figure 3.59), and showed that by distorting the grid it was possible to transform one into another. Some of

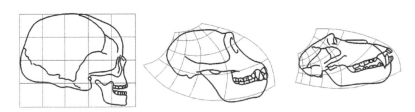

FIGURE 3.58
D'Arcy Thompson's mesh relationship between the skulls of a human, chimpanzee, and baboon.

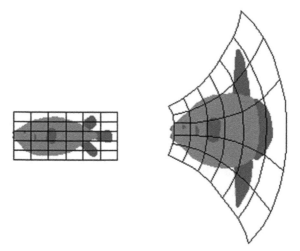

FIGURE 3.59
A mesh relationship between the puffer fish and the mola mola. (After D. W. Thompson, 1917, *On Growth and Form*, Cambridge University Press, Cambridge, UK.)

Thompson's methodology has subsequently been discredited, but the basic idea of tracking the continuous transformation of shape with aging or evolution has continued, with somewhat more rigor. One change has been the introduction of grid deformation with more uniformity, with techniques such as thin-plate splines and conformal mapping (illustrated in Figure 1.24). But the major trend has been to abandon grids altogether in favor of treating the distances between pairs of landmarks as the measure of "shape."

One area of application is the study of ontogeny, which uses landmarks (and/or grids) to study how different parts of the body of an animal or human, or any organism, or a part of one such as leaves, bones, or organs, even a cell, grow as the organism matures. Another is phylogeny, the study of how evolution has altered the shapes of body parts of different species. Both are concerned with determining how form or shape varies, as defined by the motions of various landmarks. Anyone interested in the general topic should begin by reading Gould (1966, 1977) and Alberch et al. (1979).

The most famous example of phylogenetic modification of shapes is that of Darwin's finches. Charles Darwin recognized that the variation in beak shapes of finches in the Galapagos Islands represented adaptation to different types of diets and used this as one key example in his explanation of evolution as a response to natural selection. There are a dozen species of Darwin's finches; Figure 3.60 shows examples of the differences in their beak morphology.

Recently, both the geometry of the shapes and the genetic factors that control them have been investigated by Campàs et al. (2010). They found that the insect-eating finches with bills adapted for probing and grasping, as well as their ancestor (the bird that presumably arrived on the Galapagos Islands

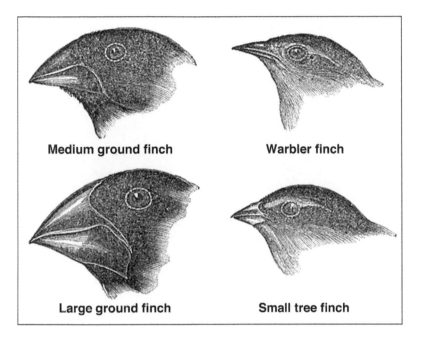

Medium ground finch

Warbler finch

Large ground finch

Small tree finch

FIGURE 3.60
Differences in beak shapes: the two finches at the left eat seeds and have crushing beaks; the two at the right eat insects and have bills for probing and grasping.

and gave rise to the various finch species), have beak shapes that can be matched by scaling transformations alone. Similarly, the seed-eating finches with bill shapes adapted for crushing have beak shapes that can be matched by scaling transformations alone. And even more remarkably, the latter are related to the former by a single shear transformation. This applies to both the bill and the underlying bone structure. Furthermore, the researchers found that the expression of two genes, *Bmp4* and *Cadmulin*, is quantitatively correlated in each bird species with the modifications in beak shape.

The study of the finch beaks uses the full outline of the shape, but many similar efforts use only the location of the landmark points. The study of the distances between landmark points and of the covariances between these distances is called "morphometrics." The object is not to describe the forms themselves, but by abstracting representative, meaningful, or perhaps critical dimensions, to make determinations of their associations, and perhaps discover correlations with causes or effects. The literature of the field rarely shows images or even sketches that actually represent the underlying structures. Instead, matrices of numbers, perhaps with vector diagrams of the lines between points, and sometimes a tesselation of the points, may be presented along with extensive statistical analysis.

Figure 3.61 illustrates a typical starting point. The image of a human hand, with its outline superimposed, is used to locate a few landmarks.

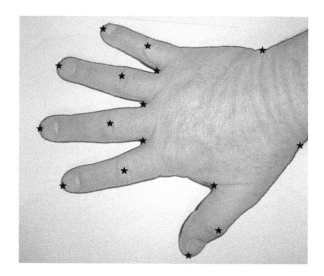

FIGURE 3.61
A human hand, with some landmark points.

The distances and directions of some of these will remain constant as the hand is moved, whereas others (for instance, the distance between the tips of different fingers) will not. In this view the knuckles and other interior points are not visible; most of the landmarks are along the periphery of the shape.

Performing a least-squares alignment of the points in this image with another image of the same hand might be useful if the residual differences indicated a repositioning of the fingers. (But if only one finger had moved, the least squares approach would indicate some shift for all of them.) Comparing the image to the same hand at a different age might provide information about the relative growth of different bones. Comparing this image to one of a different person's hand might reveal systematic differences in the relative sizes of the fingers. And comparing this hand to that of a chimpanzee might correlate in some way with how humans and chimps use their hands. An example in Chapter 4 (Figure 4.42) demonstrates the use of landmarks for identification of individual animals from their footprints.

The coordinates of the various landmarks are used to create a matrix of distances between each pair of points (and sometimes also the angles of the lines connecting them). For pure shape characterization, the distance matrices are normalized so as to be scale-free by expressing all of the distances as ratios. Sometimes weighting is added to make some points, for example, the ones that are more precisely determined, more important than others. The choices for normalization include the longest distance, a particular chosen distance, or a statistical mean of all of the distances in the matrix. The angle matrix may be similarly adjusted so as to be rotation independent by selecting a particular line as a reference or by using a statistical combination of the individual line angles as the reference.

When matrices are compared to detect changes in orientation or size (for instance to track changes as growth occurs), normalization is not used. Instead, the distances (and directions) between the corresponding landmark points are calculated and combined for alignment and to monitor changes. Various methods are used to align the sets of points, such as minimizing the sum of squares of distances, or minimizing the sum of absolute values, and so on. Each of these methods produces a somewhat different result. Analyzing the set of displacements is also performed in many different ways, depending on the subject and the researcher (Cootes et al., 2005; Ramanathan et al., 2009).

The most commonly used method for statistical shape analysis is generalized Procrustes analysis (GPA). For a series of points with coordinates (x_i, y_i), or in three dimensions (x_i, y_i, z_i), translational variations are removed from the object by finding the mean values:

$$\bar{x} = \frac{\sum x_i}{n}$$

$$\bar{y} = \frac{\sum y_i}{n}$$

(3.8)

The means are then subtracted from all of the (x, y) coordinates. Next the scale component is removed by scaling the object so that the sum of squared distances from the points to the origin is 1. The size, s, is calculated as

$$s = \sqrt{\sum \left\{ (x_i - \bar{x})^2 + (y_i - \bar{y})^2 \right\}}$$

(3.9)

All of the coordinate points are then divided by s:

$$x_i' = \frac{(x_i - \bar{x})}{s}$$

$$y_i' = \frac{(y_i - \bar{y})}{s}$$

(3.10)

Adjusting the values to remove any rotation is accomplished by minimizing the sum of squares of distances between points as a function of angle. Rotation by an angle ϕ gives new coordinates (u_i, v_i):

$$u_i = x_i' \cdot \cos\phi - y_i' \cdot \sin\phi$$
$$v_i = y_i' \cdot \cos\phi + x_i' \cdot \sin\phi$$

(3.11)

The angle is found that minimizes the sum of squares of distances:

$$d^2 = \sum \left\{ \left(u_i - x_i'\right)^2 + \left(v_i - y_i'\right)^2 \right\}$$

(3.12)

Other methods are also used, some more statistically robust than others, but none can guarantee "true" correspondence (Davies et al., 2008). Introductions and summaries for these procedures can be found in Goodall (1991), and Dryden and Mardia (1998).

Human Faces

One interesting use of the vectors representing growth is the "aging" of facial appearance. This is often used with photographs of missing children, to predict their appearance at different ages (Figure 3.62). It has also become important for security screening, to predict the changes in adult faces with age (Albert et al., 2007).

Multiple landmark points are used to model facial shape as indicated in Figure 3.63. Vector distortions of the shape are then applied, along with changes to skin texture, hair color, and so forth. Suo et al. (2010) use a model with 90 landmarks to generate synthesized aged images as shown in Figure 3.64 and Figure 3.65.

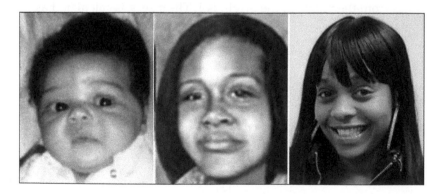

FIGURE 3.62
(See color insert.) Simulated aging of an infant's face. The image at left is the baby picture of Carlina White at age 19 days, before she was kidnapped from a hospital crib. The center image shows the simulated aged appearance. At right is a photograph of Carlina at age 23, after she recognized herself in the picture and tracked down her real parents. (National Center for Missing and Exploited Children.)

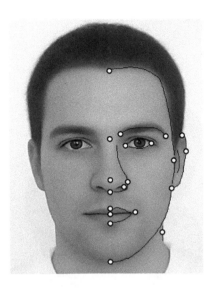

FIGURE 3.63
Illustration of a few landmark lines and points on one side of a face (note that some these
depend on facial expression as well as head "shape").

There have been some notable successes obtained by this approach. One
is recognizing the changes in facial "shape" due to fetal alcohol syndrome,
which some physicians learned to recognize but it was not easy to communicate to others until the characteristics were isolated and compiled (Figure 3.66).
Notice, however, that the characteristics are described by adjectives such as
"low," "underdeveloped," "flat," and "thin," and are not quantitative. These
are best understood as syntactical elements rather than shape measurements.

Another example is relating the occurrence of avian keratin disorder, in
which birds' beak shapes become deformed and affect feeding and preen-
ing, to environmental pollutants such as organochlorine compounds. A
current effort is underway to use landmarks to represent the footprints of
individual animals in various endangered species such as polar bears and
white rhinos (see Figure 4.49). The difficulty, of course, is the need for an
expert to select the key landmarks and then to recognize and locate them
in each footprint.

Lestrel (2000) suggests that while angles and ratios of distances are technically "size-independent" they are problematic as shape descriptors. Partly this
is because ratios do not usually exhibit normal statistics, yet virtually all of
the statistical comparison tools that are applied to landmark data assume and
are valid only for normally distributed values. For example, the assumption
that the measurements have uniform and normally distributed uncertainties
or perturbations in all directions makes it possible to test for differences in
vector distances using a t-test extended to two dimensions (a t^2-test).

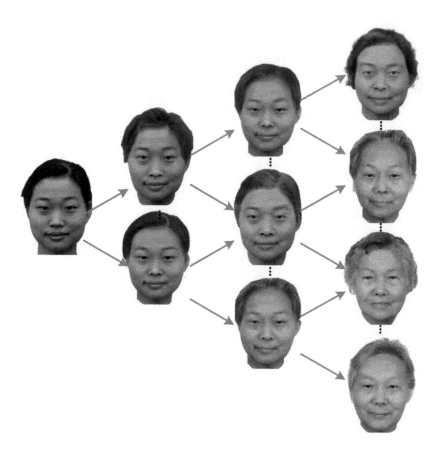

FIGURE 3.64
Aging one face. At each time interval, additional uncertainties increase the range of plausible faces in each age group. (From J. Suo et al., 2010, *IEEE Transactions on Pattern Analysis and Machine Intelligence* 32(3):385–401.)

Another objection raised is that the points oversimplify the structures and lose or ignore much detail, which is both the strength and the weakness of the landmark approach.

Just as there are few visual tools to help the analyst comprehend the "meaning" of the landmark displacements, so there are few parameters that summarize the results. One, called the "centroid size" of the structure represented by the landmarks, is simply the square root of the sum of squares of all pairwise distances between points. This has the dimension of length, but relating it to a physical size is sometimes difficult.

One approach to facial recognition is based on ratios of distances between landmark points such as the corners of the eyes and mouth, tips of the ears, chin and nose, and so forth, as illustrated in Figure 1.17. These landmarks can often be located by computer algorithms, based either on edge processing or cross-correlation.

20-30   30-40   40-50   50-60   60-80

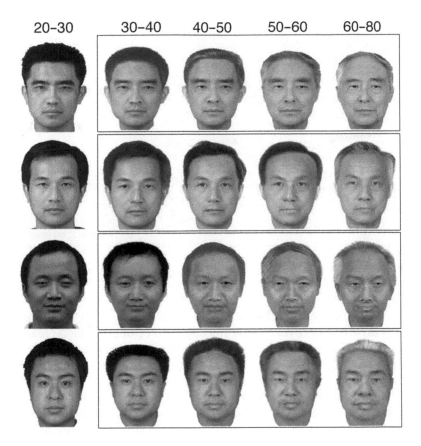

FIGURE 3.65
Aging simulations. The leftmost column shows original images, with the synthesized aged images shown for different age groups. (From J. Suo et al., 2010, *IEEE Transactions on Pattern Analysis and Machine Intelligence* 32(3):385–401.)

Fingerprint identification based on landmarks such as ends, breaks, and bifurcations in the ridge pattern (Figure 1.18) may be found automatically for an ideal fingerprint, but typical prints found at crime scenes are incomplete, distorted, or on less than ideal surfaces, and usually require human marking to locate the minutiae. Once located, the distances and directions from the central core of the print are used as landmark values to search the database.

For both facial recognition and fingerprint identification, the result of the search consists of a small number of closest matches that are shown to a human who must decide whether one of them actually matches the original image.

Another technique for identification that was originally proposed by Bertillon (1890) uses the shape of ears. These have the advantage that they do not change with age (except for becoming larger in size and some lengthening of the ear lobe). Limited use has been made in criminal trials of matching ear prints found at crime scenes to a suspect. Another area of potential use

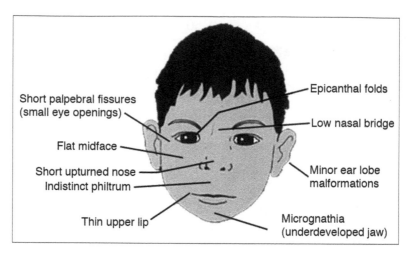

FIGURE 3.66
Indicators for fetal alcohol syndrome in the face of a young child.

is surveillance images, in which ears are often visible, even when someone has made an effort to cover or alter the appearance of other facial features.

One established method for characterizing ear shape uses a transparent template that is placed over the image and aligned with key points (Iannarelli, 1989). The ratios of radial distances at a series of fixed angles then become the basis for identification. This is a classic manual landmark procedure, but recent work has centered on automatic methods. Image processing can locate and extract the ear from profile images (Ibrahim et al., 2010). Once the outline of the ear and the principal internal structures have been isolated, it is possible to use many points to align the images, adjusting for scale, orientation, and perspective (essentially a generalized Procrustes alignment, as described earlier). A variety of statistical analysis procedures are used to search a database for matches. Although the existing data sets are small compared to those for fingerprints, high rates of success are reported for both positive matches and minimization of false matches (Hurley et al., 2007).

Other Methods

Like the landmark approach, a histogram of the lengths of all possible chords that can be drawn between any two points on the periphery of a feature is invariant to translation and rotation (Smith & Jain, 1982; You & Jain, 1984). If normalized with respect to the maximum chord length, the histogram is also invariant to scale. The histogram of the angles of the chords is invariant to scale and translation, and can be shifted modulo 360° to compensate for

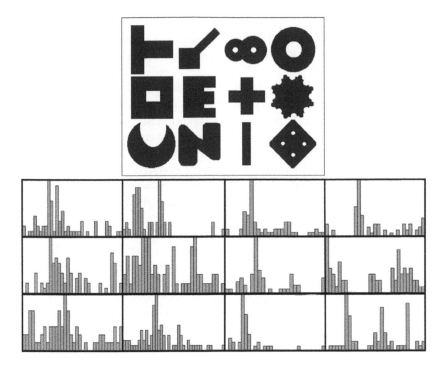

FIGURE 3.67
Twelve shapes with normalized histograms of their intercept lengths.

rotation. For small features, the finite pixel grid causes changes in the histograms with scale, translation, and rotation. For large features, the number of chords rises sharply and makes the construction time consuming. The method is not widely used, and although it uses the full outline of the shape, it cannot be used to reconstruct the original shape.

Rather than using a complete set of all possible chords, a much simpler procedure measures the intercept lengths by superimposing a grid of parallel lines. Figure 3.67 shows plots of the intercept lengths obtained by rotating the parallel lines in 10-degree steps. The scales of each histogram have been normalized to the longest value. The 12 features shown in the figure are the same ones from Figure 3.17, which have the same measured formfactor. The histograms of intercept lengths are obviously distinct for these shapes, but it is not clear how they should be used to quantitatively describe or recognize each shape.

Curvature Scale Space

Another approach to dealing with arbitrary shapes uses the local curvature along the periphery to locate landmarks, and then uses the locations to form a vector representing the shape. Searching a database for shapes similar to a target is then carried out by comparing the vectors. This method, called curvature scale space (CSS), finds the significant points along the contour by a process of progressive smoothing (Mokhtarian & Mackworth, 1986; Mokhtarian et al., 1996).

First, the pixels along the periphery are sampled to select 200 equally spaced points, with coordinates plotted against distance along the path. The inflection points along this profile are marked. Then a Gaussian smooth is applied to the graph, which removes some of the detail and with it some of the inflection points. Figure 3.68 shows the procedure. As the standard deviation of the Gaussian is increased, a graph is constructed showing the position of the inflection points along the profile (shown at right in the figure). As the points of inflection move together and finally merge, they produce a series of maxima on the graph.

A list of the maxima recording the position along the path and the height at which each occurs (the standard deviation of the Gaussian that eliminated it) then describes the shape. Noise is dealt with by ignoring maxima below 20% of the highest value. Because all shapes are sampled down to the same number of points, the method is scale independent. Shifting the graph horizontally (which amounts to adding or subtracting a constant from the location of the maxima) or using only the relative position of peaks allows for rotation independence. Traversing the graph left to right or right to left handles mirror images.

It is not obvious how the CSS representation of shapes can be used for classification or as a tool for quantitatively measuring the differences between shapes, but it produces impressive results in searching for similar shapes. The presence and location of the maxima are syntactical elements. Given a target shape, the CSS contour list is constructed and the maximum value found. The magnitude of the value corresponds to the depth of the deepest indentation or greatest protrusion present on the original periphery. The database is scanned for shapes whose maxima lie within 20% of this value. Then the next highest maximum and its distance along the periphery from the first is located, and those values are similarly used to scan the candidate points. The process continues until the number of remaining candidates is below a defined limit, and those shapes are presented to the viewer.

An interactive Web site, http://www.ee.surrey.ac.uk/CVSSP/demos/css/demo.html, demonstrates the procedure using a database of 1100 shapes, all obtained by scanning images of sea animals, mostly fish. Many of the images show considerable noise along the boundary, and some are views with unusual perspective or angles of view. The goal is not to identify the fish, per se (and indeed, no data are provided on the identity of any of the images), but rather to create an interesting array of shapes, a few of which are shown in Figure 3.69.

The demonstration allows the viewer to select a shape from the contents of the database, and then shows an array of the nearest matches, as shown in Figure 3.70. The search time for the 1100 CSS contour codes is less than a second.

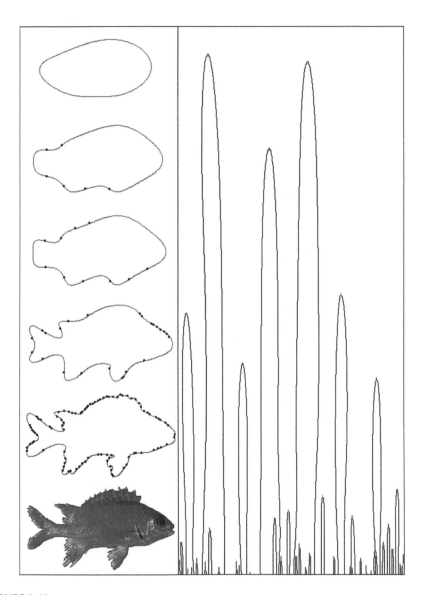

FIGURE 3.68
Progressive smoothing of a boundary (at left) reduces the number of inflection points (marked). Plotting the location of the inflection points along the boundary (horizontal axis) versus the standard deviation of the Gaussian smooth (vertical axis) produces the graph at right, whose maxima are used as the CSS code for the shape. (From P. L. Mokhtarian, S. Abbasi, J. Kittler, 1996, *Proceedings of the British Machine Vision Conference*, 53–62.)

FIGURE 3.69
Some of the shapes in the SQUID demo database.

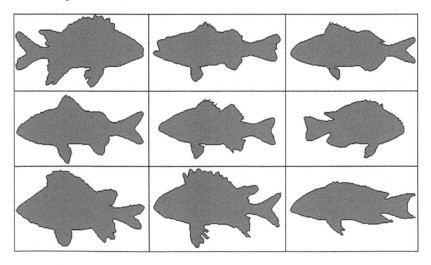

FIGURE 3.70
Shape search results. The shape at the upper left is the target. The other shapes are shown in decreasing order based on the CSS search criteria.

The identification of a sequence of inflection points along the boundary is an example—a particularly successful one—of the syntactical approach to shape description. The implementation is also a good demonstration of a class of operations called Query by Image Content (QBIC). Most of the current database search programs for images (e.g., Google® Images) use text labels or words from the accompanying documentation as the basis for the search. The goal of QBIC is to use the actual contents of the image. Most currently deployed systems, such as the one using IBM's software (Niblack, 1993; Faloutsos et al., 1994; Flickner et al., 1995; DeMarsicoi et al., 1997) for the Hermitage Museum in St. Petersburg, Russia (http://www.hermitage-museum.org/fcgibin/db2www/browse.mac/category?selLang=English) can deal with colors, textures, or a few simple shapes such as squares and circles. The SQUID (Shape Queries Using Image Databases) system based on CSS seems to be capable of dealing with complex natural shapes directly.

Some Additional Approaches

Image analysis of paintings has also been used to identify artist's distinctive patterns and shapes of brushwork (Johnson et al., 2008), and the shapes of picture elements that are used (Kirsch & Kirsch, 1988). Methods based on statistical texture analysis or wavelet analysis to isolate and identify characteristics of individual artists have also been reported (Keren, 2003; Hughes et al., 2010).

As an example of another approach to shape identification, a neural net was applied to the silhouettes of the tropical reef fish shown in Figure 3.71. The dimensionless ratios in Equations 3.2 to 3.4 were measured. The application of principal components analysis indicated that formfactor, solidity, and radius ratio are the most important variables for distinguishing the various species. Figure 3.72 plots the data, which show considerable overlap between species. A neural net solution (as discussed in Chapter 6) with just these three measurements as inputs, using a randomly selected two-thirds of the data for fitting and the remainder for validation, has an overall error rate of 10%, better than might be expected from examination of the scatterplot. Adding the rest of the dimensionless ratios produces a neural net that makes no errors, but with very large positive and negative weight values that suggest an over-trained fit that might not be robust if applied to a larger set of images.

Another approach to searching a database of images has been proposed by Xi et al. (2007). Shapes are first unrolled as radius from the centroid versus angle, and the resulting "time series" divided into a small number of equal segments. The mean value of each segment is calculated and this reduced set of points is further quantized into small number of discrete values. From that a symbolic representation is created, a process called SAX (Symbolic Aggregate approXimation). The resulting motifs (Figure 3.73) are compared to determine the most probable rotational alignment of the shapes and the most probable matches. For the shapes that survive the screening process,

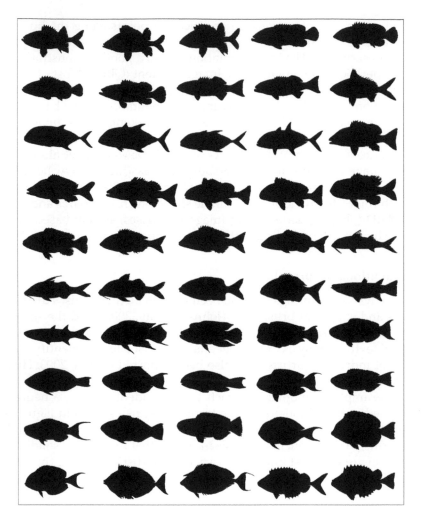

FIGURE 3.71
Silhouettes of tropical reef fish.

the profiles are compared based on the summed Euclidean distance between
them (Figure 3.74). This is calculated as

$$E.D. = \sqrt{\sum_{i=1}^{n}\left(t_{1i} - t_{2i}\right)^2} \qquad (3.13)$$

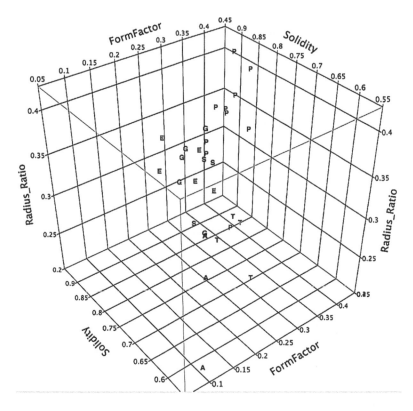

FIGURE 3.72
Scatterplot showing measured shape factors for the various fish in Figure 3.71 (A = Goatfish, E = Emperor, G = Grouper, P = Parrotfish, T = Trevally).

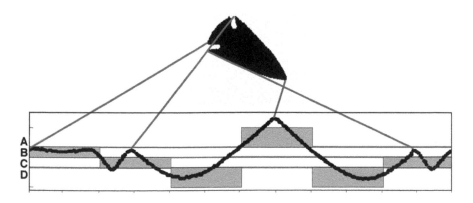

FIGURE 3.73
Generating the SAX motif from a shape. The unrolled profile is quantized into steps, which generate a symbolic sequence (BCDADC in the example) used to screen potential matches.

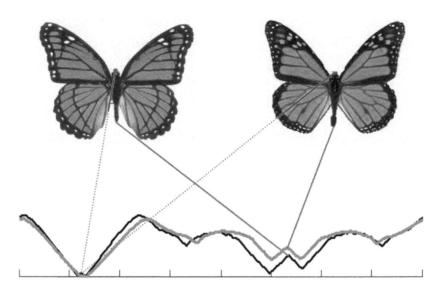

FIGURE 3.74

Comparing two butterflies from different families (left: *Limenitis archippus*, right: *Danaus plexippus*) that are similar in shape (also in color and pattern). The unrolled profiles are used to compute the Euclidean difference. (From X. Xi et al. (2007) in W. Jonker, M. Petkovic (Eds.) *SIAM Conference on Data Mining, Lecture Notes in Computer Science*, 4721:249–260, Springer Heidelberg.)

FIGURE 1.10
Brightness, color, size, and orientation illusions: (e) the red lines in the two spirals are usually seen as orange and magenta rather than identical; (f) the girl's right eye is usually seen as being cyan rather than the same neutral gray as the left eye.

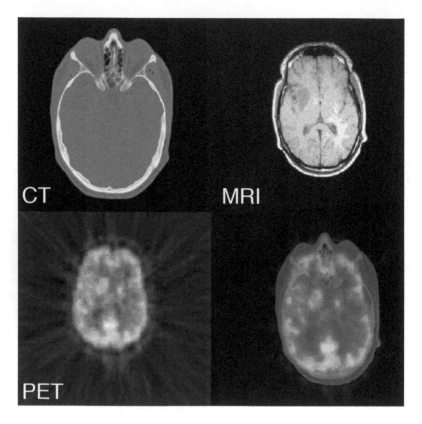

FIGURE 1.21
Images from computed tomography (CT), magnetic resonance imaging (MRI), and positron emission tomography (PET), and the merged result using color channels.

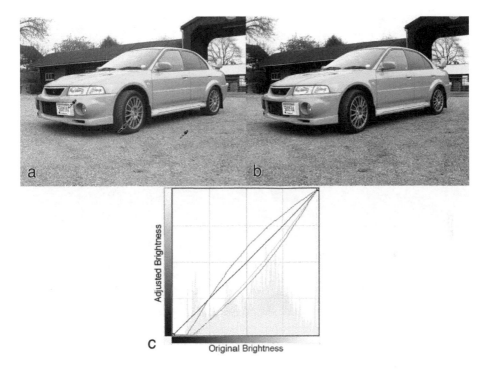

FIGURE 2.3
Neutral color adjustment: (a) original image, showing locations selected as defining neutral black, gray, and white; (b) corrected image showing that the car is blue, not green; (c) the transfer functions applied to the red, green, and blue channels to make the adjustment.

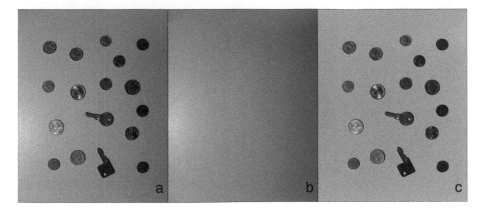

FIGURE 2.11
Subtracting a measured background: (a) the original image; (b) the background with the same lighting and the objects removed; (c) the result of removing the background.

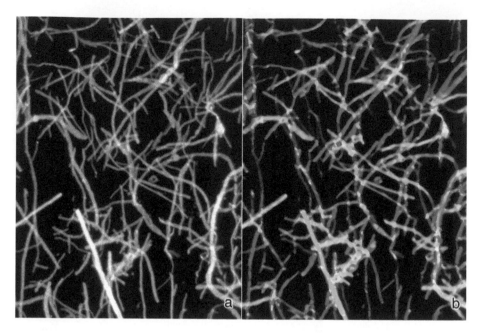

FIGURE 2.30
Image of cellulose fibers used in papermaking, with a colored representation in which the hue corresponds to the gradient direction (perpendicular to the fiber orientation) at each point along a fiber.

FIGURE 2.34
Thresholding a color image: (a) original; (h) principal components analysis applied to the original color image, with values along each axis displayed in the red, green, and blue channels.

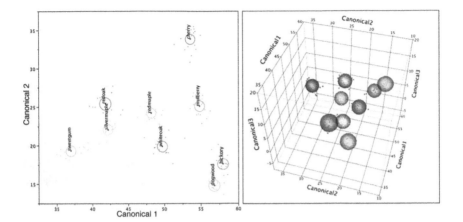

FIGURE 3.23
Classification of leaves by shape: successful grouping of the leaves based on three dimensionless shape parameters.

FIGURE 3.62
Simulated aging of an infant's face. The image at left is the baby picture of Carlina White at age 19 days, before she was kidnapped from a hospital crib. The center image shows the simulated aged appearance. At right is a photograph of Carlina at age 23, after she recognized herself in the picture and tracked down her real parents. (National Center for Missing and Exploited Children.)

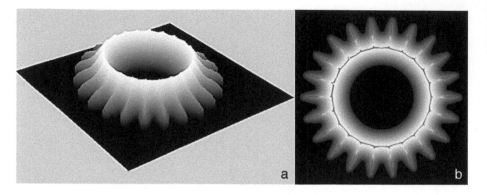

FIGURE 4.1
The EDM and MAT of the image of the gear in Figure 3.45: (a) the Euclidean distance map displayed to show the value of each pixel (the distance to the nearest point on the boundary) as an elevation; (b) the EDM with the MAT superimposed, with colors indicating the values along the midlines.

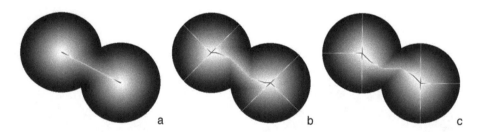

FIGURE 4.2
Distance maps for a shape composed of two overlapped circles, with the distance maps and superimposed MATs (shown in color) produced by different structuring elements: (a) Euclidean; (b) square; (c) diamond.

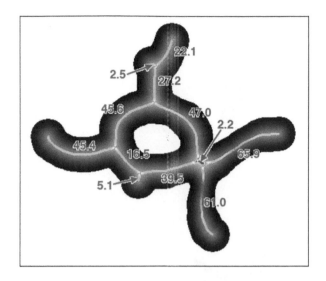

FIGURE 4.4
Pruning the MAT. The integrated value for each branch is shown (in arbitrary units). Short or narrow terminal branches and internal branches, with values shown in red, may be removed.

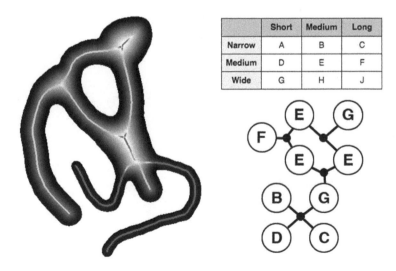

	Short	Medium	Long
Narrow	A	B	C
Medium	D	E	F
Wide	G	H	J

FIGURE 4.5
Reducing a MAT to a graph. The branch descriptions are simplified as shown in the table, and their linkage is represented by the nodes in the graph.

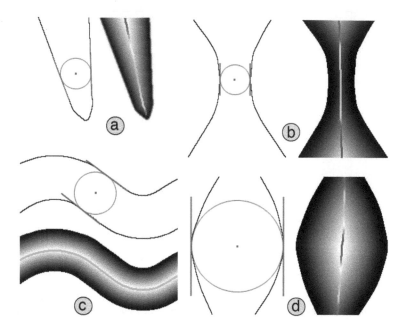

FIGURE 4.9
An illustration of the four types of shocks, as defined by the boundary curvature and the medial axis transform: (a) 1-shock; (b) 2-shock; (c) 3-shock; (d) 4-shock.

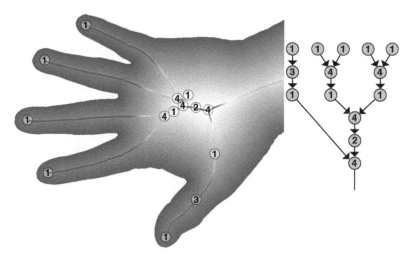

FIGURE 4.10
Silhouette of the hand from Figure 3.61, with its MAT and the graph connecting the shocks.

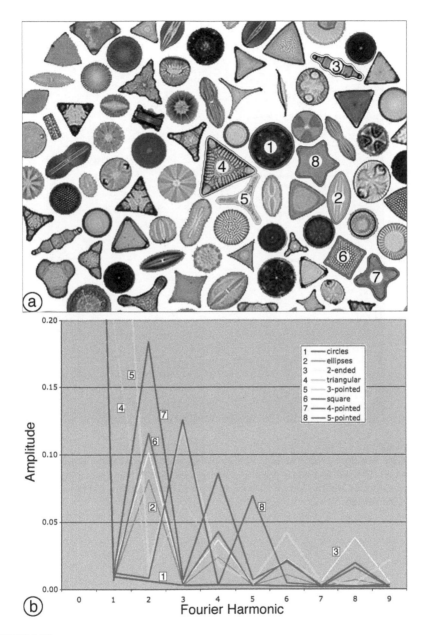

FIGURE 4.23
Some diatom shapes as discussed in the text, and plots of the amplitude of the Fourier harmonic terms for the boundaries averaged over multiple similar diatoms.

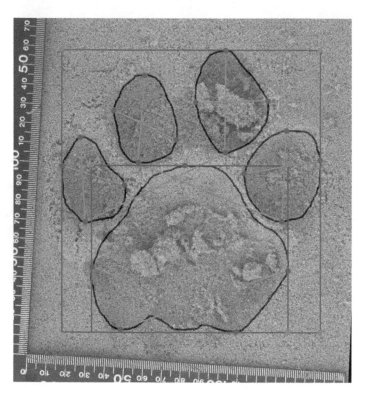

FIGURE 4.42
Landmarks on a tiger print. The red primary points are extrema on the outlines of each toe and heel pad. The blue secondary points mark intersections between several lines (green) joining primary points and the corners of bounding boxes (magenta) around the entire print and around the heel pad.

FIGURE 4.44
Outlines of five footprints from one tiger with the primary landmark points marked to show the amount of scatter.

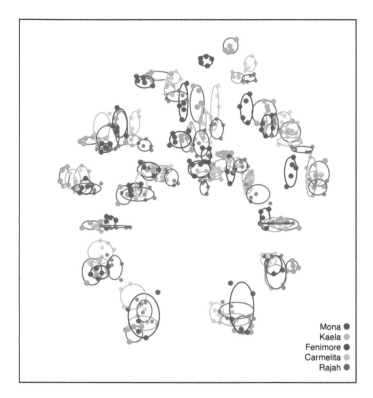

FIGURE 4.45
Superposition of the points and ellipses for each of the five sets of five tiger prints.

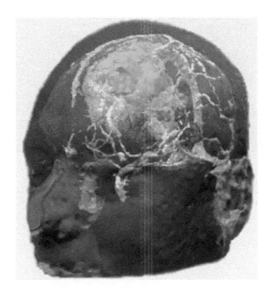

FIGURE 5.2
Three-dimensional volumetric visualization of a brain tumor.

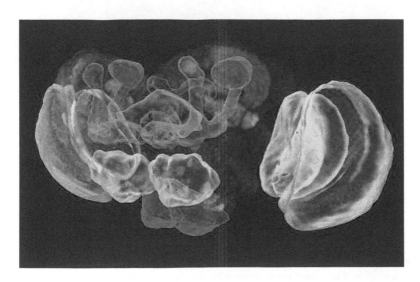

FIGURE 5.3
Reconstruction of a fruit fly brain with transparent surfaces. (Data from K. Rein, Department of Genetics, University of Würzburg; visualization by M. Zoeckler using Amira software.)

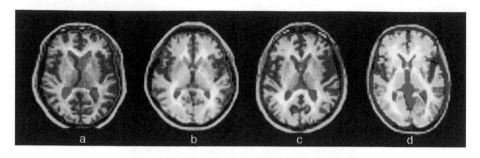

FIGURE 5.6
Aligned sections through 3D reconstructions of human brains from different imaging modalities. The green channel is an MRI image; the red channel is a PET image using fluorodeoxyglucose, and the blue channel is a PET image using 6-fluoro-L-dopa. (a) control; (b, c) two patients with Parkinson's disease; (d) patient with multiple system atrophy. (Adapted from M. Ghaemi et al., 2002, *Journal of Neurology, Neurosurgery and Psychiatry* 73:517–523.)

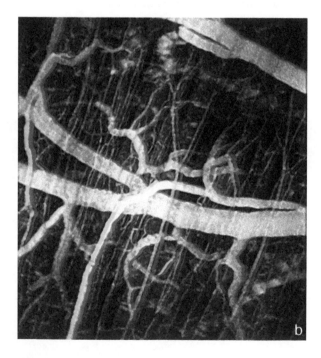

FIGURE 5.8
Stereo view of multiple focal plane images from a confocal light microscope showing light emitted from fluorescent dye injected into the vasculature of a hamster and viewed live in the skin: (b) color stereo for viewing with glasses (red filter on left eye, green or blue on right). (Courtesy of Chris Russ, University of Texas, Austin.)

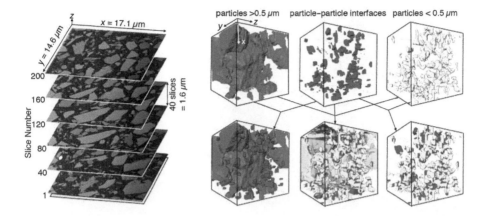

FIGURE 5.12
A stack of serial section images using FIB surface machining, and the reconstruction using different colors for particles of different sizes and the various interfaces present. (Courtesy of L. Holzer, Swiss Federal Laboratories for Materials Testing and Research.)

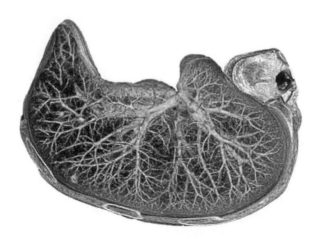

FIGURE 5.13
Volume-rendered image of branching vasculature in the lung. (Courtesy of TeraRecon, Inc., San Mateo, California.)

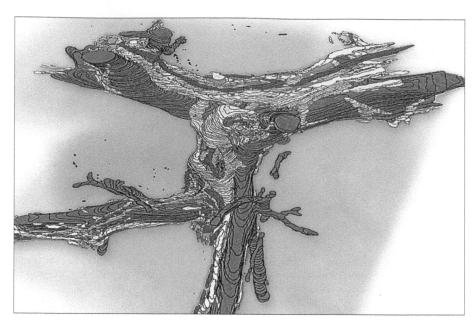

FIGURE 5.15
Visualization of vein (blue), arteries (red), bile ducts (green), and lymphatics (yellow) recon-
structed from serial sections through dog liver. (Courtesy of Dr. John Cullen, College of
Veterinary Medicine, North Carolina State University.)

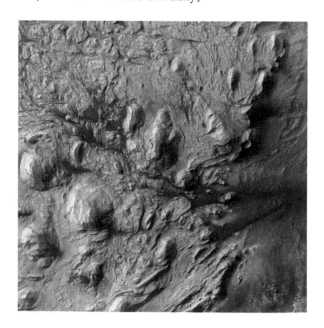

FIGURE 5.27
Stereo view of the corner of Gale Crater on Mars.

FIGURE 5.28
Stereo view of a fly's head. The scanning electron microscope pictures were recorded by rotating the specimen by 7 degrees.

FIGURE 5.34
Scanning laser rangefinder measurement of the interior of a room: (top) the panoramic image recorded by a digital camera as the unit rotates; (bottom left) the cloud of points whose distances from the unit are measured by the rangefinder; (bottom right) the resulting 3D model of the surfaces combining coordinate positions determined by the rangefinder data with image data from the camera. (Courtesy of Doug Schiff, 3rdTech Inc., Durham, NC.)

FIGURE 5.41
Elevation map of Mars. The elevation data have been color coded and shaded to emphasize the relief. (Courtesy of NASA and USGS.)

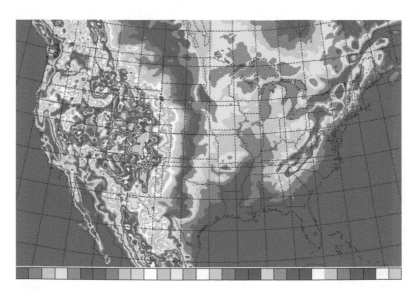

FIGURE 5.43
Elevation map of the United States, colored in steps of 100 meters.

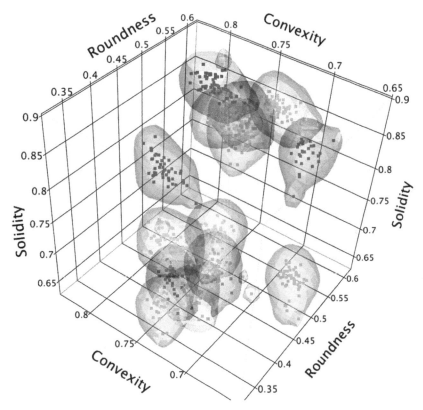

FIGURE 6.87
Scatterplot of the measured values, color coded by class with nonparametric surfaces surrounding each cluster.

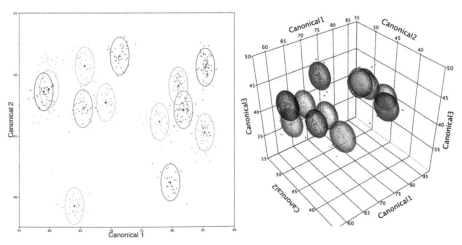

FIGURE 6.88
Linear discriminant results using three dimensionless ratios.

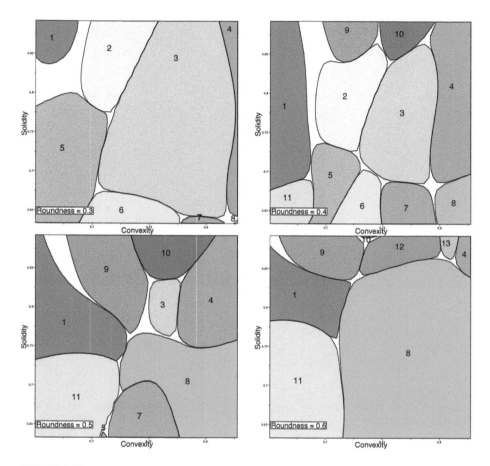

FIGURE 6.90
Contours for solidity versus convexity at different levels of roundness, showing the decision boundaries generated by the neural net model. Numbers identify the same region in the various slices.

4

Two-Dimensional Measurements (Part 2)

Chapter 3 summarizes shape descriptors that attempt to extract just a few numbers that may be used to characterize, and perhaps to compare or match, various shapes. One of the common threads in all of those parameters is their incompleteness. There are infinitely many visually different shapes that produce identical values for formfactor, or fractal dimension, or the number of loops in the skeleton. This chapter instead deals with shape descriptors that are complete, that is, they contain enough information to fully reconstitute the original shape and are therefore unique.

Although these descriptors are generally somewhat more compact than the pixel array they describe, they are not small. The problem with all of them is that they are complex, contain many numbers, and are consequently difficult to analyze or compare. The analysis of the data usually selects a small subset suitable for some purpose, such as distinguishing between classes of objects, but the subset is no longer a complete descriptor that can regenerate the original shape.

The Medial Axis Transform (MAT)

The medial axis transform (MAT) is a close cousin of the skeleton. As described in Chapter 2, the skeleton is the locus of centers of inscribed circles. This is the ridge line in the Euclidean distance map (EDM), and in some software implementations the skeleton is constructed from the EDM, rather than by a sequential pixel-by-pixel conditional erosion. If the radius of the inscribed circle, which is equivalently the value of the EDM, is assigned to the pixels in the skeleton, the MAT results. Figure 4.1 shows the results for the same image of a gear from Figure 3.45. The MAT shows the location of the ridges with the values of the EDM displayed in color. The importance of the MAT is that it is a complete shape descriptor, meaning that it can be used to reconstruct the shape of the original feature (Vermeer, 1994).

Although everything in the previous paragraph is exactly true for continuous images and functions, some practical problems arise when the original image is digitized onto a finite pixel grid. The ridges in the EDM are not necessarily single-pixel-wide continuous lines. A simple example arises when the local inscribed circle has a diameter that is an even number of pixels.

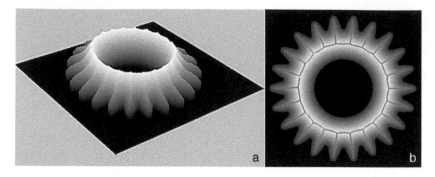

FIGURE 4.1
(See color insert.) The EDM and MAT of the image of the gear in Figure 3.45: (a) the Euclidean distance map displayed to show the value of each pixel (the distance to the nearest point on the boundary) as an elevation; (b) the EDM with the MAT superimposed, with colors indicating the values along the midlines.

That places the center midway between two pixels, both of which will have the same distance value assigned. In forming the skeleton, it is usual to randomly or systematically select one or the other, and to do so in a way that maintains the continuity of the skeleton line. Assigning the EDM value to this pixel then underestimates, if only slightly, the dimension of the feature at that point. Other similar problems can create plateaus of uniform values with irregular shapes in the centers of features. If integers rather than floating point values are used in the construction of the EDM, steps are created in the EDM and consequently in the MAT.

For nearly 40 years the MAT has been an intriguing tool for analyzing and computing with form, but it is one that is notoriously difficult to apply in a robust and stable way (Katz & Pizer, 2003). A tiny change to an object's boundary can cause a large change in its MAT. So can a single-pixel hole in the interior of a shape. There has also been great difficulty in using the MAT to decompose an object into a hierarchy of parts reflecting the natural components and a hierarchy that humans perceive, or to use the MAT as the basis for comparison of shapes.

There is a substantial body of literature on the MAT (and on the skeleton), but the majority of the papers are concerned with algorithms for efficiently generating a satisfactory approximation for an image represented on a pixel grid. There are fewer papers that discuss the use of the MAT for shape description or recognition. One thread of analysis deals with the use of the MAT of relatively simple geometric forms, often with considerable symmetry, to aid in decomposing the shape into a set of standard circles or polygons. The purpose is primarily for constructing suitable representations to aid in the machining and manufacture of parts, but a related interest is in human vision and the visual description of shapes as being composed of a collection of simplified forms.

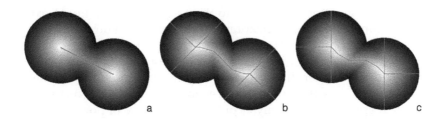

FIGURE 4.2
(See color insert.) Distance maps for a shape composed of two overlapped circles, with the distance maps and superimposed MATs (shown in color) produced by different structuring elements: (a) Euclidean; (b) square; (c) diamond.

The critical step in decomposing a complex shape into a set of component parts is finding those points along the MAT that are not redundant. A redundant point on the MAT is one whose corresponding circle does not add any points (or pixels) to the feature that are not already specified by a larger circle from some other point on the MAT. For example, the filled-in "8" shape in Figure 4.2 consists of two slightly overlapped circles. Only the two maximum points in the MAT in Figure 4.2a are needed to reconstruct it. The circles from other points along the MAT are redundant and do not add any additional information. The interpretation is that the original shape can be decomposed into the two circles.

Once a distance map has been constructed, the MAT is extracted from it and the interpretation of nonredundant points is carried out as for the circles. Unfortunately, there is no shortcut method for locating the nonredundant points, and an exhaustive search is the only sure approach.

When this approach is used to decompose features into other primitives, such as squares or polygons, the MAT is constructed not as the locus of centers of circles but as the locus of centers of those shapes. For some shapes, such as squares and diamonds, there are straightforward ways to generate a distance map. Because they require only integer processing on a pixel array, these non-Euclidean distance maps were in use well before the isotropic EDM came into widespread use. However, if the structuring element used does not correspond well to the shape or its parts, poor results are obtained, as shown in Figure 4.2. The MAT can still be used to reconstruct the original shape, but all of the points must be used and the MAT does not lead to any decomposition or simplification.

For complex shapes in either two or three dimensions, there are various approximations and procedures for pruning some of the branches of the MAT so that the reconstruction preserves "most" of the information that is considered visually significant (Ogniewicz, 1994; Attali & Montanvert, 1997; Shaked & Brukstein, 1998; Tam & Heidrich, 2002, 2003). Figure 4.3 shows an example. While the reconstructed shapes from pruned MATs are often visually close approximations to the originals, the ability to extract a simplified numerical characterization of the shape is not demonstrated.

Some of the other applications of the MAT are to medical images (e.g., using the MAT to map locations along images of chromosome structures), analysis of printed letter shapes (e.g., for type design, and for detecting forgeries in Chinese signature blocks), and even generating patterns for embroidery. When applied to the analysis of the morphology of pore structures in porous media, local minima in the 3D MAT are interpreted as the radii of throats that constrict and limit flow.

For the purpose of this book, which is to find shape descriptors that can provide a few numbers to use in comparison or recognition tasks, the MAT is not a good general candidate. In spite of its completeness as a representative of the original shape, it does not easily reduce to a suitable set of numerical values for statistical treatment.

Syntactical Analysis with the MAT

The MAT does offer a natural way to isolate primitive elements useful for syntactical analysis (introduced in Chapter 3). Each internal and terminal branch of the MAT is defined by its length (which is also the length of the corresponding branch of the skeleton) and by a measure of its magnitude (which are the values of the EDM along that skeleton). This may be represented by either the maximum value, which is the maximum width of the corresponding portion of the original feature, the minimum, or the mean value. The metric properties of the MAT branch must be combined with the topological order of their connections to each other, which are determined by the identities of the branches that meet at each of the nodes. There may be either three or four branches originally joined at a node. Each branch also possesses the topological property of being either a terminal branch or an internal one that meets a node at both ends.

Taken together, these metric and topological properties define each branch and become the alphabet and the grammar that describe the feature. Although some of these properties are metric, it must be understood that syntactical description is inherently a qualitative and not a quantitative shape description.

In many cases, as illustrated for the three-dimensional object in Figure 4.3, it is useful to prune the MAT by eliminating "insignificant" primitives. One principle of syntactical analysis is that the number of primitives cannot be too large. Discarding terminal branches that are either short or narrow (or both) is one way to simplify the MAT. This corresponds to removing bumps around the periphery of the original feature. Another useful simplification method is to find short internal branches and eliminate them by merging the nodes at both ends. In both cases, the determination of the cutoff values for the elimination of branches is essentially arbitrary but can be made somewhat less so by establishing a threshold as a percentage of the largest or longest branch in the MAT.

| Original Shape | Medial Axis | Pruned Medial Axis | Heavily Pruned Medial Axis | Surface Reconstructed from Heavily Pruned Axis |

FIGURE 4.3
Illustration showing the pruning of the medial axis transform of a 3D object, and the resulting reconstruction. (From R. Tam, W. Heidrich, 2003, *Proceedings of IEEE Visualization*, 481–488.)

Figure 4.4 illustrates these operations. The decision for removal is based on the integrated value of the MAT along each branch (the product of the length and average width), which is shown in the figure. The surviving branches may be further grouped, for instance using a 3 × 3 array of classes of long, medium, and short by wide, medium, and thin, again using percentages of the maximum values present. This produces an alphabet of nine types. The syntactic analysis of the features then applies the rules, or grammar, to connect the various primitives into graphs as shown in Figure 4.5.

The syntactic representation of a graph can be done in several ways. One is to represent the nodes connecting the branches by rules such as {1 = branch; 0 = return; 2 = joins}. Using these rules, the graph in Figure 4.5 generates the sequence F2E1G0E1E201G1B0D0C. However, depending on the beginning point and the path taken through the graph, many other sequences are

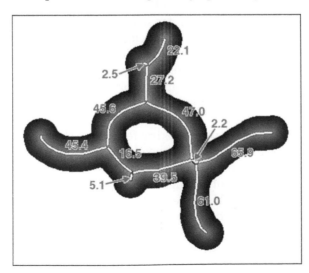

FIGURE 4.4
(See color insert.) Pruning the MAT. The integrated value for each branch is shown (in arbitrary units). Short or narrow terminal branches and internal branches, with values shown in red, may be removed.

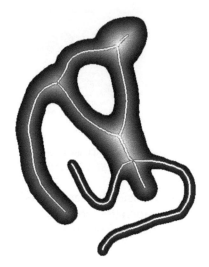

	Short	Medium	Long
Narrow	A	B	C
Medium	D	E	F
Wide	G	H	J

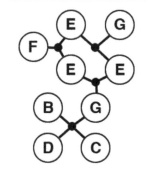

FIGURE 4.5
(See color insert.) Reducing a MAT to a graph. The branch descriptions are simplified as shown in the table, and their linkage is represented by the nodes in the graph.

possible, and direct comparison of the sequences cannot be used to match the graphs. Many other representations can be created, some constructed top-down while others are context-free or nondeterministic.

Graph theory is a rich field of mathematics, which goes far beyond the scope or needs of this text. A graph, consisting of nodes or vertices, and edges or branches (which may include an associated direction) can be represented by list structures like the example shown or by matrix structures. There are several different forms of each, with various advantages for analysis, and different storage requirements. Graph theory has many other applications, organizing information and knowledge that have nothing to do with images or shapes.

Comparing features consists of looking for isomorphisms between graphs. In this search it is usual to give more weight to the topology, that is, the order of linkages, than to the identity of the primitives. In other words, if the order of the branches is correct but one figure has a wide and short branch where another has a medium and long one, that is a smaller difference (and the two features are considered to be a closer match) than if the branches are connected in a different order. Most of the procedures for comparing, matching, or fitting graphs consider only the topology of the network and do not use information about the metric properties of the branches.

A complete topological match between two graphs is called graph isomorphism. Partial matches, including matching one graph with a portion of another (subgraph isomorphism), are illustrated in Figure 4.6. Finding complete or partial matching is computationally expensive and depends on

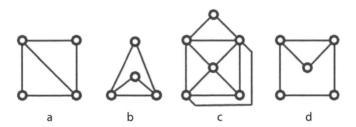

FIGURE 4.6
Isomorphisms and matching: a is isomorphous with b and has subgraph isomorphism with c.
The partial match of a with d depends on ignoring an additional node.

heuristic rules. There is a rich literature devoted to the topic of grammars, and to the matching, partial matching, or fitting of one graph with another (Harary, 1969; Ballard & Brown, 1982; Fu, 1982; Bollabas, 2002; Sonka et al., 2008). One of the uses of syntactical analysis of images is to construct a graph of the relationship between objects and parts of objects in three-dimensional scenes, as part of image understanding and robotics vision.

As a an example of the complexity of pattern comparison based on graph theory, Figure 4.7 shows the patterns of window panes in several houses near one of the author's. Matrix notation listing the connections between the nodes that are numbered in the figures is shown Table 4.1. It is also possible to enter the width, length, or any other property of each edge into the cells. The process of matching one matrix to another, or to a part of another,

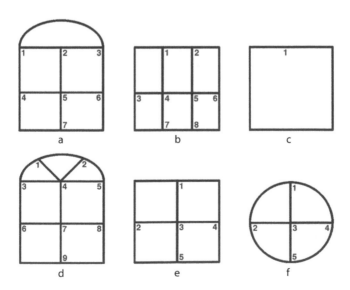

FIGURE 4.7
Shapes of window panes with numbered nodes, as discussed in the text.

TABLE 4.1

Matrix Representation for Each of the Window Pane Shapes in Figure 4.7, Based on the Node Numbering Shown

A	1	2	3	4	5	6	7
1	•	1	1	1	0	0	0
2		•	1	0	1	0	0
3			•	0	0	1	0
4				•	1	0	1
5					•	1	1
6						•	1
7							•

B	1	2	3	4	5	6	7	8
1	•	1	1	1	0	0	0	0
2		•	0	0	1	1	0	0
3			•	1	0	0	1	0
4				•	1	0	1	0
5					•	1	0	1
6						•	0	1
7							•	1
8								•

C	1
	•

D	1	2	3	4	5	6	7	8	9
1	•	1	1	1	0	0	0	0	0
2		•	0	1	1	0	0	0	0
3			•	1	0	1	0	0	0
4				•	1	0	1	0	0
5					•	0	0	1	0
6						•	1	0	1
7							•	1	1
8								•	1
9									•

E	1	2	3	4	5
1	•	1	1	1	0
2		•	1	0	1
3			•	1	1
4					1
5					•

F	1	2	3	4	5
1	•	1	1	1	0
2		•	1	0	1
3			•	1	1
4				•	1
5					•

requires permuting the rows and columns, because the node numbering scheme is arbitrary.

A few observations can be made visually:

- The simple loop pattern in C will match many areas in the other windows, but only if the fact that those other areas have nodes along their periphery is ignored. It may be helpful to consider C as a single edge or branch with one virtual node where the ends meet.
- E and F are topologically identical, that is, isomorphous.
- E is subgraph isomorphous with several areas in A, B, and D.
- A and D are visually closely similar, with the addition of branches and nodes in D.

Any topological matching scheme must reflect these same visual characteristics (Malsburg, 1988).

Because the information from the MAT or skeleton typically includes the order of the connections, and perhaps the dimensions of each branch, but not their relative orientations, it is well suited to some kinds of flexible shape matching, as shown by the various poses of the mannequin in Figure 4.8.

FIGURE 4.8
Various poses for a jointed mannequin, with the EDM and skeleton superimposed. The dimensions and order of connections of the branches exhibit only minor changes.

The Shock Graph

The practical difficulties of finding isomorphisms in graphs are reduced somewhat by constructing a directed graph from the MAT. The medial axis transform as described earlier encodes complete information about the topology of an object and its local shape. Siddiqi and Kimia (1995) proposed a method, which they call the shock graph, in which the skeleton (or MAT) of an object is encoded with values representing interesting features of the local boundary curvature of the object at each point rather than with the width. This method has the advantage of encoding information about a region of the boundary centered on the point of interest, averaging out the effects of noise or discretization error.

The transformation adds a term to Blum's evolution equation (Blum, 1967, 1973) for the medial axis transformation that measures the local boundary curvature. Because the resulting transform has the same form as a reaction-diffusion equation and the interesting points correspond to a shock in the diffusing system, the term "shock" has been applied to these points.

Siddiqi and Kimia (1995) define four classes into which each point on the skeleton falls on the basis of the tangents to the inscribed circle centered at that point and the sum of the two boundary curvatures. Figure 4.9 illustrates these four classes of points. Their definition of the various types of shocks is based on the boundary curvature and is appropriate to continuous geometric boundaries. For a pixelated image, the MAT provides a more readily interpreted characterization as shown in the figure.

A 1-shock can be seen as a sequence of points where the tangents to the inscribed circle at that point are not parallel. These points typically lie toward the terminal ends of branches and protuberances. The MAT has values that monotonically diminish toward the ends. Higher shocks all have parallel tangents but a difference in the local boundary curvatures. A 2-shock

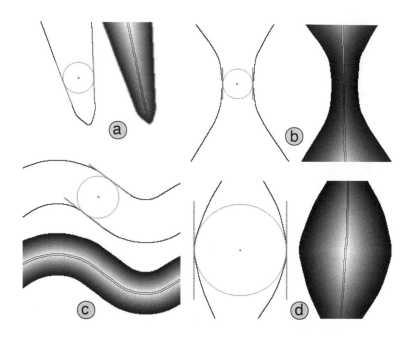

FIGURE 4.9
(See color insert.) An illustration of the four types of shocks, as defined by the boundary curvature and the medial axis transform: (a) 1-shock; (b) 2-shock; (c) 3-shock; (d) 4-shock.

represents a point where the sum of the two curvatures is negative, that is, both boundaries curve away from the point. This might represent a point where two thicker regions touch or represent a watershed boundary. The MAT at the 2-shock reaches a local minimum value.

A 3-shock represents a sequence of points where the sum of the curvatures is zero within some defined tolerance. In these regions, the boundaries are locally parallel, although their orientation may vary from one point to another. The MAT is characterized by uniform values. A 4-shock represents a point where the local curvature is positive, that is, toward the central point. These can be thought of as isolated thicknesses or "lumps" in the object. The MAT at the 4-shock reaches a local maximum value. Nodes, where branches of the skeleton or MAT join, are usually 4-shocks (in rare cases a 2-shock is possible).

Each of the shocks has direction information associated with it pointing "uphill" to greater MAT values (i.e., greater width of the original feature). Thus, a 1-shock points toward increasing values, a 2-shock points away from the minimum value, a 3-shock simply connects whatever is on either end, and a 4-shock has directions pointing in toward the maximum value. These directions allow construction of a simplified graph that omits the details of the width and length of the various branches but preserves the order, as illustrated in Figure 4.10.

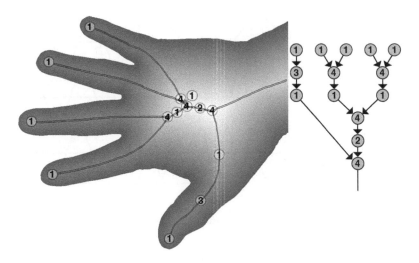

FIGURE 4.10
(See color insert.) Silhouette of the hand from Figure 3.61, with its MAT and the graph connecting the shocks.

Comparing the graphs from different objects or different views of an object is performed by starting at top-level nodes and finding and grading each subtree for its similarity. The overall similarity scores have been shown to be useful for classifying shapes (Pizer et al., 1999; Siddiqi et al., 1999).

Fourier Shape Descriptors

Of all the various qualitative and quantitative shape description methods, Fourier coefficients are arguably the most useful (or at least the most often used) for statistical classification techniques and at the same time one of the least accessible to human recognition or interpretation. The method is often called harmonic analysis or spectral analysis, and provides potentially complete information about the shape of a feature (Schwartz & Shane, 1969; Ehrlich & Weinberg, 1970; Zahn & Roskies, 1972; Beddow et al., 1977; Persoon & Fu, 1977; Flook, 1982; Kuhl & Giardina, 1982; Kaye et al., 1983; Barth & Sun, 1985; Ferson et al., 1985; Bird et al., 1986; Diaz et al., 1989, 1990; Rohlf, 1990; Verscheulen et al., 1993; Lestrel, 1997; Haines & Crampton, 2000; Bowman et al., 2001; Pincus & Theriot, 2007).

The basic idea starts with a simple observation: starting anywhere on the border of an object, it is possible to proceed around it, recording various kinds of information (direction of the edge, coordinates, etc.), and eventually return to the starting position. (Of course, this considers only the outer edge

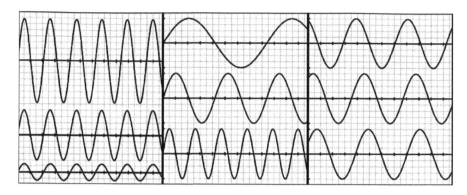

FIGURE 4.11
Variations in sinusoid amplitude, frequency, and phase, respectively.

and ignores any interior holes.) Continuing the process around the object again will produce the same results, and if plotted this would generate an endlessly repeating graph. Any continuous (about which, more later) and repeating function is a natural candidate for Fourier analysis, which decomposes the graph as a series of sinusoids. The amplitudes of those sinusoids are often useful as a relatively short list of numerical values that can be used to describe the object's shape, for example, in statistical classification applications. Both the amplitude and phase are needed to reconstruct the shape.

$$F(u) = \int_{-\infty}^{+\infty} f(x) \cdot e^{-2\pi i u x} dx$$

$$e^{-2\pi i u x} = \cos(2\pi u x) - i \cdot \sin(2\pi u x) \tag{4.1}$$

$$f(x) = \sum_i amplitude_i \cdot \sin\left(frequency_i \cdot 2\pi x + phase_i\right)$$

The Fourier series expansion may be written either with sines and cosines, or equivalently with exponential notation, as shown in Equation 4.1, in which $f(x)$ is an original function in either the spatial or time domain and $F(u)$ is the Fourier transform in terms of frequencies u. For a digitized image, the integral equation becomes a summation, and instead of being infinite it is limited at high frequencies by the spacing of the pixels. The role of the amplitude and phase are summarized in Figure 4.11.

There are several possibilities for selecting the information to plot as a function of position along the edge, as described later. First, it may be helpful to consider the difficulties that humans seem to have with this approach. Fourier analysis does involve math, which puts some people off, but that is a minor issue. It is clear that the lower indices in the Fourier series represent variations that occur more gradually around the object's periphery, whereas

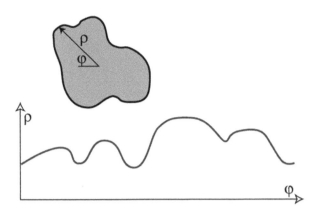

FIGURE 4.12
Unrolling a shape as radius versus angle.

higher index terms represent more rapid, short range or local fluctuations. But this may not be easily related to visual observation of groups of natural objects. Finding from statistical analysis that one set of objects produces a greater amplitude in, say, the seventh term than another group does not generally help a human observer to recognize what that represents in a collection of naturally varying objects. It is that disconnect between the numbers and the visual appearance that has limited acceptance of the method, now that computing power makes it practical to apply the procedures.

Shape Unrolling

The first approach to Fourier analysis plots radius (usually from the object's centroid) as a function of angle. As shown in Figure 4.12, this amounts to "unrolling" the shape.

The Fourier series expansion for the profile can be written as

$$\rho(\varphi) = a_0 + a_1 \cos(\varphi) + b_1 \sin(\phi) + a_2 \cos(2\varphi) + b_2 \sin(2\varphi) + \ldots \quad (4.2)$$

but it is more common to write this in terms of amplitude c and phase δ, as

$$c_k = \sqrt{a_k^2 + b_k^2}$$
$$\rho(\varphi) = \sum c_k \cdot \sin\left(2\pi k\varphi - \delta_k\right) \quad (4.3)$$

The constant term c_0 in the Fourier series for this plot is just the mean radius of the shape, one of the most commonly used measures of size for an irregular object. The c_1 term is a measure of the elongation or aspect ratio of the shape, c_2 is a measure of triangularity, and so on. The phase

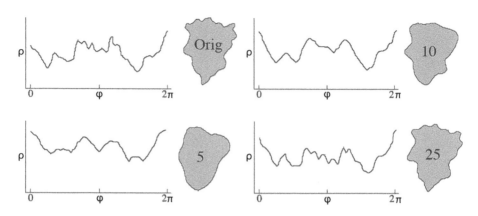

FIGURE 4.13
Unrolled shape for a natural particle (an extraterrestrial dust particle), and the recreation of
that shape using a limited number of terms (5, 10, and 25, respectively) in the Fourier series.
(From B. H. Kaye, 1989, *A Random Walk through Fractal Dimensions*, VCH Verlagsgesellschaft,
Weinheim.)

terms are often overlooked or not used for statistical shape matching, but
contribute strongly to the visual judgment of shape. For any natural object,
the ability to discern the contribution of the individual components to
shape is difficult.

One of the very nice attributes of Fourier analysis is that the magnitudes
of the terms in the series generally decrease very rapidly with frequency, so
that the list of numbers that can adequately describe the shape is manage-
ably small. When digitized images of objects are used, there is also of course
a finite limitation in the resolution of the shape. Figure 4.13 shows that for
the shape of a natural particle, the number of terms needed to reconstruct it
is not too large.

It is important to emphasize that although most of the analysis of shapes
based on harmonic analysis uses only the magnitudes of the various terms
in the Fourier series expansion, these by themselves are not adequate to
reconstruct the original shape. The phase angles are also required.

If the magnitude of the series coefficients drops with frequency such that a
log–log plot produces a straight line, and the phase angles are random, then
the shape is fractal and the slope of the plot is related to the fractal dimension
as shown in Equation 4.4 (Meloy, 1977; Feder, 1988; Russ, 1994). As pointed
out in Chapter 3, the dimension is for the boundary line, and so can vary
between 1.0 (a line specified by Euclidean geometry, in this instance a circle) and 1.999... (a line that is so irregular that it wanders all over the plane,
whose topological dimension is 2.0). Measurement of fractal dimension on
a log–log plot of Fourier amplitude versus frequency is particularly appropriate for measuring surfaces, and is used in Chapter 5 for that purpose.
However, there are more convenient ways to measure fractal dimensions of
planar features as shown in Chapter 3.

$$Fractal\ Dimension = \frac{4 - |Slope|}{2} \tag{4.4}$$

One objection to plotting radius versus angle is that portions of the profile that are very irregular are undersampled, relative to smooth areas. A more natural method plots radius versus distance along the perimeter. Although this requires more calculation, it is more generally used. Using the distance along the perimeter allows interpolating a fixed number of points at equal distances along the boundary, which may be helpful when comparing results for different size objects. It also allows choosing a number of points that is a power of 2, which simplifies using a fast Fourier transform (FFT) as noted later.

There is another, major problem with shape unrolling. As shown in Figure 4.14, if the outline is reentrant or undercut, the radius-versus-angle curve becomes multivalued. Selecting a convention of always using the minimum (or the maximum) radius at each angle eliminates the multiple values, but makes the plot discontinuous. Fourier analysis depends upon the function being continuous.

This limitation encourages other approaches to the unrolling instead of plotting radius from the centroid. One of the common ways of encoding an object in image analysis software uses "chain code," introduced in Chapter 2 in Figure 2.45. As shown in Figure 4.15, this is a series of digits that record the path from pixel to pixel along the perimeter. Plotting the string of digits produces a graph that also repeats for every circumnavigation and might be subjected to Fourier analysis. But because of the discontinuity between direction 1 and direction 8, this graph is not suitable. The solution is to use differential chain code. Instead of plotting the absolute direction of each link in the chain, changes in direction are plotted. If the chain continues in the same direction, a value of zero is assigned. Values of +1 or +2 represent turns to the right, while –1 and –2 represent turns to the left.

This convention has the further advantage that it ignores the overall orientation of the object, producing the same set of differential values if the object is rotated with respect to the pixel grid as shown in Figure 4.16. At least, that

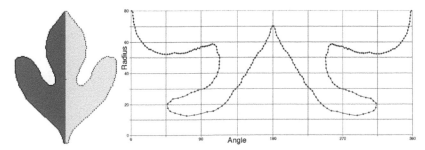

FIGURE 4.14
The reentrant shape of a leaf and the resulting multivalued plot of radius versus angle.

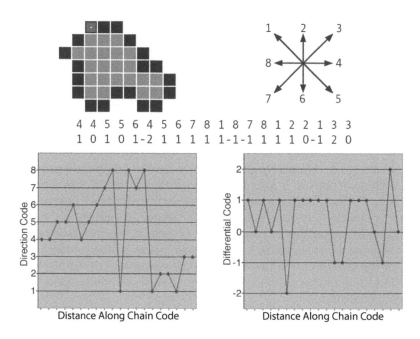

4 4 5 5 6 4 5 6 7 8 1 8 7 8 1 2 2 1 3 3
1 0 1 0 1-2 1 1 1 1 1-1-1 1 1 1 0-1 2 0

FIGURE 4.15
Chain code and differential chain code for a shape, proceeding clockwise from the marked pixel and following the dark edge pixels.

is approximately true. Aliasing of the border of the shape with the pixel grid produces one-pixel-high deviations that can be a problem for any chain code representation of shapes using finite pixels, but this is a minor issue for those aspects of shape that are large compared to the pixel size and are not significantly affected by the individual pixels.

There are remaining difficulties for performing Fourier analysis using differential chain code. First, the links in the chain are not all of the same length and so the values are not uniformly spaced, as required for the mathematics. The diagonal links are $\sqrt{2}$ longer than the orthogonal ones. A simple and reasonably effective way to handle this is to repeat the digits for orthogonal links 5 times and those for diagonal ones 7 times. The expanded differential chain code produces a string of numbers that

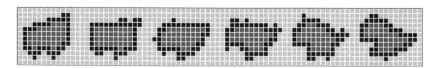

FIGURE 4.16
Details of a shape as it is rotated on a pixel grid in 10-degree steps, showing aliasing of the pixels, and some boundary pixels that must be traversed more than once in a chain-code representation.

is much longer, but because $\frac{7}{5}$ is fairly close to $\sqrt{2}$ it makes the spacings close to uniform.

The set of digits can be used to calculate a discrete Fourier transform of the shape, but as a practical matter the use of an FFT is much preferred (it can be hundreds of times faster). The FFT (Cooley & Tukey, 1965; Bracewell, 1989) works by a divide-and-conquer strategy that requires the number of values in the original function or graph to be a power of 2 (64, 128, 256, 512, etc.). There are adaptations of the fast transform that do not have the power-of-2 restriction (Frigo & Johnson, 2005; Johnson & Frigo, 2007), but the most commonly used forms that produce straightforwardly understood frequency terms have this requirement.

The expanded differential chain code is unlikely to have a number of digits that meet this requirement, and in any case different objects will have varying lengths. The usual solution is to interpolate a fixed number of values along the graph, producing a sample of the values that maintain enough resolution to characterize the important details of the shape but is always an exact power of 2 in length. Of course, the number of points must be large enough to produce an adequate number of terms in the expansion (i.e., the minimum number of points is twice the number of terms). Fourier analysis is not the only tool available for analysis of the boundary. Wavelets have also been used (Antoine et al., 1996) and produce similar results, as shown later.

When the Fourier transform coefficients are used for shape classification, it is important to have enough coefficients in the expansion to represent important details. In some cases the high frequency terms have very small values yet still produce significant refinements of the shape's boundary. In the example of Figure 4.17, a red oak leaf requires 100 coefficients to reproduce the finest points along the leaf edge, but the first 20 coefficients capture enough of the leaf shape to distinguish this from another oak species such as a white oak. This is one of the leaves from the set used in Chapter 3 (Figure 3.23) that are analyzed later for comparison to those results.

This method has been used for a variety of applications, including classifying sediment particles that have traveled down different river beds or have been eroded in different settings, soil particles, and sand grains (Schwartz & Shane, 1969; Ehrlich & Weinberg, 1970; Van Nieuwenheuise et al., 1978; Dowdeswel, 1982; Ehrlich et al., 1984; Thomas et al., 1995; Bowman et al., 2001; Santamarina & Cho, 2004), characterizing the shape of lunar craters (Eppler & Ehrlich, 1983), and correlating diatom shape with environmental factors such as water temperature (Healy-Williams & Williams, 1981; Belyea & Thunell, 1984; Healy-Williams et al., 1997). Neto et al. (2006) apply the technique to plant species identification based on leaf shape.

Other applications have been to assess the fertility of bull sperm (Ostermeier et al., 2001), classify cells and cell nuclei (Diaz et al., 1989, 1990, 1997), evaluate metastatic potential in tumors (Partin et al., 1989; Delfino et al., 1997),

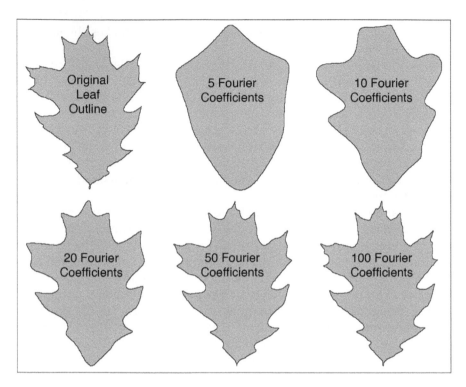

FIGURE 4.17
Reconstruction of a leaf shape using an increasing number of terms in the Fourier expansion.

track evolution in cephalopods (Canfield & Astey, 1981), classify fossil shells (Crampton, 1995), distinguish between different strains of mice based on the shape of vertebrae (Johnson et al., 1985; Johnson, 1997) or mosquitoes based on wing shape (Rohlf & Archie, 1984), and track the development of human skeletal morphology (Lestrel & Brown, 1976; Lestrel, 1997; Lestrel & Roche, 1986; Jacobshagen, 1997; Ohtsuki et al., 1997). The method has even been applied to recognition of hand-printed letters (Granlund, 1972) and the analysis of the shapes of buildings to compare the complexity of various architectural designs (Psarra & Grajewski, 2001).

Most of these studies have used statistical techniques to identify a few of the amplitude coefficients in the Fourier series that provide the characterization important in each individual case. It also turns out that in some cases the amplitudes can be plotted against frequency to further simplify and reduce the volume of information. For example, as mentioned earlier, for a fractal shape this plot is a straight line and the slope of the plot gives the fractal dimension. Interpreting the shape of a graphical representation of numbers derived from an object's shape can be thought of as a second-order interpretation of shape.

Complex Coordinates

Fourier analysis deals naturally with complex numbers. In performing the mathematics using expanded differential chain code, the digits form the real part of the input and the imaginary part is set to zero. There is an alternate way to proceed that begins with a complex input. The (x, y) coordinates of points along the object periphery are expressed as $(x + iy)$ to produce a complex function suitable for Fourier analysis (Kuhl & Giardina, 1982). Figure 4.18 illustrates this. The complex profile has projections on the real

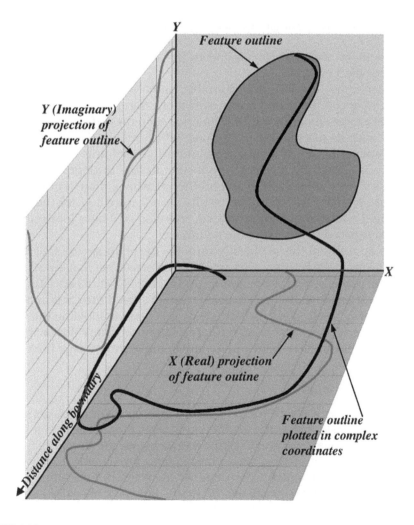

FIGURE 4.18
Combining the x- and y-coordinates of the object outline to form a complex number profile in three-dimensional space.

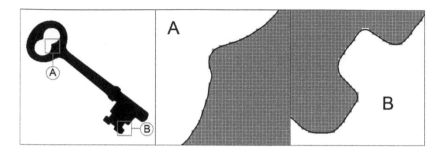

FIGURE 4.19
Example of the super-resolution boundary produced by interpolation through the boundary
pixels of a feature; enlarged details show the pixelation of the image.

and imaginary axes that correspond to the x- and y-coordinates of points
along the object outline.

In order to use the FFT, it is desirable, as noted earlier, to interpolate a
fixed number of uniformly spaced points along the boundary. This approach
produces a more smoothly varying function than the differential chain code
method with its discrete integer values, which makes the high frequency
terms better behaved. If a super-resolution line defining the boundary of
a feature is created by interpolating a smooth line through the boundary
pixels, as shown in Figure 4.19, in order to measure a more accurate perimeter (and one that gives the same results for a feature and for the hole that
inverting the image contrast produces), this also facilitates locating a set of
uniformly spaced points along the boundary.

It may appear that this artificially smooths the boundary so that high-frequency terms in the Fourier series would be incorrect, but that is not true. The
number of points along the boundary used for the FFT, and hence the highest
frequency terms that can be calculated, cannot exceed the limit imposed by
the number of pixels along the boundary. It is these that define the shape of
the feature, and the interpolated super-resolution boundary does not add any
new information to that shape or lose any of the original shape information.

Vector Classification

When Fourier (or wavelet) coefficients are calculated for an object's profile,
they can be considered as the vector coordinates representing that shape in
a high-dimensionality space. Classification schemes for objects such as particulates typically use these vector coordinates for statistical analysis. The
methods are fundamentally the same as the statistical classification techniques used with other measures of shape, but the dimensionality of the
vector space is typically much greater, making it more difficult to visually
detect clusters or patterns in the data.

Standard statistical tests such as stepwise regression, discriminant analysis, or principal components analysis can be applied to determine which of

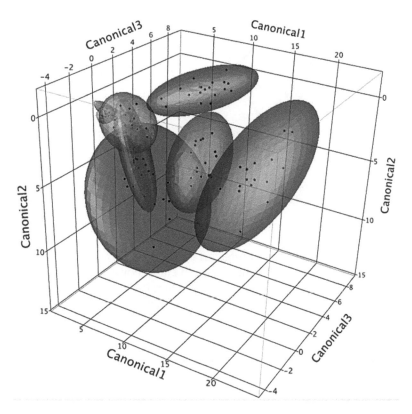

FIGURE 4.20
Discriminant analysis results for the leaf samples from Figure 3.23 using six Fourier coefficients.

the terms are useful for feature classification or recognition, as described more fully in Chapter 6. In a significant number of cases, this approach proves to be successful. The identification of particles in sediments, distinguishing the seeds from various closely related plant species, and the discrimination of normal from potentially cancerous cells in Pap smears are a few of the successes of this approach.

Using the same leaf images shown in Figure 3.23, the magnitudes of the first 30 Fourier coefficients were determined, and normalized Fourier terms obtained with the Shape program (Iwata & Ukai, 2002). The data set was analyzed using the same discriminant analysis procedure used with the classical ratio parameters illustrated in that figure. The minimum set of Fourier coefficients needed to produce classifications without any errors is coefficients 2, 4, 6, 7, 10, and 14. Figure 4.20 shows the groupings in parameter space. Comparison with the results shown in Figure 3.23 using dimensionless ratios as shape descriptors shows that the class separation is not as good, in spite of using twice the number of parameters (and ones that are probably more difficult to comprehend).

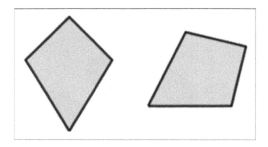

FIGURE 4.21
Two identical shapes. The one at the left has a vertical axis of symmetry, which is more readily recognized than the one at the right.

In this plot, the clusters of points can be seen with the ellipsoidal regions representing the two-sigma limits for each class. For a case such as this one, in which more than three parameters are used, it is not possible to show the full *n*-dimensional parameter space; the plot shows the three most significant canonical parameter axes.

Symmetry Estimation Using the Boundary

Human visual inspection of objects is sensitive to symmetries. Mirror or reflection symmetry is particularly noticed when the axis of symmetry is vertical or horizontal, as illustrated in Figure 4.21. Rotational symmetry of order *k* results when an object can be rotated by 360°/*k* and matches its original form. For small values of *k* (typically less than 10) this is readily recognized. Of course for real objects both types of symmetry may be imperfect.

The symmetry of a shape is encoded in the Fourier transform of its boundary. This symmetry can be measured in several ways. Zahn and Roskies (1972) state that the value of R_k calculated with Equation 4.5 approaches zero for objects with increasing *k*-fold rotational symmetry:

$$R_k = \sum_{i \bmod k > 0} a_i \tag{4.5}$$

In this equation, the term a_i is the coefficients of the real terms of the Fourier transform of the boundary, and the summation is over those terms that are not multiples of the rotational symmetry being tested. For example, for twofold symmetries, the sum would be only over the odd terms and for sixfold symmetry, the sum would exclude *i* = 6, 12, 18, etc.

Reflection symmetry can be measured from the phase angles using Equation 4.6,

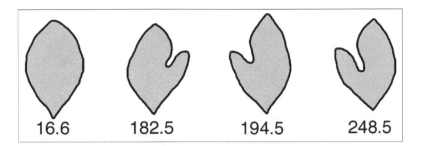

FIGURE 4.22
Outlines of sassafras leaves, with the measured asymmetry parameter. (From D. T. Kincaid, R. B. Schneider, 1983, *Canadian Journal of Botany* 61:2333–2342.)

$$X = \sum_{i,j} w_{i,j} \cdot \left| \left(i * \delta_j - j * \delta_i \right) - \tfrac{\pi}{2} \left(i * - j * \right) \right| \tag{4.6}$$

where $w_{i,j}$ is a weighting factor equal to the minimum of $\{\delta_i, \delta_j\}$, and i^*, j^* are the Fourier indices i and j divided by the greatest common divisor of the two. For objects with a high degree of reflection symmetry, X is small. The implication of Equation 4.6 is that for objects that have reflective symmetry, the associated phase angles for the nonzero Fourier terms are integer multiples of π.

Kincaid and Schneider (1983) define a normalized asymmetry parameter that is a function of the coefficients of the imaginary components of the Fourier transform. A shape is symmetrical about its principal axis only if all of the Fourier terms are real. Using the notation from Equation 4.2, the asymmetry parameter is calculated as the normalized sum of all the imaginary terms:

$$A.P. = \sqrt{\frac{\sum b_k^2}{Area}} \tag{4.7}$$

This provides an easily computable single number that can be used to rank order shapes based on their asymmetry or to be used in a classification model where symmetry is important. Figure 4.22 illustrates this for leaf shapes.

Equation 4.5 indicates that a shape having k-fold rotational symmetry has zero harmonic amplitudes for all indices that are not integral multiples of k. This is illustrated in Figure 4.23 for several shapes of diatoms. The circular shape (#1) has no higher order terms. The elongated shapes (#2, #3) have peaks for terms 2, 4, 6, etc. (as do the square and four-pointed shape, which also of course have twofold symmetry). The triangular and three-pointed shapes (#4, #5) have peaks in terms 3 and 6. The square and four-pointed

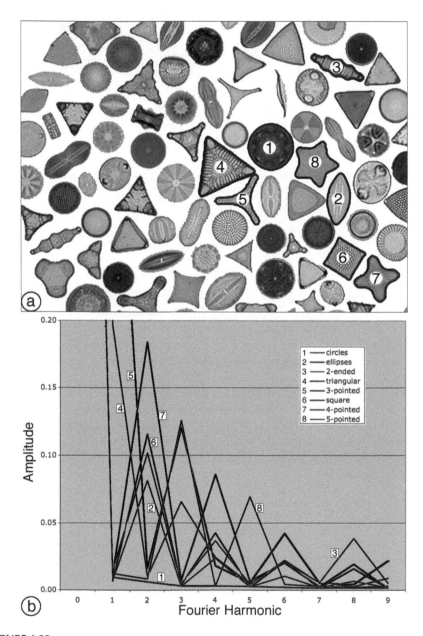

FIGURE 4.23
(See color insert.) Some diatom shapes as discussed in the text, and plots of the amplitude of the Fourier harmonic terms for the boundaries averaged over multiple similar diatoms.

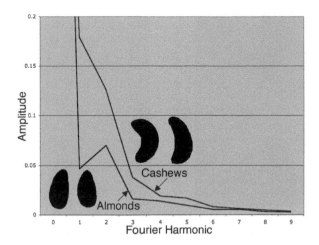

FIGURE 4.24
Plots of the average amplitudes of Fourier harmonic terms for almonds and cashews.

shapes (#6, #7) have peaks in terms 4 and 8. The five-pointed shape (#8) has a peak in term 5, but also one in term 3 due to a slight irregularity in form that has partial matches with equilateral triangles.

It is also possible to detect partial or approximate symmetry in this way. Figure 4.24 shows a plot of the averaged harmonic amplitudes for the boundaries of several dozen almonds and cashews (which are illustrated in Figure 6.20). The almonds are pointed at one end and rounded at the other, but have a greater approximate 180° rotational symmetry than do the cashews (both have approximate mirror symmetry, about different axes). The plot shows a drop in the first harmonic amplitude for the almonds, but not the cashews, corresponding to the much greater (but imperfect) rotational symmetry.

Symmetry Estimation Using the Interior Pixels

The preceding section discusses symmetry estimation based on the boundary of an object. Information about rotational and reflectional symmetries of an object is also encoded into the 2D Fourier–Mellin transform of the object's full area (Shen et al., 1999; Derrode & Ghorbel, 2004). Rather than performing the Fourier transform in the commonly encountered row-and-column order, the image is processed in polar rho–theta coordinates, applying the Mellin transform to each line. The combined Fourier–Mellin transform is given in Equation 4.8 and is well suited to extracting rotational and mirror symmetries because the transform-space image is invariant to translation and scale:

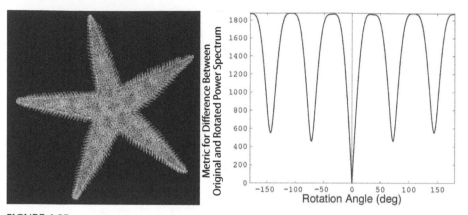

FIGURE 4.25
An image with fivefold symmetry, and a plot showing the difference between the Fourier-Mellin spectrum and a rotated copy. (From S. Derrode, F. Ghorbel, 2004, *Signal Processing* 84(1):25–39.)

$$M(k,s) = \frac{1}{2\pi} \int_0^\infty \int_0^{2\pi} r^{s-1} f(r,\theta) \cdot e^{-ik\theta} d\theta\, dr \qquad (4.8)$$

In these cases, as with the boundary, objects that are k-fold symmetric have Fourier coefficients that are zero (or near zero) at multiples of k. Derrode and Ghorbel (2004) calculate a metric E_{dd} based on the summed Euclidean difference between the Fourier–Mellin transform image and the transform of a rotated copy and show that it drops to mark positions of rotational symmetry. Figure 4.25 shows an example of fivefold rotational symmetry characterized by this approach.

Similarly, Shen et al. (1999) define a generalized complex moment $GC_{p,q}$ as the higher moments of the Fourier transform, which is also computed in rho–theta polar coordinates. In this notation, q is the index of the associated Fourier coefficient and p is the order of the moment, such that $GC_{0,q}$ are the Fourier coefficients. Using this form, the greatest common divisor of the orders of the first three nonzero moments (i.e., the values of q) yields the number of rotational-symmetric planes in the object. The generalized complex moments themselves are a selection of the Fourier–Mellin transform given in Equation 4.8, where the exponent s is selected to be an integer rather than a complex number.

Wavelet Analysis

Fourier analysis represents the boundary of a feature as a summation of sinusoids of increasing frequency. Analysis of the frequencies, and comparison of the amplitudes of certain frequencies to those present in other shapes, provides one important tool for comparison. But the frequency resolution in the Fourier series tells nothing about where along the boundary the frequencies occur or if indeed they are localized. Wavelet analysis is able to represent the boundary profile in a way that does localize the various frequencies that may be present. It does not, however, take advantage of the repetitive nature of boundaries around features; many of the applications of wavelet analysis have been to nonrepeating profiles, such as time series of weather or financial data.

Wavelet analysis works by defining a localized shape (the "mother wavelet") that is shifted to all positions along the signal (the boundary) and the values convolved together to obtain a measure of similarity. Then the wavelet shape is compressed so that it becomes shorter and the procedure is repeated. As this process is continued a two-dimensional spectrum is obtained of the amplitude versus position for each wavelet size.

The profile function $f(x)$, which may be the unrolled radius as a function of angle or position, or the differential chain code, is represented by a double summation over values of s (scaling) and t (translation) of a two-dimensional spectrum of amplitude coefficients c (a function of both s and t) times the scaled and translated copies of the mother wavelet $g(x,s,t)$ as shown in Equation 4.9:

$$f(x) = \sum_{s=0}^{S} \sum_{t=0}^{T} c_{s,t} \cdot g(x,s,t) \tag{4.9}$$

The simplest of the functions used as a mother wavelet is the Walsh function, better known as the Haar wavelet, which satisfactorily demonstrates the basic characteristics of the procedure. As shown in Figure 4.26, it is simply a step function, and the most convenient way to scale the mother wavelet is by factors of 2.

As for the Fourier transform, the wavelet transform has a "fast" version and is easiest to implement when the profile has a length that is a power of 2. To illustrate the procedure, Figure 4.27 shows a simple shape composed of 33 pixels, with a plot of the differential chain code around the boundary; the chain code consists of 16 links. The process of constructing the wavelet transform is shown in Figure 4.28. At each step the average of the values at various scales along the boundary is computed, and then the difference between that average profile and the detail or difference profile obtained at the preceding step. Summing up all of the difference profiles and the final average profile reconstructs the original boundary code.

Just as the Fourier analysis of the boundary results in both amplitude and phase information, the wavelet analysis results in amplitude and position data. Both are needed to reconstruct the original boundary, but for

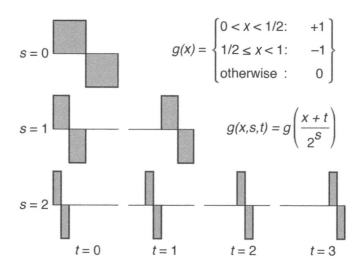

$$g(x) = \begin{cases} 0 < x < 1/2: & +1 \\ 1/2 \leq x < 1: & -1 \\ \text{otherwise}: & 0 \end{cases}$$

$$g(x,s,t) = g\left(\frac{x+t}{2^s}\right)$$

FIGURE 4.26
The Haar wavelet family. The mother wavelet (top line) is scaled (s) and translated (t) to form a complete set of basis functions, which reconstruct the original signal being modeled when multiplied by corresponding amplitude coefficients and summed.

statistical comparison purposes the amplitude values averaged for the spectrum may be used. Alternatively, for some purposes such as locating "interesting" points along the boundary, just the positions of maximum power may be used. The curvature scale space (CSS) method shown in Figure 3.68 is an example of the latter, in which inflection points along the boundary are located with a mother wavelet having the shape of the first derivative of a Gaussian. Positions where this passes through zero are the interesting points, which are plotted as a function of the scale of the daughter wavelets.

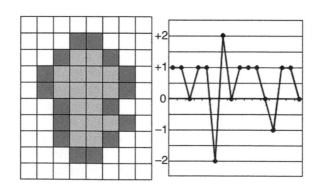

FIGURE 4.27
A small shape with its differential chain code.

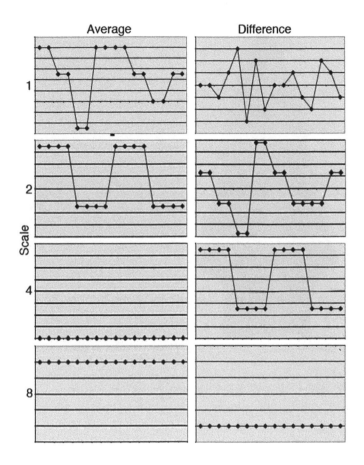

FIGURE 4.28
The progressive steps in the Haar wavelet decomposition of the shape in Figure 4.27.

The wavelet transform is also used for filtering data. Just as the first few coefficients in the Fourier series contain much of the information about the shape, and have values that may be useful for classification or comparison, so the initial (larger scale) amplitude values in the wavelet transform are the most significant. To illustrate, Figure 4.29 shows a shape (one of the arrow points from the set shown in Figure 4.52) with 256 points along the boundary. The boundary is represented by the radii from the centroid to these points, plotted as a function of angle (i.e., the shape is unrolled). The wavelet spectrum for the shape contains the same number of values as the profile, but as can be seen in Figure 4.30 most of them are very small, and the first portion of the transform contains most of the power.

There are many other mother wavelets in addition to the Haar function, and although they are somewhat more complicated in shape, they often are able to extract the major shape details more compactly. Figure 4.31 shows the boundary profile from the arrow point in Figure 4.29 with its reconstruction

FIGURE 4.29
An arrow point, with 256 boundary locations used for wavelet analysis.

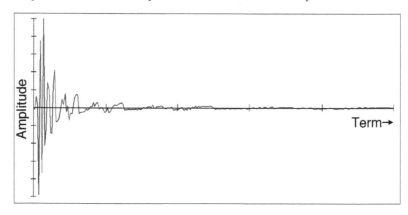

FIGURE 4.30
Amplitude spectrum of the Haar wavelets calculated for the boundary of the arrow point in
Figure 4.29.

omitting the 75% least significant of the Haar wavelet terms. Much better
results can be obtained with the Daubechies-4 mother wavelet (Daubechies,
1992), as shown in Figure 4.31. This is perhaps the most widely used mother
wavelet but certainly not the only one. Some of the other common wavelets
are shown in Figure 4.32. Each of these, used in a continuous or discrete

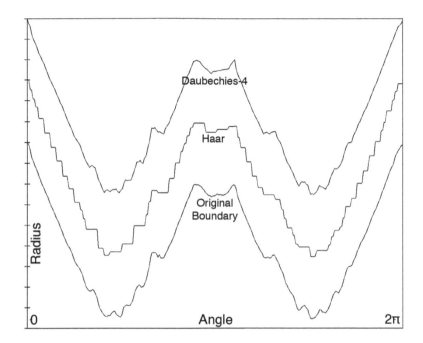

FIGURE 4.31
Comparison of the reconstructed boundary to the original using just the most significant 25% of the wavelet transform values for the Haar and Daubechies-4 mother wavelets.

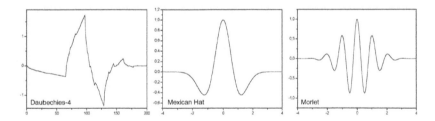

FIGURE 4.32
Mother wavelets: Daubechies-4, Mexican hat (second derivative of a Gaussian), and Morlet (damped cosine function; the wavelet is complex with a damped sine function for the imaginary part).

transform, is capable of extracting somewhat different types of information about the shape of a profile.

One way to present the spectrum from a wavelet transform is shown in Figure 4.33 (Torrence & Compo, 1998). The same boundary data for the unrolled arrow point shown in the previous figures are analyzed using Haar, Daubechies-4, Mexican hat, and Morlet mother wavelets. The density represents the relative amplitude of the various terms in the series expansion of Equation 4.9 as a function of position and scale. Obviously, they are quite

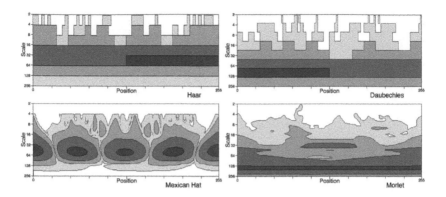

FIGURE 4.33
Wavelet spectra using Haar, Daubechies, Mexican hat, and Morlet mother wavelets on the arrow point shape from Figure 4.29 (calculated using the online IDL software at http://ion. researchsystems.com/cgi-bin/ion-p?page=wavelet.ion)

different representations of the same initial shape. The interpretation of such spectra and the desire to extract a few meaningful numeric descriptors are challenging tasks. Some use has been made of principal components analysis (PCA) but this has primarily been applied to time series not to object or feature boundaries.

Wavelets are particularly useful for locating corners or points of high local curvature (Costa & Cesar, 2009). Figure 4.34 shows an example. The unrolled star shape (radius from the centroid as a function of position along the boundary) has five exterior and five interior corners, which are evident in the wavelet spectrum.

The wavelet analysis (using the Mexican hat wavelet) of the differential chain code for the boundary is less obvious (Figure 4.35) because of the large number of +1 and –1 steps that arise from the pixelation of the shape. The

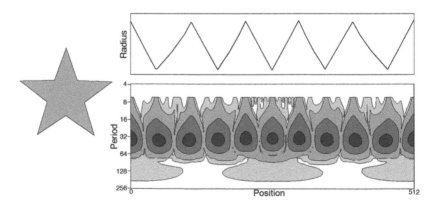

FIGURE 4.34
An unrolled star shape and its wavelet spectrum using the Mexican hat mother wavelet.

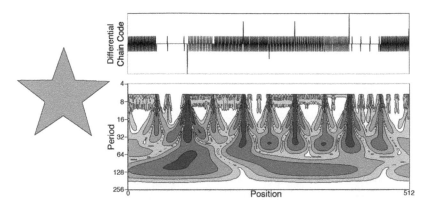

FIGURE 4.35
Differential chain code and its wavelet spectrum for the same star shape as Figure 4.34.

corners are detected because clusters of steps in the chain code produce local curvature that is strongly selected by the compressed versions of the wavelet that cover a corresponding distance along the profile. The fact that the chain code happens to begin at one of the points on the star causes the wavelet analysis to miss the high curvature at that location; this is a consequence of the fact (noted earlier) that the wavelet analysis, unlike Fourier analysis, does not take into account the repetitive nature of the outline.

The principal challenge to extracting a meaningful shape description lies in the choice of mother wavelet. The selection depends on the characteristics or features of interest. The Morlet wavelet gives better frequency resolution, whereas the Mexican hat produces better spatial localization. The latter wavelet can be used to locate "corners," as illustrated in Figure 4.36. The arrow point has several features with localized curvature, such as the tip, the serrations on each edge, and the corners of the base. Each of these has a different scale, as measured by length along the boundary.

The spectrum based on the Mexican hat wavelet applied to the unrolled shape shows the correlations at large size scales, represented by the four darkest regions (marked A) of high power at scales between 32 and 64 pixels (Figure 4.36). These represent the tip, the cutting surface, and the base, and the spectrum shows that these regions of high power extend upward to shorter scales as well. This is especially true for the corners of the base (B). At the next smaller scale, starting at about 16 pixels, are the indentations above the base and the rear corners of the cutting edge (C). The serrations along the sides of the blade (D) primarily involve the finer scales.

This example illustrates both the power and the difficulty of wavelet analyses. Although it is possible to analyze and to measure objects in this fashion, it is not obvious that these measurements are easier, more precise, or preferable to other, simpler methods.

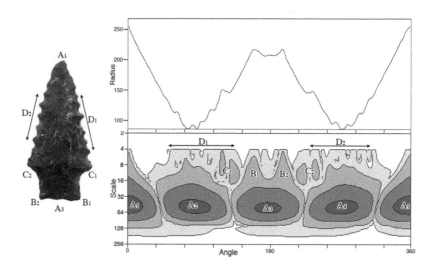

FIGURE 4.36
Corresponding locations on the arrow point and the wavelet spectrum, which begins at the point A_1 and proceeds clockwise around the periphery.

For a more complex shape, the spectrum is more difficult to interpret. Figure 4.37 illustrates an example of a reentrant shape for which unrolling of radius versus angle cannot be used. The differential chain code (2048 points) along the periphery of a chrysanthemum leaf is shown.

Most of the values in the code are either 0 (continuing in the same direction) or ±1 (45-degree turns to the left or right). Individually, these have little to do with the actual shape of the leaf, but result from the pixelation of the image. It is the sequence of code values over a range of positions that represents a significant interior or exterior indentation or lobe, some of which are quite small. The higher power at longer periods (longer physical scales)

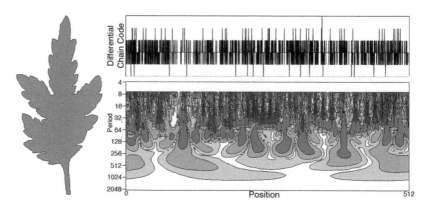

FIGURE 4.37
Differential chain code and wavelet spectrum for a leaf shape using the Mexican hat wavelet.

marks the major indentations and lobes on the leaf, but these are not necessarily easy to detect, to characterize in terms of size or curvature, or to accurately locate on the leaf periphery from the wavelet spectrum.

It is worth noting again the similarity of the spectra shown here to the CSS code representation in Figure 3.68. In that example, the plot is inverted with larger spatial distances—more smoothing—at the top of the graph. The features detected are not points of maximum curvature (for which the Mexican hat, the second derivative of a Gaussian, can be used) but rather points of inflection (for which the first derivative of a Gaussian is appropriate).

Moments

The moments of a feature composed of pixels are calculated as

$$\mu_{i,j} = \sum_{x,y} x^i y^j \cdot v \tag{4.10}$$

where v is the value of the pixel at location (x, y), if the pixel values have meaning. These are the moments about the x and y axes, but, as shown next, they can be used to calculate a set of moments about the principal axis of the feature. When $i = j = 0$, and $v = 1$ (in other words, a binary feature in which all the pixels are identical), this gives the area. Similarly, μ_{10} and μ_{01} are the centroid coordinates of the feature.

Higher moments describe the shape of the feature by measuring the distribution of pixels with respect to the x- and y-axes. Like the medial axis transform and Fourier analysis methods, with enough moments it is possible to exactly reconstruct the feature shape (Teague, 1980; Liao & Pawlak, 1996). Unlike the Fourier and wavelet analysis methods, which depend only on the coordinates of the periphery of the feature, moments use all the interior pixels as well.

Usually the meaning of the values is simplified by using central moments, measuring the (x, y) coordinates from axes that pass through the centroid. In addition, the moments are generally transformed to normalized central moments by calculating

$$\eta_{i,j} = \frac{\mu_{i,j}}{\mu_{00}^{\frac{i+j+2}{2}}} \tag{4.11}$$

One use for moments is to determine the principal axis or orientation angle of a feature. This can be calculated based on the covariance of the moments.

In implementing Equation 4.12, note that some methods for calculating the inverse tangent give results with a 180° ambiguity, which is converted to a potential 90° error in the angle.

$$\theta = \frac{1}{2}\tan^{-1}\left(\frac{2 \cdot \mu_{11}}{\mu_{20} - \mu_{02}}\right)$$

(4.12)

A set of seven invariant moments has been shown (Hu, 1962; Flusser & Suk, 2006) to remain constant with translation, rotation, and size (at least within the constraints of the pixel representation). These meet the definition of pure shape factors and therefore can be considered as a way to measure shape. The first one is analogous to the moment of inertia around the centroid of the feature.

$$m_1 = \eta_{20} + \eta_{02}$$

$$m_2 = \left(\eta_{20} - \eta_{02}\right)^2 + 4 \cdot \eta_{11}^2$$

$$m_3 = \left(\eta_{30} - \eta_{12}\right)^2 + \left(3 \cdot \eta_{21} - \eta_{03}\right)^2$$

$$m_4 = \left(\eta_{30} + \eta_{12}\right)^2 + \left(\eta_{21} + \eta_{03}\right)^2$$

$$m_5 = \left(\eta_{30} - 3 \cdot \eta_{12}\right) \cdot \left(\eta_{30} + \eta_{12}\right) \cdot \left[\left(\eta_{30} + \eta_{12}\right)^2 - 3 \cdot \left(\eta_{21} + \eta_{03}\right)^2\right]$$

$$+ \left(3 \cdot \eta_{21} - \eta_{03}\right) \cdot \left(\eta_{21} + \eta_{03}\right) \cdot \left[3 \cdot \left(\eta_{30} + \eta_{12}\right)^2 - \left(\eta_{21} + \eta_{03}\right)^2\right]$$

(4.13)

$$m_6 = \left(\eta_{20} + \eta_{02}\right) \cdot \left[\left(\eta_{30} + \eta_{12}\right)^2 - \left(\eta_{21} + \eta_{03}\right)^2\right]$$

$$+ 4 \cdot \eta_{11} \cdot \left(\eta_{30} + \eta_{12}\right) \cdot \left(\eta_{21} + \eta_{03}\right)$$

$$m_7 = \left(3 \cdot \eta_{21} - \eta_{03}\right) \cdot \left(\eta_{30} + \eta_{12}\right) \cdot \left[\left(\eta_{30} + \eta_{12}\right)^2 - 3 \cdot \left(\eta_{21} + \eta_{03}\right)^2\right]$$

$$+ \left(3 \cdot \eta_{12} - \eta_{30}\right) \cdot \left(\eta_{21} + \eta_{03}\right) \cdot \left[3 \cdot \left(\eta_{30} + \eta_{12}\right)^2 - \left(\eta_{21} + \eta_{03}\right)^2\right]$$

To demonstrate the invariance of these values, the leaf shape from Figure 4.17 was used, with rotation by different amounts, side-to-side reversal, and scaling to different sizes. The results are shown in Table 4.2. The slight variations in the calculated values arise because of the pixel aliasing that results from the rotation and scaling.

TABLE 4.2

Values for Moments from Equation 4.13 Calculated for the Original Leaf in Figure 4.17

	Original	25° Rotation	45° Rotation	Side-to- Side Reflection	75% Scaling	66% Scaling
m_1	1.851 E–01	1.851 E–01	1.851 E–01	1.851 E–01	1.851 E–01	1.8513E–01
m_2	4.100 E–03	4.100 E–03	4.100 E–03	4.100 E–03	4.100 E–03	4.100 E–03
m_3	3.434 E–04	3.434 E–04	3.432 E–04	3.434 E–04	3.431 E–04	3.434 E–04
m_4	6.768 E–06	6.774 E–06	6.764 E–06	6.768 E–06	6.746 E–06	6.786 E–06
m_5	1.184 E–10	1.188 E–10	1.178 E–10	1.184 E–10	1.179 E–10	1.198 E–10
m_6	2.881 E–07	2.885 E–07	2.876 E–07	2.881 E–07	2.873 E–07	2.895 E–07
m_7	–3.040 E–10	–3.044 E–10	–3.039 E–10	–3.040 E–10	–3.024 E–10	–3.049 E–10

While these seven moments are invariant to size and orientation, and are consequently pure shape factors, they are not sufficient to reconstruct the original figure and are not easily interpreted in terms of the visual shape of an object. The same set of leaf images introduced in Chapter 3 and used earlier with the Fourier coefficients can be measured to show the ability of these moments to classify and identify objects. Figure 4.38 and Table 4.3 show the results.

The final separation of classes is good, but requires all of the invariant moments and a high dimensionality space to accomplish the discrimination. Moments, which depend on all of the interior pixels within the feature, have difficulty distinguishing cherry from dogwood, which are both compact shapes whose most obvious difference is the smoothness or irregularity of the periphery. The Fourier coefficients shown in Figure 4.20, on the other hand, are derived from the boundary and separate these two shapes easily. This is the same distinction between methods that characterize shape using the interior or boundary that was shown for the convexity and solidity parameters in Chapter 3.

Moments have been applied to feature measurement, classification, and recognition in a few applications, including handwritten numerals (Pan & Keane, 1994), fingerprints and palm prints (Noh & Rhee, 2005), measurement of cell shape (Dunn & Brown, 1986), the density patterns in human chromosomes (Hilditch & Rutovitz, 1969), and the distribution of chromosomes within the dividing cell (Barton & David, 1962).

Although the seven moments shown in Equation 4.13 are invariant to translation, rotation, and scaling, they are not invariant to general affine transformations or to perspective distortion. There is an additional set of moments, derived from the second- and third-order moments as shown in Equation 4.14 (Flusser & Suk, 1993) that also remain invariant under general affine transformations.

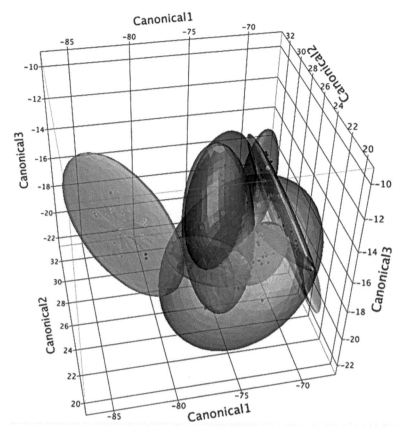

FIGURE 4.38

Discriminant analysis results for the leaf samples from Figure 3.23 using invariant moments (the three most significant canonical axes are shown).

TABLE 4.3

Comparison of Actual and Predicted Leaf Types Using Invariant Moments

Actual Class:	Predicted Class:								
	Cherry	Dogwood	Hickory	Mulberry	Red Maple	Red Oak	Silver Maple	Sweet Gum	White Oak
Cherry	16	1	0	1	0	0	0	0	0
Dogwood	0	14	0	0	0	0	0	0	0
Hickory	0	0	16	0	0	0	0	0	0
Mulberry	0	1	0	16	0	0	0	0	0
Red Maple	0	0	0	0	19	0	0	0	0
Red Oak	0	0	0	0	0	12	0	0	0
Silver Maple	0	0	0	0	0	0	18	0	0
Sweet Gum	0	0	0	0	0	0	0	16	0
White Oak	0	0	0	0	0	0	0	0	16

$$I_1 = \frac{\mu_{20} \cdot \mu_{02} - \mu_{11}^2}{\mu_{00}^4}$$

$$I_2 = \frac{\mu_{30}^2 \cdot \mu_{03}^2 - 6 \cdot \mu_{30} \cdot \mu_{21} \cdot \mu_{12} \cdot \mu_{03} + 4 \cdot \mu_{30} \cdot \mu_{12}^3 + 4 \cdot \mu_{21}^3 \cdot \mu_{03} - 3 \cdot \mu_{21}^2 \cdot \mu_{12}^2}{\mu_{00}^{10}}$$

$$I_3 = \frac{\mu_{20} \cdot \left(\mu_{21} \cdot \mu_{03} - \mu_{12}^2\right) - \mu_{11} \cdot \left(\mu_{30} \cdot \mu_{03} - \mu_{21} \cdot \mu_{12}\right) + \mu_{02} \cdot \left(\cdot \mu_{30} \cdot \mu_{12} - \mu_{21}^2\right)}{\mu_{00}^7}$$

$$I_4 = \frac{1}{\mu_{00}^{11}} \left(\mu_{20}^3 \cdot \mu_{03}^2 - 6 \cdot \mu_{20}^2 \cdot \mu_{11} \cdot \mu_{12} \cdot \mu_{03} - 6 \cdot \mu_{20}^2 \cdot \mu_{02} \cdot \mu_{21} \cdot \mu_{03} \right. \tag{4.14}$$

$$+ 9 \cdot \mu_{20}^2 \cdot \mu_{02} \cdot \mu_{12}^2 + 12 \cdot \mu_{20} \cdot \mu_{11}^2 \cdot \mu_{21} \cdot \mu_{03} + 6 \cdot \mu_{20} \cdot \mu_{11} \cdot \mu_{02} \cdot \mu_{30} \cdot \mu_{03}$$

$$- 18 \cdot \mu_{20} \cdot \mu_{11} \cdot \mu_{02} \cdot \mu_{21} \cdot \mu_{12} - 8 \cdot \mu_{11}^3 \cdot \mu_{30} \cdot \mu_{03} - 6 \cdot \mu_{20} \cdot \mu_{02}^2 \cdot \mu_{30} \cdot \mu_{12}$$

$$\left. + 9 \cdot \mu_{20} \cdot \mu_{02}^2 \cdot \mu_{21}^2 + 12 \cdot \mu_{11}^2 \cdot \mu_{02} \cdot \mu_{30} \cdot \mu_{12} - 6 \cdot \mu_{11} \cdot \mu_{02}^2 \cdot \mu_{30} \cdot \mu_{21} + \mu_{02}^3 \cdot \mu_{30}^2 \right)$$

Moments have several potential advantages as shape measurements in some applications. Because they depend on all of the pixels within the shape, not just those along the periphery, they are attractive for images in which determining the exact border of the feature is difficult. Furthermore, moments are not restricted to shapes that consist of contiguous blocks of pixels. Some shapes consist of separated regions, and even the number of those regions may not be fixed.

Table 4.4 shows the invariant and affine-invariant moments for the simple shapes in Figure 4.39. Moments offer another tool for detecting mirror symmetry in shapes, in addition to the Fourier harmonics described earlier. There are a few obvious interpretations, such as the zeros for the even-symmetric circle, square, hexagon, and ellipse (but not for the four-pointed or six-pointed stars). Some of the values are extremely small and are nonzero as the consequence of making up the shapes with finite pixels. Other than that, few straightforward interpretations of relationships between the visual interpretation of the shape and the numbers provided by the moment calculations are evident. This does not mean that moments may not be quite useful for computer classification and recognition.

Example: Tiger Footprints

As an example comparing moments to other shape descriptors from Chapter 3, a problem of current practical interest is the identification of footprints of individual endangered animals. This can potentially offer a noninvasive method for monitoring populations of animals and tracking their movements. Images of multiple footprints (always the left hind foot) for several

TABLE 4.4A

Invariant Moments for the Shapes in Figure 4.39

Shape	Inv. Mom. 1	Inv. Mom. 2	Inv. Mom. 3	Inv. Mom. 4	Inv. Mom. 5	Inv. Mom. 6	Inv. Mom. 7
1	0.159	0	0	0	0	0	0
2	0.166	0	0	0	0	0	0
3	0.192	1.68 E–08	4.55 E–03	1.46 E–09	–1.82 E–15	–1.42 E–13	3.30 E–15
4	0.161	5.06 E–09	4.17 E–09	8.75 E–11	–4.54 E–20	–3.06 E–16	–2.79 E–20
5	0.160	2.09 E–07	0	0	0	0	0
6	0.261	4.29 E–02	0	0	0	0	0
7	0.229	6.64 E–09	4.29 E–09	3.62 E–10	4.52 E–19	2.95 E–14	1.94 E–28
8	0.303	9.45 E–07	4.15 E–02	4.28 E–07	–5.69 E–11	–4.16 E–10	–4.43 E–12
9	0.208	4.10 E–08	6.20 E–08	7.78 E–09	9.29 E–17	–9.62 E–13	1.43 E–16
10	0.200	5.32 E–07	2.02 E–08	3.06 E–09	2.41 E–17	2.12 E–12	1.12 E–18
11	0.417	4.27 E–02	1.68 E–02	1.36 E–03	–5.98 E–06	–1.58 E–04	2.59 E–06
12	0.374	3.26 E–03	2.73 E–06	1.97 E–06	–2.99 E–12	–5.07 E–08	3.49 E–12
13	0.393	3.96 E–03	5.50 E–03	1.50 E–03	–2.48 E–06	–7.21 E–05	3.57 E–06
14	0.457	2.26 E–03	9.82 E–04	3.80 E–04	–1.87 E–07	1.79 E–05	1.37 E–07
15	0.445	1.97 E–02	8.86 E–02	5.64 E–07	–5.53 E–11	5.82 E–08	–1.13 E–10

TABLE 4.4B

Affine-Invariant Moments for the Shapes in Figure 4.39

Shape	Aff. Inv. 1	Aff. Inv. 2	Aff. Inv. 3	Aff. Inv. 4
1	6.33 E–03	0	0	6.95 E–37
2	6.94 E–03	0	0	8.33 E–38
3	9.25 E–03	–3.23 E–07	–5.47 E–05	4.05 E–06
4	6.54 E–03	–2.32 E–19	–4.13 E–11	2.21 E–12
5	6.42 E–03	0	0	5.44 E–34
6	6.33 E–03	0	0	2.20 E–36
7	1.31 E–02	–1.92 E–19	–5.62 E–11	6.45 E–12
8	2.29 E–02	–2.69 E–05	–7.86 E–04	1.44 E–04
9	1.09 E–02	–2.36 E–17	–7.08 E–10	7.07 E–11
10	1.00 E–02	–3.18 E–18	–2.50 E–10	2.02 E–11
11	3.29 E–02	–1.46 E–06	–2.80 E–04	1.14 E–04
12	3.42 E–02	9.48 E–13	–1.78 E–08	2.08 E–08
13	3.76 E–02	7.23 E–07	–6.71 E–05	3.09 E–05
14	5.17 E–02	5.02 E–08	–2.20 E–05	1.21 E–05
15	4.46 E–02	–1.22 E–04	–2.47 E–03	8.99 E–04

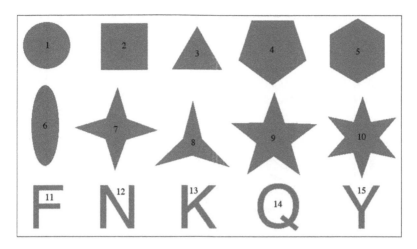

FIGURE 4.39
Some simple shapes whose moments are listed in Table 4.4.

tigers were provided by Zoe Jewell (Wildtrack.org). As shown in Figure 4.40, the prints from a single animal vary, even to the extent that in some cases the toes touch and change the number of separate regions. The question is whether the prints of one animal are more like themselves than like those of another.

A set of images of five footprints from each of five different tigers was used to compare techniques. Using the seven invariant moments plus the four affine-invariant moments it is possible to use linear discriminant analysis to separate the individual animals, as shown in Figure 4.41. Although some of the individual prints can be confused with that from another animal, with the values from a small number of prints the separation is sufficient for identification. (In practice, a minimum of six to eight prints along a single continuous track is always collected in order to identify a specific animal.)

Using landmark techniques (described in Chapter 3), much better results are obtained. A semiautomatic system for identifying individual tigers begins by identifying 21 primary landmarks along the outlines of the imprint

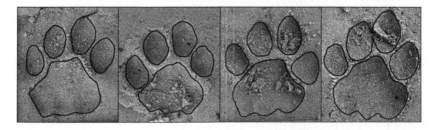

FIGURE 4.40
Tiger prints (left to right): Carmelita, Kaela, and two prints from Rajah.

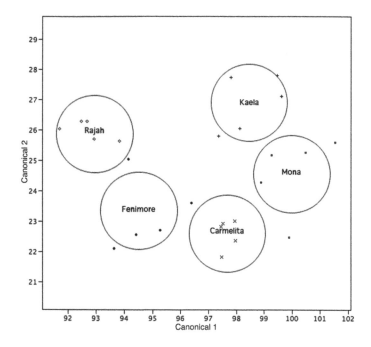

FIGURE 4.41
Linear discriminant results for five prints each from five different animals, using the invariant and affine-invariant moments.

of toes and heel pad. After the image is manually rotated to a standard alignment based on the posterior edge of the lateral and medial toes, the outline of
each imprint is marked with human assistance. Then for each toe, the points
with maximum and minimum x and y coordinates, respectively, are located
(4 toes × 4 points each = 16 points). An additional five points are located on
the outline of the heel pad imprint: the minimum and maximum x coordinates, the topmost y coordinate, and the bottommost y values on either half
of the pad. These primary points are marked in red in Figure 4.42.

An additional 12 secondary or constructed landmarks are created from
these. As indicated in Figure 4.42, six points are defined by the intersections
of lines joining specific pairs of the primary points, and another six correspond to corners of the bounding boxes around the entire footprint and the
heel pad. From the resulting 33 total landmark coordinates, a set of distance
and angle measurements between various pairs of points produces a total of
more than 120 measurements, which are then used for discriminant analysis
using SAS JMP 9 (results provided by Sky Alibhai, Wildtrack.org).

This method was used with the same set of five footprints each from five
different tigers to produce the results shown in Figure 4.43. For this particular set of data, the statistical analysis chose variables 1, 4, 13, 40, 67, 75, 82,
97, 100, and 112 to construct the canonical variables. Complete separation

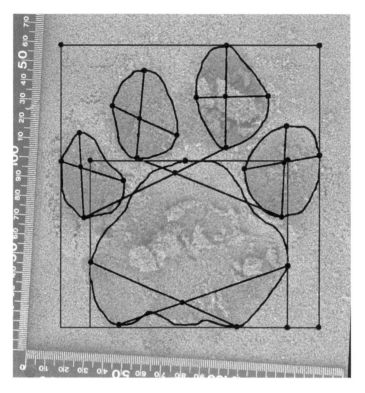

FIGURE 4.42
(See color insert.) Landmarks on a tiger print. The red primary points are extrema on the outlines of each toe and heel pad. The blue secondary points mark intersections between several lines (green) joining primary points and the corners of bounding boxes (magenta) around the entire print and around the heel pad.

is achieved so that an animal can be identified from a single print. This is impressive, but in practical terms it is necessary to decide whether the results are robust. One approach to evaluating this is by examining the landmarks from the multiple images to see if the points for each animal are consistent, and if the scatter of the locations within each animal's prints is less than the difference between the animals.

(It is important to note that the actual FIT procedure [Alibhai & Jewell, 2008] developed for animal identification from footprints in the wild does not rely simply on discriminant analysis, and that rather than five prints from five animals, the models for identification are built on multiple sets of prints from multiple tracks and deal with populations 10 times greater in size. The purpose of using a more limited set of data here is to demonstrate methods and approaches to data interpretation. The FIT method explicitly uses both size and shape variables; the emphasis here is primarily on shape.)

Figure 4.44 shows the outlines of the five footprints from one tiger, with the primary landmarks for each print and ellipses marking the variation

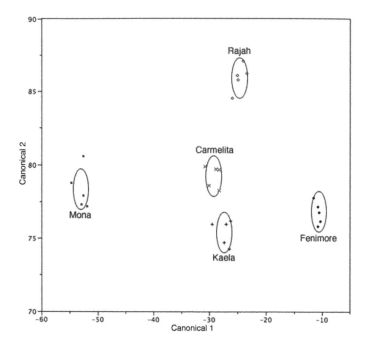

FIGURE 4.43
Linear discriminant results for the same prints as in Figure 4.41, but using landmark values.

among the five prints. Repeating this procedure for each of the tigers, and superimposing the points and ellipses as shown in Figure 4.45, shows that the clusters of points for each tiger's prints are as large and often larger than the differences between them. This suggests that the apparently good discrimination shown in Figure 4.43 may not be robust.

The fact that statistical analysis of more than a hundred measurements can select a few that successfully discriminate between a small number of individuals does not necessarily mean that those measurements constitute a meaningful representation of "shape," or, as one researcher involved in the program has remarked, "It is more of an engineering solution than a scientific discovery." However, as has been pointed out (Law, 2011), for endangered species management the number of individual animals in an area may be very small (tens, not hundreds) and distinguishing them based on footprints may be a practical approach, better than invasive monitoring with radio collars or attempting to identify them from sightings or cameras set up along game paths.

Further analysis shows that the landmark method, at least in this instance, is not dealing just with shape as it is dealt with in this book. The various landmark points in Figure 4.42 are used to construct a table of pairwise distances and angles between them. If ratios of the distances were used, in order to eliminate scale as a factor, or if a full generalized Procrustes analysis had

FIGURE 4.44
(See color insert.) Outlines of five footprints from one tiger with the primary landmark points marked to show the amount of scatter.

been performed, which would likewise have removed scale, then the variables would reflect only the shape of the object. But in this case it is clear that the actual difference between the prints from the various tigers is primarily one of size. (The moments used to generate Figure 4.41 are invariant to size and orientation, and so are "pure" shape descriptors.)

Figure 4.46 shows a scatterplot of the measured areas of the prints (the sum of the heel pad and the four toes) and the vertical and horizontal dimensions (X- and Y-Feret's diameters) of the bounding box around the entire print. The five clusters fully separate the individual animals. If the same discriminant analysis approach is used with size measurements (the net area as described earlier, the area within the bounding polygon, and the dimensions of the bounding box), the result is complete separation and the ability to identify each print from each animal (Figure 4.47). A single-layer neural net with three nodes (as described in Chapter 6) is also able to successfully identify each print using these three size inputs.

If, in addition to the size measurements, the dimensionless ratios introduced as shape descriptors in Chapter 3 are used for discriminant analysis, the results are further improved as shown in Figure 4.48. The practical

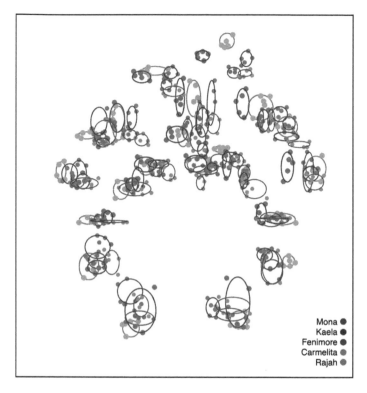

Mona ●
Kaela ●
Fenimore ●
Carmelita ●
Rajah ●

FIGURE 4.45
(See color insert.) Superposition of the points and ellipses for each of the five sets of five tiger prints.

question is whether adequate results can be obtained in the field, with minimum need for expert interpretation of the prints and location of the landmarks. There is no reason to exclude size variables in the quest for a solution; in fact many papers that apply landmark methods specifically refer to "size and shape" in presenting results.

The issue with landmark methods is not their ability to provide discrimination, but that they are based primarily on points selected by an expert lying along the outer boundary of a structure. As previously noted, perimeter-based measurements, including harmonic Fourier coefficients and landmarks along the periphery of a shape, are highly sensitive to "missing" information, which leads to practical drawbacks in their use. This is particularly evident with animal tracks such as those in Figure 4.49.

The nonuniformity of the textures of the substrate, as well as the indistinctness of the track margins, make thresholding or outlining these features difficult. The task illustrated for the tiger prints is but one example of the desire to classify many images of animal tracks, and information that is missing or uncertain along the periphery makes harmonic analysis or fractal dimension a poor choice. It also complicates the landmark method, since the landmarks

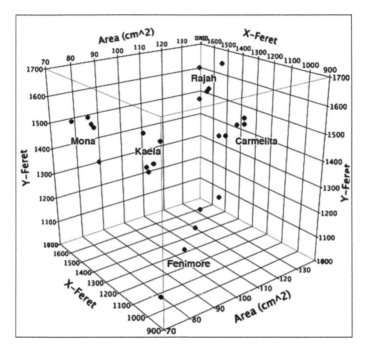

FIGURE 4.46
Scatterplot of the measured area and dimensions of the bounding box for the tiger prints.

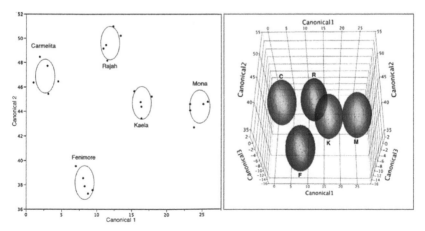

FIGURE 4.47
Discriminant analysis for the tiger prints, using only size measurements.

of importance for each species are quite different, and their location may not be clearly shown in the print and must be determined by expert judgment.

An area-based descriptor such as invariant moments, or the dimension-less ratios based on both periphery and interior pixels, suffers less from the missing information but may not offer as efficient a means of identification

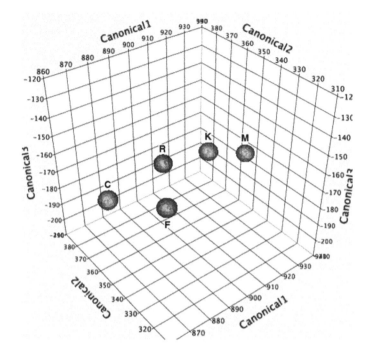

FIGURE 4.48
Discriminant analysis results using shape descriptors (dimensionless ratios) in addition to measures of size.

if proper landmarks can be robustly identified. Also, most of the methods discussed in this and the preceding chapter do not readily handle shapes consisting of multiple unconnected parts, as is the case for these tracks.

Invariant moments do not capture all of the details of feature shape. A complete set of moments, which does fully represent the shape and can be used to reconstruct it, can also be calculated. For a feature that has a periphery consisting of N points or vertices, a set of $(2N - 1)$ moments is theoretically sufficient to reconstruct the shape. For a shape without indentations, the vertices of the bounding polygon can be used to calculate a complete set of moments, requiring a relatively small value of N. For an irregular feature defined by pixels, the number N may be as high as the number of boundary pixels, which results in a very large required number of moments (but recall that for the complete set of Fourier descriptors the real and imaginary terms are also equal in number to the boundary points).

In practice, the calculations by which the shape can be recovered from the moments, and the sensitivity of the reconstruction to noise in the original image and the presence or absence of a few pixels in the representation, make this theoretical possibility one of limited use. In terms of the classification scheme suggested in the Introduction, the full set of moments is an example of a characterization that is able to reconstruct the original shape.

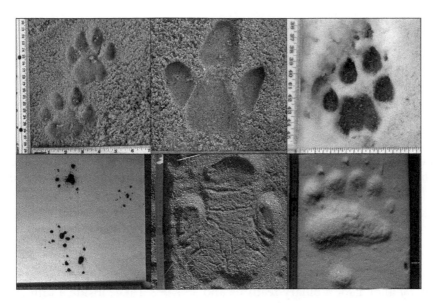

FIGURE 4.49
Animal footprints used for identifying individuals in wild populations (from upper left: cougar, tapir, cheetah, wood mouse, white rhino, polar bear). The images are not printed at the same scale. Tracks may be recorded in sand, soil, and snow. (Images courtesy of Sky Alibhai and Zoe Jewell, WildTrack.org.)

The invariant moments, on the other hand, cannot reconstruct the details of the shape. Both are based on all of the pixels within the feature, as opposed to using just the periphery.

For comparison purposes, consider harmonic analysis, which can also be used to reconstruct a shape. The fractal dimension, which may be obtained from a plot of the amplitude of the Fourier coefficients as a function of frequency, describes the boundary "roughness" but cannot reconstruct the shape. But harmonic analysis and fractal dimension use information from the periphery of the feature to define shape, while moments are calculated using all of the pixels and are thus strongly influenced by the interior. That distinction is the second axis of the classification scheme described in the Introduction.

Zernike Moments

While the moments described by Equation 4.10 and Equation 4.11 are complete, they are not efficient (both in terms of the number of terms and the computation needed to reconstruct a shape, and because they are redundant), and their meaning is difficult to interpret. Zernike moments have several advantages for shape representation. They use complex polynomials that form a complete set of orthogonal functions, meaning that each is independent of the others, and they automatically produce translation, rotation, and scale independence. The Zernike basis functions are

$$V_{nm}(\rho,\theta) = R_{nm}(\rho) \cdot e^{im\theta}$$

$$R_{nm}(\rho) = \sum_{s=0}^{\frac{n-m}{2}} (-1)^s \cdot \rho^{n-2s} \cdot \frac{(n-s)!}{s! \cdot \left(\dfrac{n+|m|}{2} - s\right)! \cdot \left(\dfrac{n-|m|}{2} - s\right)!} \qquad (4.15)$$

where n is a positive integer (or zero), m is an integer such that $|m| < n$ and $(n - |m|)$ is even, and (ρ, θ) are polar coordinates. The moments are calculated as

$$A_{nm} = \frac{n+1}{\pi} \sum_{x,y} f(x,y) \cdot V_{nm}^{*}(\rho,\theta) \qquad (4.16)$$

where the * indicates the complex conjugate; f is the value of the pixel at location x,y (0 or 1 for a binary feature); and $x^2 + y^2 \le 1$ (i.e., inside a unit circle around the feature, which gives scale independence).

As illustrated in Figure 4.50, reconstruction of a shape from the moments is a simple additive process starting with the lowest n,m terms and proceeding upward until the desired quality (reproduction of the original) is achieved, analogous to the use of Fourier harmonics for reconstruction. But, of course, with the moments it is possible to have shapes that consist of separated parts and internal gaps, which Fourier or wavelet analysis of the boundary cannot handle.

Khotanzad and Hong (1990) demonstrate that the use of Zernike moments is useful both for efficiently coding and reproducing shapes, and also for comparing shapes for similarity using classifiers such as the k-nearest-neighbor method described in Chapter 6. Belkasim et al. (1991) use them to identify handwritten numerals. Retrieval of similar shapes from a database using Zernike moments is shown in Lin and Chou (2003) and Toharia et al. (2007). Biometric applications such as fingerprint (Qader et al., 2007), hand (Amayeh et al., 2006), and face (Ono, 2003) recognition have also been proposed, and the ability to identify sex from handprints demonstrated (Amayeh et al., 2008).

FIGURE 4.50
Reconstruction of a shape using Zernike moments. The original shape (the letter E) is at left. The following images show the results obtained by adding together the first 2, 4, 6, 8, 10, and 12 moments. (From A. Khotanzad, Y. H. Hong, 1990, *IEEE Transactions on Pattern Analysis and Machine Intelligence* 12(5):489–497.)

Cross-Correlation

Although it does use all of the pixels within the feature, and consequently is relatively insensitive to minor changes or noise in the boundary definition, cross-correlation does not provide a useful tool for comparing shapes. The relative magnitude of the cross-correlation peak and the peak shape and width are dependent on too many factors, including the relative areas of the object and the target and their rotational alignment, to be readily interpretable. Figure 4.51 shows the silhouettes of several of the snowflake images

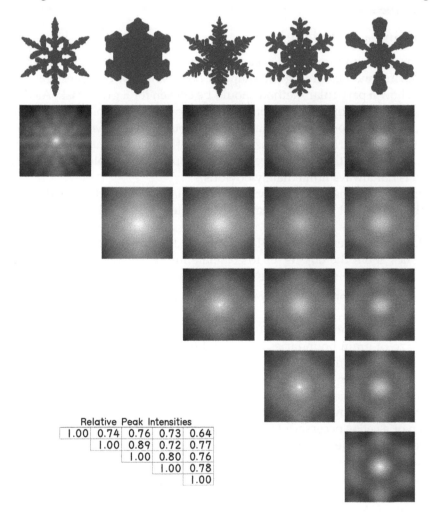

Relative Peak Intensities				
1.00	0.74	0.76	0.73	0.64
	1.00	0.89	0.72	0.77
		1.00	0.80	0.76
			1.00	0.78
				1.00

FIGURE 4.51
Silhouettes of five snowflakes, with their cross-correlation images and the relative magnitude of each central peak.

from the more extensive set shown in Figure 6.26, with their cross-correlation images and the relative values of the peak heights. The silhouettes have their major axes aligned and adjusted to the same dimension. The 1.0 values that occur along the matrix diagonal result from matching each image to itself, but although the peak values are equal, the shapes of the peaks are quite different. The other values vary between 64% and 89% but do not correspond to the visual impression of similarity between the various shapes.

Choosing a Technique

This chapter and the previous one have illustrated a variety of techniques for measuring and classifying the shape of objects. It now becomes reasonable to ask how a particular method should be chosen for a given task. As digital imaging has become more sophisticated, the number of pixels constituting an image has grown dramatically. As computational power has increased, techniques requiring an increasing degree of complexity have become practical. Because of this, the choice of an appropriate descriptor for the task at hand is not always immediately obvious.

The choice of shape descriptor for a particular task should always be driven by the purpose of the measurement. Most often, a task will involve either

1. the measurement of similar things to determine population statistics about the objects, or to correlate minor changes in shape with some other information (e.g., changes in history or performance); or
2. the measurement of many dissimilar things in order to classify them.

Examples of the first type of measurement include measuring the shape and size of grain boundaries in a metal to understand structure–property relationships, or measuring the modification ratio of extruded plastic fibers to understand process variation. In biology, the description of a species must include information about the range of variation (in shape as well as in size). In medicine, knowing the natural range of variability in the shape of an organ or a cell is necessary to spot deviations that may indicate disease.

Examples of the second task include using topological characteristics for character recognition, or a convexity measurement as a means of quality assurance to detect foreign objects in powders, seeds, or foods. Classification of new insect species is often based on significant differences in the shape of body parts. Analysis of the curvature of an archaeological collection of potsherds can indicate the different sizes of the vessels that were in use.

One way to make a good choice of descriptor for a particular application is to start with the classification scheme suggested in the Introduction (Figure

I.2). Most shape description methods are either complete or reductive, and either based on the periphery of an object or include its interior. A further consideration is the desired or practical number of measurements. An imaging system designed to measure the output of a running production line can tolerate less complexity in measurement and calculation than a scientist measuring a relatively small number of samples in the lab. Some methods involve human interaction, which is costly and time consuming, as well as raising concerns about consistency from one operator to another or for one person over time. If the resolution needed to achieve adequate measurement precision for a given method is high, the cost of the cameras and the computer power required increase, although this is becoming less important.

Methods that cannot reconstruct the shape are reductive. As shown in Chapter 3, these discard all but a limited set of information about the shape. These descriptors have the advantages of highlighting those aspects of the overall shape for which the technique is designed, so that appropriate statistics can be easily computed. These include invariant moments and the classical shape descriptors based on dimensional measurements such as formfactor and aspect ratio. These descriptors are often the initial choice because of their simplicity, ubiquity of implementation in commercial software packages, and their ability to be understood or perhaps recognized intuitively.

With reductive techniques, it is often not possible to choose a small (or even a large) set of descriptors that separate different classes of objects correctly. This is most often the case when the classes are visually similar to each other, such as distinguishing grains in different types of sediments, or histological classifications. In these cases, it is necessary to retain more information about the objects. This will almost invariably require the use of statistical techniques to gain any insight from the measurements. Unfortunately, in some instances the statistical analysis may tend to obscure the visual "meaning" of the descriptors.

The choice between descriptors that use only the periphery of an object and ones that use the entire area of an object is more complex. Many of the perimeter-based measurements, such as fractal dimension, are largely scale-invariant within the limits of precision. This allows them to be used effectively for classification across multiple images. Typically, the periphery of an object will have a higher density of data about an object's shape, and measurements based on the perimeter may be preferred as giving greater efficiency in calculating a result.

Example: Arrow Points

Consider the following example, in which a total of 38 images of arrow points were thresholded and measured. Only substantially intact points were used, not broken fragments. Five recognized styles are represented (Figure 4.52), each with variations from one point to another due to the individual creator's technique, material variations, and so forth. In particular,

FIGURE 4.52
From left to right: Caraway, Hardaway, Kirk serrated, Morrow Mountain, Dalton. (Courtesy of
Dr. Billy Oliver, Office of North Carolina State Archaeology Research Center.)

the length-to-breadth ratios vary significantly within each type, and the
edge roughness or serrations are quite variable.

Dimensionless ratios were used to classify the objects, with poor results.
Using only the dimensionless ratio shape factors, it was not possible to
distinguish all the individual points or to completely separate the groups.
Invariant moments fared somewhat better (Figure 4.53), separating the individual points but with considerable scatter.

In contrast, using just three terms from the harmonic analysis separates all
of the arrow points (Figure 4.54), and analysis using thirty terms gives very
tight clusters for each set (Figure 4.55.). The three terms used to achieve the
classification shown in Figure 4.54 are the 4th, 6th, and 10th, and their combinations to create the canonical variables are shown in Equation 4.17. It is a
plausible speculation that the lower frequency terms are primarily related to
the gross shape of the point, whereas the higher frequency terms represent
variations due to individual knappers and in the materials used. However,
no hard data are available to support that notion.

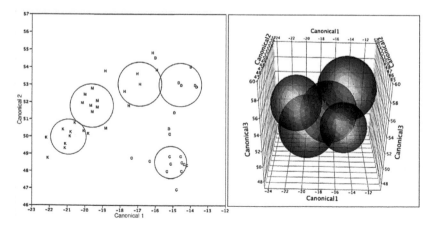

FIGURE 4.53
Classification of arrow points by invariant moments.

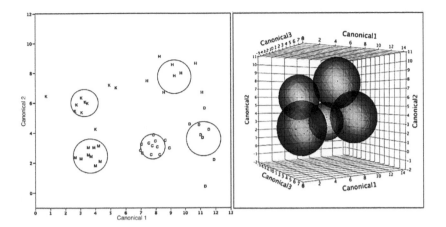

FIGURE 4.54
Classification of arrow points by harmonic analysis using three terms.

$$Canonical\ 1 = +66.294 \cdot FFT_4 + 97.744 \cdot FFT_6 - 74.393 \cdot FFT_{10}$$
$$Canonical\ 2 = -12.374 \cdot FFT_4 + 54.315 \cdot FFT_6 + 308.87 \cdot FFT_{10} \qquad (4.17)$$
$$Canonical\ 3 = +25.329 \cdot FFT_4 + 95.309 \cdot FFT_6 + 124.135 \cdot FFT_{10}$$

The use of Fourier series in this particular case is powerful and successful, but turns out not to be the only way to accomplish the task. Another classification scheme can be constructed using various size measurements (area, perimeter, convex area, length, breadth, etc.) in addition to the dimensionless shape descriptors, as shown in Figure 4.56. Shape is an important component of object recognition but obviously not the only one.

Even for simple geometric shapes, selecting a measurement that efficiently captures the shape characteristic of greatest utility is important.

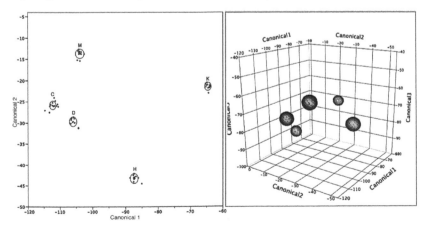

FIGURE 4.55
Classification of arrow points by harmonic analysis using the full 30-term Fourier expansion.

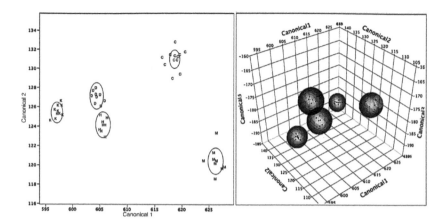

FIGURE 4.56
Classification of arrow points by size and dimensionless ratio shape descriptors.

The flags of many nations include stars; Figure 4.57 reproduces a few of them. Human vision quickly recognizes the number of points in each star as a defining topological characteristic. Another distinguishing feature is the degree of indentation of each point, which, for instance, distinguishes the Jordanian and Australian seven-pointed stars. The topology of each star is easily measured by counting the number of end points in the skeleton of the shape, as shown in the figure. The ratio of inscribed diameter to circumscribed diameter corresponds to the depth of the indentations. Those two measured variables completely describe the shapes of the stars, regardless of the size of the flag.

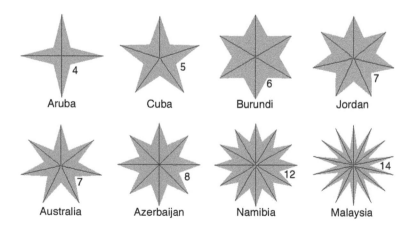

FIGURE 4.57
Star shapes from selected national flags, with the superimposed skeleton and number of end points in each.

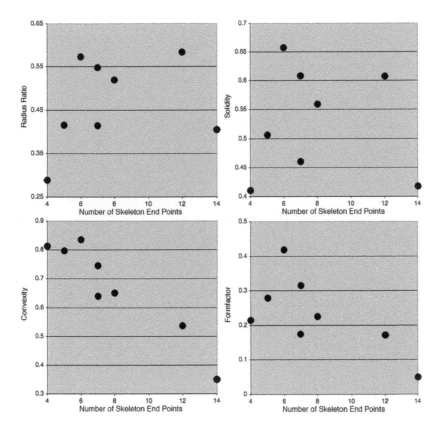

FIGURE 4.58
Graphs of the radius ratio, solidity, convexity and formfactor for each star in Figure 4.57 plotted against the number of skeleton end points. The convexity plot shows the highest correlation but still does not serve to distinguish the shapes.

Other dimensionless ratios such as the radius ratio, formfactor, solidity, and convexity do not provide a useful measurement of these shapes (Figure 4.58). Combining those dimensionless ratios in a complex logic net can distinguish these eight shapes. But it would not be able to accommodate a general case with other stars having different numbers of points or different indentation depths.

5

Three-Dimensional Shapes

The direct application of the two-dimensional shape measurements described in the previous chapters to three-dimensional objects is often difficult, incomplete, or ambiguous. Reducing a three-dimensional object to a two-dimensional shape that can be measured using the various procedures illustrated does not produce a meaningful result in most cases. A projected image of the outline or silhouette of a three-dimensional object does not fully reveal the shape and may even be quite misleading (see Figure 1.5). Even with multiple views there will often be missing information. Indentations in the surface are covered up in the projected outline, which presents a smoother appearance as hills cover valleys. For a particle with a fractal surface, projection produces a lower fractal dimension (and indeed the projected outline may not even be fractal).

Randomized section images represent objects only in a statistical sense, and do not show the details or the variation that may be present. A section image produces an outline that is essentially a single profile measurement across the surface, as might be obtained with a scanning profilometer. A common parameter extracted from profilometer traces across engineering surfaces is R_z, the range between the lowest and highest points. But the probability that the trace will pass through the actual lowest or highest points on a surface is extremely low, and the actual range observed depends on the length of the scan and potentially on its orientation, so there is no fixed relationship between the measured value and the actual surface roughness.

Slightly more meaningful parameters such as R_a, the average deviation from the mean profile, or R_q, the standard deviation of the elevation values, are also used, but these also depend on the length of the traverse. They require a fitting procedure to determine the mean profile, which may or may not be a straight line. Figure 5.1 shows a typical elevation profile measured on a ground metal surface. Note the difference between the lateral dimensions and the scale of the height measurements, as illustrated in the figure. The profiles are blind to any undercut regions, and the depth and steepness of slopes that can be followed are limited by the shape and length of the tip. The measurement of surfaces is discussed later.

When full three-dimensional measurements are performed on objects, such as a complete voxel array, some of the two-dimensional shape descriptors can be extended by analogy to three dimensions. For example, the use of dimensionless ratios such as formfactor ($[4\pi \cdot Area]/Perimeter^2$) to describe a departure from a circular shape is similar in approach to comparing a

FIGURE 5.1
Profilometer trace on a ground metal surface. The total length of the profile is 1.5 cm, but the vertical range of measurement is just tens of micrometers.

dimensionless ratio based on the surface area (S) and volume (V) of an object to that of a sphere. Equation 5.1 illustrates a possible relationship:

$$Sphericality = \frac{\left(36\pi\right)^{\frac{1}{6}} \cdot V^{\frac{1}{3}}}{S^{\frac{1}{2}}} \tag{5.1}$$

As for the two-dimensional shape factors, there is practically no limit to the number of such formally dimensionless ratios that can be proposed nor to the variety of names that may be proposed for them. And also, as for the two-dimensional case, there is no reason to expect that any of these will represent the particular variations of shape that may be recognized or important in any specific situation.

Acquiring Data

An additional difficulty that affects all three-dimensional measurements, and especially any shape descriptors that depend on how well the surface of the object is defined, is the comparatively poor resolution that many 3D imaging methods provide. Voxels, the volume elements that comprise a 3D representation in the same way that pixels comprise one in 2D, are usually larger in both an absolute and relative sense than pixels. For some techniques this is due to technological limitations; for example, computed tomography using X-rays is typically limited to a resolution of about a millimeter for medical imaging. Specialized instruments and the use of synchrotron sources to image small objects can achieve 3D resolution of about 1 µm, whereas a light microscope examining a 2D section can easily resolve better than 1 µm, and electron microscopes have a resolution of about 1 nm. In addition, many 3D imaging techniques, such as tomography, rely on reconstruction methods that are sensitive to noise. This makes defining the position of surfaces more difficult and less accurate.

It is not the purpose of this text to fully describe or compare all of the various methods for acquiring 3D data sets, which vary widely. But the consequences of the different methods for the characteristics of the voxel array do deserve mention. Some produce a series of planar images, which may require alignment to each other. For example, traditional serial sectioning of biological tissue captures images of individual slices cut by a microtome, using transmission light or electron microscopy. The slices are picked up and viewed in random rotational orientations, and furthermore are compressed somewhat by the cutting process. If fiducial marks (e.g., embedding wires in the block before sectioning) are available, the alignment and distortion removal are greatly simplified; otherwise alignment of the features themselves from slice to slice can result in cumulative errors and bias.

Sequential removal of surfaces, whether by grinding, ion beam erosion, shovel, or bulldozer (e.g., in a quarry or archaeological site), avoids the alignment problem but may cause different rates of removal in different locations, if the material hardness varies. This results in a distortion of the voxel information. Optical sectioning, as in the confocal light microscope, is the most satisfactory in terms of distortion but generally has somewhat poorer resolution in the depth direction than laterally. Also, resolution may vary with depth, and, of course, many materials are not transparent.

Many tomographic reconstruction methods (CT, MRI, PET, etc.) also produce images from one planar array of voxels at a time, and these may similarly have different spacing from plane to plane than the resolution within the plane. The use of noncubic voxels makes processing and measurement much more difficult. Many of the processing algorithms shown in Chapter 2 depend on the local neighborhood around each point (pixel or voxel) and become more difficult to apply if the distance varies with direction. Volumes can still be determined by counting voxels, but surface area, length, and so forth are not readily obtained and may also be directionally biased.

For these practical reasons, and also perhaps because the principal use of volumetric imaging remains visualization rather than measurement (Ruthensteiner et al., 2010; Ziegler et al., 2010), few currently available programs offer much in the way of measurement, and particularly shape measurement. This is not true for dedicated surface measurements, which are discussed later. Their importance in industry requires precise characterization, and there are many instruments and algorithms in use.

Tomographic reconstruction is capable of providing direct measurement of an array of cubic voxels if the raw data are collected appropriately. This may be done, for instance, by rotating the specimen through a range of angles and capturing projected cone beam images, which are then combined in the reconstruction. Another method that is inherently three dimensional is holography, but the result is an image, not an array of voxels. It has been speculated but not demonstrated that information about object shape can be extracted directly from a hologram without reconstructing an image or generating a fringe pattern (Seebacher et al., 1998; Sang et al., 2011).

Finally, 3D imaging places significant demands on computer memory and speed, and even with modern computers it is often necessary to limit the size of the 3D voxel arrays to obtain practical results. For example, a 3000 × 4000 pixel image with 2 bytes per pixel occupies 24 Mbytes of memory; that same memory will hold only a 300 × 200 × 200 voxel image. Applying a 2D convolution with a kernel consisting of a 7 × 7 array of weights requires 49 multiplications and additions for each pixel; a 7 × 7 × 7 array applied to a voxel image requires 343 multiplications and additions for each voxel.

In testing pixel neighbors to determine whether they are connected, there are 8 neighbors to be visited in 2D, but 26 in 3D as shown in Figure 2.60. If the presence or absence of each of the 8 neighbor positions in 2D is used as an index into a table to determine the outcome of operations such as erosion or dilation, the table occupies just $2^8 = 256$ bytes and provides an extremely efficient means for application. But in 3D where there are 26 neighbors, the table would require $2^{26} = 6.7 \cdot 10^7$ entries, and so other, slower algorithms for carrying out the operations must be devised.

Measuring the projected or caliper diameters of a shape to find a maximum dimension or to locate the corners of a bounding polygon (2D) or polyhedron (3D) presents another example. In 2D, a precalculated table of sine and cosine values for a small set of axis rotation angles (usually between 10 and 20) allows determining the maximum dimension with a very small potential error (with 20 angles, the worst case error is less than 0.1%). In 3D, rotations about two axes are required. Achieving the same level of accuracy would require more than 300 axis rotations, with comparison of the resulting data.

Furthermore, the addressing of voxels in memory requires more elaborate calculations to determine where in memory the neighboring voxel values are located. It is only within the past decade that desktop computers have really been able to provide practical power for reconstructing and processing voxel arrays, allowing real-time manipulations such as 3D rotation or interactive sectioning of the array. Measurement possibilities are still rather limited. It is reasonable to expect all of the technological limitations to be pushed back, but the difficulties of working in 3D will always be greater than in 2D.

Registration and Alignment

In some fields of application, such as biomedical imaging, the concern with shape is not to measure it but to compare one shape to another, for example, the deviations of the shape of an organ or cell from a canonical reference, such as a healthy organ or a cancerous cell. In most cases these comparisons are performed by visual inspection, relying on the knowledge of an experienced pathologist, for example, to know what deviations are important and which are not. This sort of human judgment is rarely quantified by measurement and is taught by example.

Registration is critical for image-based treatment planning and delivery, such as radiation treatment of tumors as in Figure 5.2 (Xing et al., 2006) and robotic surgery (Burschka et al., 2005). Although automatic registration is available, manual, visual-based image fusion using three orthogonal planar

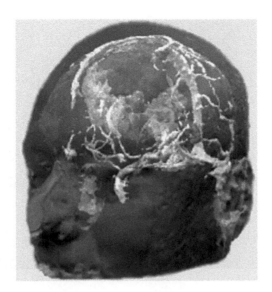

FIGURE 5.2
(See color insert.) Three-dimensional volumetric visualization of a brain tumor.

views is employed clinically to verify and adjust the automatic registration result. This can be time consuming, observer dependent, as well as prone to errors, owing to the limitations of 3D volumetric image representations (Li et al., 2008).

Medical imaging using X-ray or positron emission tomography, magnetic resonance imaging, ultrasound, endoscopy, and other technologies is concerned with establishing the shape, structure, size, and spatial relationships of anatomical features, with spatial resolutions that vary from less than 1 mm to several cm. Pathologists can generate volumetric images of tissue using confocal light microscopes with spatial resolution of about 1 μm.

The voxel arrays are examined using a variety of computer-rendering methods, which may be as simple as viewing an arbitrarily positioned section plane through the array or a sequential "movie" as a plane is moved through the volume. This requires the diagnostician to mentally "fuse" the images into a 3D representation. More elaborate presentations include volumetric views in which voxels are made partially transparent based on the measured value (which may be density, water content, or whatever property the imaging technique assesses), and assigning color to regions based on some prior knowledge about what structure is represented.

With additional processing of the array, methods similar to the edge-finding algorithms described in Chapter 2 for 2D images can find surfaces, and those surfaces can be rendered to depict structures. Figure 5.3 shows an example in which surfaces have been rendered with color coding and partial transparency to depict internal structures. With all of these presentation modes there is usually the ability to rotate the array to show views from any direction, as shown in Figure 5.4.

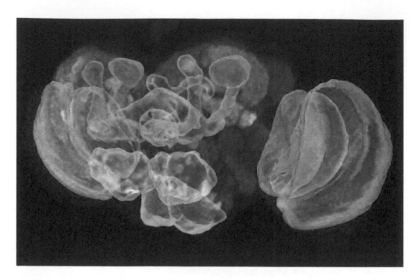

FIGURE 5.3
(See color insert.) Reconstruction of a fruit fly brain with transparent surfaces. (Data from K. Rein, Department of Genetics, University of Würzburg; visualization by M. Zoeckler using Amira software.)

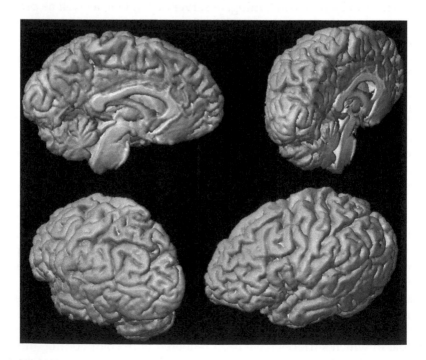

FIGURE 5.4
Selected still frames from an animation in which a reconstructed view of one-half of a brain is continuously rotated. (Image courtesy of Mark Dow, University of Oregon.)

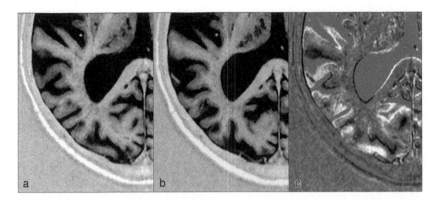

FIGURE 5.5
Aligned images from a time sequence. Sections through 3D arrays obtained by T1-weighted MRI: (a) baseline brain image of a patient with Alzheimer's disease; (b) the same person 1 year later; (c) difference between the two. (Images a and b from R. Schmidt et al., 2008, *Journal of Neurology, Neurosurgery, and Psychiatry* 79:1312–1317.)

Interactive rotation of a surface or volumetric display on a computer screen is an important and powerful aid to visual examination. Publications on paper cannot provide this capability, but electronic distribution of documents can. It is possible to incorporate 3D models that can be interactively rotated and subjected to limited types of measurement using the Adobe Acrobat® 3D Toolkit with surface geometry models created by CAD (computer-aided design) software. Converting the voxel arrays to surface models in one of the specific formats recognized by the display software remains a challenge. The resulting file can be read like any other pdf file and the images manipulated with mouse motions and clicks. For an example in which data have been converted to this format, see the file http://www.biomedcentral.com/content/supplementary/1741-7015-9-17-s1.pdf (Kumar et al., 2008; Ruthensteiner et al., 2010; Ziegler et al., 2011). The purpose of these displays is comparative visualization, not shape measurement.

Sometimes subtraction of one array from the other is used to see differences as shown in Figure 5.5. This may involve acquiring a sequence of images over time (a fourth dimension) in order to monitor dynamic processes (e.g., the opening and closing of heart valves) or disease progression. The sequential images must be aligned so that local motions or changes can be correctly seen. In many cases multiple imaging modalities are used, which provide different information about structure, and the images must be combined. This is made more complicated by the fact that the resolution of the different techniques (and hence the voxel dimensions and their alignment) differs, requiring interpolation. Figure 5.6 illustrates a typical case, combining MRI and PET.

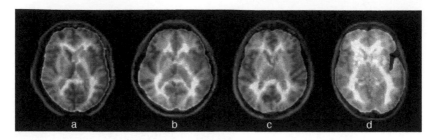

FIGURE 5.6
(See color insert.) Aligned sections through 3D reconstructions of human brains from different imaging modalities. The green channel is an MRI image; the red channel is a PET image using fluorodeoxyglucose, and the blue channel is a PET image using 6-fluoro-L-dopa. (a) control; (b, c) two patients with Parkinson's disease; (d) patient with multiple system atrophy. (Adapted from M. Ghaemi et al., 2002, *Journal of Neurology, Neurosurgery and Psychiatry* 73:517–523.)

The comparison of the recorded images to standard references, the alignment of sequential images, and the combination of different imaging modes are facilitated by registration of one 3D image onto another (Besl & McKay, 1992; Li & Hartley, 2007; Cheng et al., 2011). Once accomplished, the registration makes it possible to reach conclusions with much greater confidence than is possible using side-by-side or sequential examination of the images separately. The key requirement is establishing correspondence between reference points (landmarks) in the images. This may be done either manually or automatically.

In order to locate a sufficient number of matching points, and to avoid subjective bias, automatic methods are preferred but not always possible. Sometimes the landmarks can be externally added markers, such as pins or tattoos, but in many instances they are naturally present anatomical structures, such as branch points in blood vessels or distinctive features on bones (Maintz & Viergever, 1998). Long-term studies in which tissue changes may occur, or organ size can change, make locating correspondence points difficult; in some cases a point may vanish entirely.

For some purposes, such as alignment of images acquired with different techniques but of the same person, only translation, rotation, and scaling may be necessary. If a sequence of images is acquired with the same instrument over a brief period of time, then only a rigid body (translation and rotation) alignment is required. In principle, very few points are needed to accomplish these types of alignment: two points for rigid body motion, three for translation and scaling, and four for an affine transformation. In practice, it is desirable to have multiple landmark points and to use least squares to calculate the best fit.

The basic two-dimensional transformations illustrated in Chapter 1, Figure 1.22 are translation, scaling, rotation, and shear. It is most convenient to describe and to implement these using a matrix notation, called homogeneous coordinate form, that specifies the transformed coordinates (x', y') in terms of the original coordinates (x, y) and the translation (T_x, T_y), scaling (S_x, S_y), rotation (H), and shear (H_x, H_y). For translation, the relationship is

$$\begin{bmatrix} x' & y' & 1 \end{bmatrix} = \begin{bmatrix} x & y & 1 \end{bmatrix} \cdot \begin{bmatrix} 1 & 0 & 0 \\ 0 & 1 & 0 \\ T_x & T_y & 1 \end{bmatrix} \tag{5.2}$$

For scaling:

$$\begin{bmatrix} x' & y' & 1 \end{bmatrix} = \begin{bmatrix} x & y & 1 \end{bmatrix} \cdot \begin{bmatrix} S_x & 0 & 0 \\ 0 & S_y & 0 \\ 0 & 0 & 1 \end{bmatrix} \tag{5.3}$$

For rotation:

$$\begin{bmatrix} x' & y' & 1 \end{bmatrix} = \begin{bmatrix} x & y & 1 \end{bmatrix} \cdot \begin{bmatrix} \cos\theta & \sin\theta & 0 \\ -\sin\theta & \cos\theta & 0 \\ 0 & 0 & 1 \end{bmatrix} \tag{5.4}$$

And for shear:

$$\begin{bmatrix} x' & y' & 1 \end{bmatrix} = \begin{bmatrix} x & y & 1 \end{bmatrix} \cdot \begin{bmatrix} 1 & H_y & 0 \\ H_x & 1 & 0 \\ 0 & 0 & 1 \end{bmatrix} \tag{5.5}$$

The advantage of this matrix notation is that a sequence of operations can be combined by matrix multiplication, but the attendant problem is that the order of the operations must be clearly understood. Application of translation before rotation before scaling produces a different result than rotation before translation before scaling, and so on.

Because sections through the human body rarely line up through the same organs or structure, many if not most medical imaging registration tasks are actually three dimensional rather than two dimensional. For three-dimensional transformations, the matrices are similar but become 4×4 rather than 3×3.

Translation:

$$
\begin{bmatrix} x' & y' & z' & 1 \end{bmatrix} = \begin{bmatrix} x & y & z & 1 \end{bmatrix} \cdot \begin{bmatrix} 1 & 0 & 0 & 0 \\ 0 & 1 & 0 & 0 \\ 0 & 0 & 1 & 0 \\ T_x & T_y & T_z & 1 \end{bmatrix} \tag{5.6}
$$

Scaling:

$$
\begin{bmatrix} x' & y' & z' & 1 \end{bmatrix} = \begin{bmatrix} x & y & z & 1 \end{bmatrix} \cdot \begin{bmatrix} S_x & 0 & 0 & 0 \\ 0 & S_y & 0 & 0 \\ 0 & 0 & S_z & 0 \\ 0 & 0 & 0 & 1 \end{bmatrix} \tag{5.7}
$$

Rotation about the z-axis:

$$
\begin{bmatrix} x' & y' & z' & 1 \end{bmatrix} = \begin{bmatrix} x & y & z & 1 \end{bmatrix} \cdot \begin{bmatrix} \cos\theta_z & \sin\theta_z & 0 & 0 \\ -\sin\theta_z & \cos\theta_z & 0 & 0 \\ 0 & 0 & 1 & 0 \\ 0 & 0 & 0 & 1 \end{bmatrix} \tag{5.8}
$$

Rotation about the x-axis:

$$
\begin{bmatrix} x' & y' & z' & 1 \end{bmatrix} = \begin{bmatrix} x & y & z & 1 \end{bmatrix} \cdot \begin{bmatrix} 1 & 0 & 0 & 0 \\ 0 & \cos\theta_x & \sin\theta_x & 0 \\ 0 & -\sin\theta_x & \cos\theta_x & 0 \\ 0 & 0 & 0 & 1 \end{bmatrix} \tag{5.9}
$$

Rotation about the y-axis:

$$
\begin{bmatrix} x' & y' & z' & 1 \end{bmatrix} = \begin{bmatrix} x & y & z & 1 \end{bmatrix} \cdot \begin{bmatrix} \cos\theta_y & 0 & -\sin\theta_y & 0 \\ 0 & 1 & 0 & 0 \\ \sin\theta_y & 0 & \cos\theta_y & 0 \\ 0 & 0 & 0 & 1 \end{bmatrix} \tag{5.10}
$$

A generalized three-dimensional transformation produced by the product of these individual matrices produces a transformation matrix containing

12 coefficients (excluding the right hand column). In principle these can be determined from four landmarks (3 coordinate values times 4 points gives the necessary 12 values), but in most cases many more are used.

An alternative approach to alignment when many reference points are available is to subdivide the space as a set of tetrahedra, and apply the transformation separately to each one. This allows for flexible morphing of the data set, as described later.

Lack of a common objective coordinate system for multiple objects requires some method to establish one. Least-squares fitting is often used but may be biased if more points lie in one region than another. Other methods are also used, some more statistically robust than others, but none can guarantee "true" correspondence (Davies et al., 2008). When multiple landmarks are used, the quality of the alignment may be measured by the root mean square distance between matched points (fiducial registration error or FRE). A more meaningful measure for some purposes is the target registration error (TRE), which is the mismatch between points in the specific region of interest (which may not include any of the landmarks). The resolution of the images (voxel size) sets a baseline error limit for matching the arrays, and inherent uncertainty in the location of landmarks, particularly if they cover multiple voxels, contributes to both errors.

In many cases, rather than a predetermined set of landmark points supplied by known markers, the landmarks are determined by matching of the voxels within the images. Cross-correlation, described in Chapter 2, may be applied either within each 2D plane or, in the more general case, within the 3D voxel array to match points between the two data sets to be aligned. Usually this is done for points that are selected in one array as having high local contrast or locations of very high local brightness gradient (which usually indicates a surface). Cross-correlation is capable of providing a reference location with subvoxel accuracy, but the overall registration errors are generally at least as large as the voxel dimension.

In the more general case, such as fitting an image to a reference template, more flexible morphing as shown in Figure 1.23 may be required. This is referred to in the medical image registration literature as nonrigid transformation, and there are several different methods used. One is to allow deformation of the voxel array to have quadratic or higher order polynomial terms, in addition to the linear ones associated with translation, rotation, and scaling. A more general method creates a mesh using all of the identified matching points, and linearly deforms each small region to fit. In two dimensions, a mesh of triangles is used for this type of morphing. The equivalent mesh in 3D consists of tetrahedra. Usually there is some constraint applied to how much difference is allowed in the deformation of neighboring regions or how much local curvature is permitted for elastic surfaces (Amini & Duncan, 1992; Sun & Yang, 2011).

When transformation of any kind, even rigid body alignment, is carried out, the resulting voxel addresses are not integer values. In other words,

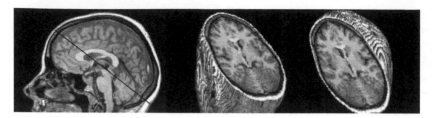

FIGURE 5.7
Presentation of section images along an arbitrary plane.

they fall between the voxels in the original array, so that some interpolation method must be used. The same choices are available as shown in Chapter 2 for 2D images (nearest neighbor, bilinear or bicubic, etc.). In a 3D voxel array, the number of neighbors that are used in the calculation is significantly increased; for example, a bicubic interpolation in 2D involves 16 neighboring pixels, whereas in 3D, 64 voxels are used. If the voxels are not cubic, but consist of planes that have different spacing between them than the spacing within each plane, the interpolation becomes more complex and usually produces results of poorer quality.

Interpolation in the voxel array is also needed to generate displays in which section planes can be viewed in various orientations, as shown in Figure 5.7. There is a wide assortment of display modes for volumetric data, including stereo viewing (Figure 5.8). Most of these require interpolation, or weighted averaging along lines through the array, or shading of surfaces, and so on. Fortunately, computer display capability (both hardware and software) has evolved to serve other markets, particularly computer games, and can be used for scientific purposes. There are even full-immersion headsets with tactile feedback (haptic) gloves that enable a high degree of interaction with 3D data sets.

Measuring Voxel Arrays

The algorithms for processing, thresholding, and measuring voxel arrays are all logical extensions of those shown in Chapter 2 for pixel arrays. However, there are a number of difficulties in implementing them. For one thing, as noted earlier, the number of neighboring voxels that must be addressed to perform convolutions or comparisons is large. For example, a 5×5 pixel neighborhood contains 25 pixels, whereas a $5 \times 5 \times 5$ voxel neighborhood contains 125. Ranking operations in particular take significantly longer as the size of the neighborhood grows. If the voxels are not cubic, so that the distances to the various neighbors must be calculated and taken into account,

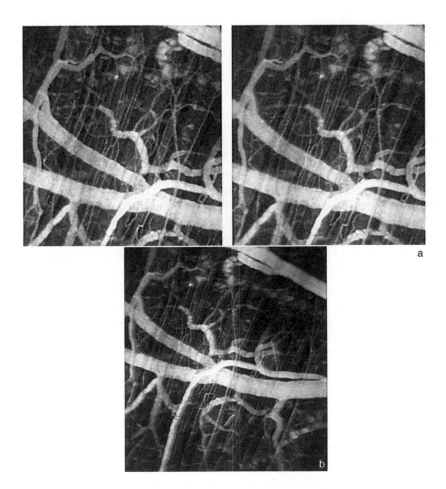

FIGURE 5.8
(See color insert.) Stereo view of multiple focal plane images from a confocal light microscope showing light emitted from fluorescent dye injected into the vasculature of a hamster and viewed live in the skin: (a) side-by-side stereo pair; (b) color stereo for viewing with glasses (red filter on left eye, green or blue on right). (Courtesy of Chris Russ, University of Texas, Austin.)

additional overhead is added. Large voxel arrays may not be held entirely in memory, requiring reading and writing to disk files, and the addressing of the individual voxels is more complex.

Locating edges in two-dimensional images can be performed by taking brightness gradients in two directions, combining them to get a gradient vector (the Sobel filter), and then perhaps locating just the local maximum values along that gradient to delineate the boundary position (the Canny filter). The same logic can be performed in three-dimensional voxel arrays by combining the gradients in three orthogonal directions and applying similar logic for locating the maximum value in the direction of the maximum gradient. Fourier transform methods for image processing, such as high- or low-pass filters and periodic noise removal, generalize directly from two to three dimensions. Thresholding, if it is based on voxel values as summarized in a histogram, is the same as for pixel arrays. If conditional logic is used to compare neighbors, the principles are the same although the programming is more involved.

More important than the implementation issues for image processing are the fundamental questions about image measurement. Areas in two-dimensional images are typically measured by counting pixels, and the equivalent operation for volumes in three-dimensional arrays is counting voxels. But what about surface areas? The same questions of treating voxels as points or volumes arise as were described for pixels, but with the added complexity that the surface elements are triangles rather than links in a chain.

When the chain is constructed around a two-dimensional feature, it is guaranteed that following it will cover all of the external periphery and return to the starting point. In three dimensions there is no procedure for sequentially following the surface of an object and adding triangles that is assured of reaching all parts of the surface, and no way to know when the task is complete. This is especially true if the object may be multiply connected. Heuristic procedures that scan the entire voxel array in line–row–column order, adding triangular facets wherever voxels are encountered, and then figuring out which object they belong to, are used instead.

As shown in Figure 5.9, there are two ways that triangular facets can be placed to connect any four voxels. Except for the cases where the four voxel centers are coplanar, the areas of the triangular facets are not the same. Consequently, there is a bias in the measured surface area. One possibility, rarely used, is to generate both triangulations and use the average of the two results.

The most commonly used method for generating a surface in a voxel array and measuring its area is the marching cubes algorithm (Cline & Lorensen, 1987; Lorensen & Cline, 1987). This works by first applying a threshold to the voxel values to determine which ones are inside and outside the structure(s) of interest. The entire array is scanned to examine each $2 \times 2 \times 2$ set of voxels.

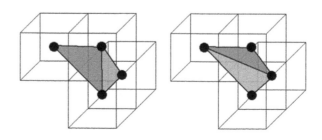

FIGURE 5.9
Two different ways to locate triangular facets joining the centers of four voxels.

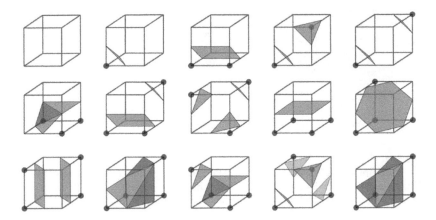

FIGURE 5.10
The 15 basic configurations of surface facets based on whether the voxel values at the corners are above or below the threshold setting.

The status of the 8 voxels in the cube (i.e., whether each is inside or outside the object) is used to construct an 8-bit index (0 to 255) into a table that contains the various reflections, inversions, and rotations of 15 basic possible configurations. The table specifies the planar surface facets that are placed in the position of the cube. This also gives the area of those facets and the orientation of the surface normals, which are used for surface rendering. Then the process is repeated for the next cube position. Since the entire voxel array is scanned, and the various facets are rendered or summed without regard to their connectivity, there is no need to find a continuous path that will cover a complex surface.

Figure 5.10 shows the basic cube configurations (each one covering the 8 voxels in the 2 × 2 × 2 array), with marks indicating the corners that are inside objects, and the resulting facets. Reflections, inversions, and rotations of these patterns cover the full set of 256 possible arrangements. The results are generally good enough for visualization purposes, provided that the size of the voxels (and consequently the size of the cubes and facets) is small enough, as shown in Figure 5.11. "Contouring" due to the abrupt change in facet orientations is visible, particularly on the smoothly rounded side of the skull. Some visual improvement can be produced by smoothing the shading of the facets, using either Gouraud or Phong shading (the former interpolates brightness values across the facet area; the latter interpolates the orientation of the normal vector).

Several problems arise when the basic marching cubes method is used for measurements. There are some ambiguous configurations of voxels that could correspond to topologically different surfaces. Resolving these cases may be attempted using more complex algorithms that take into account the configuration of the neighboring cubes. The surface that is produced is not

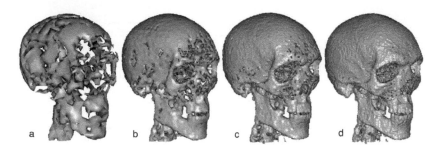

FIGURE 5.11
Surface rendering of a CT scan of a human skull, showing the facets generated by varying voxel sizes: (a) 10 mm; (b) 5 mm; (c) 3 mm; (d) 2 mm. (F. Sachse et al., 1996, *Proceedings of First Users Conference of the National Library of Medicine's Visible Human Project*, 123–126.)

everywhere continuous, and because of the faceting it tends to overestimate the actual area. This is similar to the error in perimeter produced by chain code, as shown in Chapter 2.

Figure 5.12 shows an example of the measurement of volumes and surfaces of three-dimensional objects. The voxel array was acquired as a series of two-dimensional images, revealed as material was eroded away by focused ion beam (FIB) machining. The surfaces around and between particles have been delineated, and the volumes of the particles measured. Combining volume and surface area as shown in Equation 5.1 to calculate a "sphericity" value for each particle is a possibility, subject to the usual caveats about the precision of the values and the meaning (if any) of the results.

The Morph+ software package for the processing and measurement of voxel arrays has been described by Brabant et al. (2010), with applications

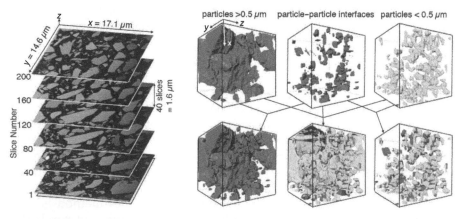

FIGURE 5.12
(See color insert.) A stack of serial section images using FIB surface machining, and the reconstruction using different colors for particles of different sizes and the various interfaces present. (Courtesy of L. Holzer, Swiss Federal Laboratories for Materials Testing and Research.)

shown to foam and network structures of materials. The program provides several 3D operations, all of which are direct extensions of 2D algorithms as described in Chapter 2. This is a representative set of basic tools available for application to cubic 3D voxel arrays:

- Neighborhood processing with median, Gaussian, and bilateral filters (the bilateral filter, in either 2D or 3D, is a smoothing operation like the Gaussian that modifies the weights applied to each pixel or voxel according to both the distance from the center and the difference in value).
- Thresholding based on voxel values, with an option for the same automatic method that is most commonly applied to 2D images, based on the *t*-test analysis of the histogram with the assumption of two populations of values.
- Processing of binary images using morphological operations such as erosion, dilation, opening, closing, watershed segmentation, hole filling, and skeletonization (the skeletonizing method produces a set of lines shown in Figure 5.14b).
- Global measurements of volume fraction, total surface area based on the marching cubes algorithm, porosity (the volume fraction of enclosed holes), the Euler number of objects as described in the next section, and the surface fractal dimension (by the box-counting method).
- Object labeling and measurement of centroid position, bounding box dimensions, volume (the count of voxels), and a shape factor named "sphericity" that is defined as the radius ratio of the largest inscribed and smallest circumscribed spheres. (Notice that no measurement of individual object surface area is provided, and consequently no shape factor based on the surface.)

The stereological measurement techniques based on two-dimensional images shown in Chapter 2 can also be extended for three-dimensional use. If a series of parallel sections is obtained (for example, with a confocal microscope) or if a full voxel representation of the structure is available, a grid of points or lines may be used for counting. A series of concentric spherical surfaces avoids the directional bias that may be present when two-dimensional sections are examined. Several current stereological techniques use three-dimensional images, particularly from the confocal light microscope, to obtain unbiased quantitative microstructural measurements. However, these provide global rather than feature-specific data.

Topology, Skeletons, and Shape Factors

For space-filling structures such as foams and single-phase metals, at equilibrium the bubbles or grains take on polygonal shapes that can be characterized by the number of faces and edges. For example, Raj (2010) measured metal and polymeric foams and counted the number of edges on each cell face. He found that 24% to 28% of the cell faces were four-sided, 50% to 57% were five-sided, and 15% to 22% were six-sided. These results are in agreement with observations on soap bubbles and with much earlier work on metal grains (DeHoff et al., 1972). Euler's rule for the topology of a polygonal shape is given in Equation 5.11, where V is the number of vertices, E the number of edges, and F the number of faces.

$$V - E + F = 2 \qquad (5.11)$$

Table 5.1 illustrates this relationship for the regular convex polyhedra, but it holds for any polyhedron. A typical equilibrium grain structure or foam contains convex polyhedra with a distribution of shapes and sizes, the smaller ones typically having fewer faces so that they can fit between and around the larger ones.

Just as for two-dimensional features, shape parameters based on topology and the local surface geometry (e.g., fractal dimension) are often useful for three-dimensional objects. A discussion of surface fractal dimension is presented next in the context of surface characterization. The skeleton represents the topological structure in two dimensions. For a branching "linear" structure (in which the lateral dimensions are small compared to the length),

TABLE 5.1

Euler's Rule for Regular Polyhedra

	Shape	V	E	F	$V - E + F$
	Tetrahedron	4	6	4	2
	Octahedron	6	12	8	2
	Cube	8	12	6	2
	Icosahedron	12	30	20	2
	Dodecahedron	20	30	12	2

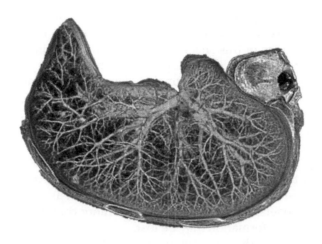

FIGURE 5.13
(See color insert.) Volume-rendered image of branching vasculature in the lung. (Courtesy of
TeraRecon, Inc., San Mateo, California.)

such as the vasculature in the lung (Figure 5.13) or the branches of a tree, the
skeleton has an unambiguous meaning.

In a more general structure, there are two kinds of skeletons that can be
created and they provide different types of information. A skeleton in three
dimensions can be constructed either by sequential erosion processes in the
voxel array, or by generating the 3D Euclidean distance map (which is done
in three dimensions similarly to the procedure in two) and locating the local
maxima (Lobregt et al., 1980; Zhou et al., 1998; Tran & Shih, 2005).

The conditional erosion method produces a set of surfaces whose voxels are
equidistant from two locations on the boundary, as indicated in Figure 5.14a.

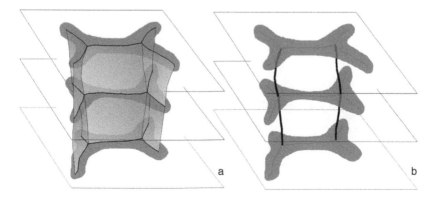

FIGURE 5.14
Two different 3D skeletons: (a) sheets (which approximately pass through the 2D skeletons in
neighboring planes); (b) lines (which approximately pass through the nodes of the skeletons in
neighboring planes).

These surfaces pass approximately through the two-dimensional skeletons that would be constructed in each parallel plane of the three-dimensional array. The advantage of this sheet skeleton is that it has information about twists in the structure and the angles between the planes.

The other type of three-dimensional skeleton is the set of lines that follow the maxima in the three-dimensional Euclidean distance map, which pass approximately through the nodes in the two-dimensional skeletons as shown in Figure 5.14b and are points equidistant from three surfaces (i.e., the centers of inscribed spheres). Combining this skeleton with the Euclidean distance map produces a medial axis transform (MAT) representation in 3D analogous to that for 2D features. The branch and end points in this line skeleton describe the topology of a network such as the vasculature in Figure 5.13, but oversimplify the structure of more complex shapes so that potentially useful information may be lost.

The line skeleton is the most frequently used. It has been used to model earthworm burrows for length, branching, and orientation (Capowiez et al., 1998; Pierret et al., 2002). It was reported that the total burrow length did not vary with season, but directions became more vertical and the number of nodes and branches increased in the spring. In another type of material and at a different scale, 3D skeletons have been used to characterize the porosity in trabecular bone (Pothaud et al., 2000). Pilgram et al. (2006) used this technique to measure size and shape variations in healthy and diseased human hearts.

Figure 5.15 shows a visualization of the vein, arteries, bile ducts, and lymphatics in dog liver (Livingston et al., 2011). The medial axis transform is

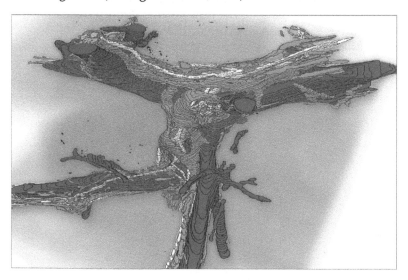

FIGURE 5.15
(See color insert.) Visualization of vein (blue), arteries (red), bile ducts (green), and lymphatics (yellow) reconstructed from serial sections through dog liver. (Courtesy of Dr. John Cullen, College of Veterinary Medicine, North Carolina State University.)

particularly useful for structures of this type. The values along the medial axis of each structure are the radii of inscribed spheres, and so the variation in values measures the variation in cross-section. In addition, the count of the number of nodes per unit length (the reciprocal of the average length of a branch in the skeleton) measures the amount of branching present.

The three-dimensional MAT determined from X-ray tomographic recon-struction provides an important characterization of porosity in sandstones, soils, and manufactured products such as catalytic converters in automobiles and fluid bed reactors. Minima in the MAT measure the size of the pore throats linking one pore opening to another. These control the potential flow rates of oil, water, or gases.

The "sphericity" defined in Equation 5.1 is not a standard measure of three-dimensional shapes. There are no standard definitions or names for dimensionless ratios, producing the same opportunity for confusion, as is the case for 2D images. In practice, few current software packages provide much measurement capability for 3D voxel arrays at all. Some measure total volume, and a few report total surface area (often without defining how it is determined, but in most cases probably by the marching cubes method).

For those systems that generate the line skeleton, measurement of its length by summing the distances from neighbor to neighbor may be carried out. There are three types of links possible: a distance of 1 for face-sharing neigh-bors, 1.414 for edge-sharing neighbors, and 1.732 for corner-sharing neighbors.

For individual objects, the volume is the most commonly reported mea-surement. The maximum caliper dimension may be determined by projec-tion on rotated axes in the same way as the 2D case shown in Chapter 2. The inscribed and circumscribed sphere radii mentioned earlier can be calcu-lated. But the topology of an object (number of loops, ends, and nodes) is dif-ficult to determine, and generally the number of measurement parameters is small compared to that available for 2D images.

Projections (Silhouettes) of Shapes

From even these few measurements of size, it is possible to construct for-mally dimensionless ratios that may be useful as shape descriptors in selected cases. The classification of sand grains provides an example. Sand is not a mineral. Sand grains are defined as particulates in the size range from $\frac{1}{16}$ to 2 mm, without regard to the type of mineral(s) present; finer particles are called silt, larger ones are gravel or pebbles. Sands are not limited to the familiar quartz, but include basalt (black sand) and olivine (green sand), both found, for example, on the beaches of Big Island, Hawaii. In Bermuda the pink sand contains bits of sea shells. Red sand at Rapida Island, Galapagos, is formed from coral, but the red sand in the Empty Quarter of Saudi Arabia is quartz stained by iron oxide. The white sand at White Sands, New Mexico, is gypsum. There is a famous (and well guarded!) beach in Namibia with diamond sand. Many sands contain a mixture of grains of various minerals.

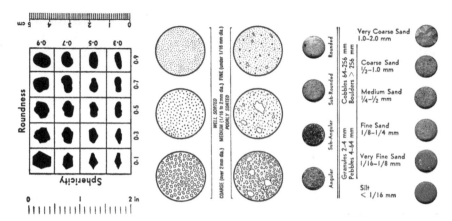

FIGURE 5.16

A pocket field guide for classification of sand grains, prepared and distributed by Sigma Gamma Epsilon fraternity at Kent State University. (Adapted from W. C. Krumbein, L. L. Sloss, 1963, *Stratigraphy and Sedimentation*, Freeman, San Francisco, CA.)

The shapes of sand grains arise from many sources. Gypsum sand particles are cubic, reflecting the underlying crystal structure. But most sand grain shapes are the result of weathering processes, in which water and wind act differently (and continuously flowing water has a different effect compared to the back-and-forth flow of a tidal beach). Interaction with other particulates, including larger pebbles and cobbles, also affects shape. The guide to classification of sand grains shown in Figure 5.16 does not rely upon digitized two- or three-dimensional imaging. Rather, it is intended as a chart or field guide for visual comparison. The two axes labeled *Sphericity* and *Roundness* are definitely shape factors, one dealing (apparently) with the degree of elongation or departure from an equiaxed shape, and the other related to the local smoothness or angularity of the surface. Since manual and visual examination of physical grains allows for turning them over, these silhouettes are compared to multiple views of the grains. Of course, the names used conflict with other definitions and names that are also in use.

There have been attempts to introduce some shape factors for sand and other rock particles based on measurement rather than visual comparison. Heywood (1954) specifically attempted to combine surface area and volume to calculate the roundness and sphericity values for the silhouette shapes in Figure 5.16. Other combinations include the ratio of the radius of curvature at corners to the mean radius of the particle (Wadell, 1932), which is related to the angularity of the corners. The ratio of volume to the smallest circumscribed sphere (Krumbein, 1941) is a measure in three dimensions similar to the "extent" defined in Chapter 3 for a two-dimensional feature. But no widely accepted or standardized methods are in use (Barrett, 1980).

The silhouettes in Figure 5.16 are projections of the three-dimensional shape of sand grains. The extent to which they can be used depends on factors such as the random or controlled orientation of the grains, and the fact that they are mostly convex, or at least approximately so. Projected outlines are not suitable for an accurate measurement of either size or shape; randomized cross-sections can be used to determine statistically unbiased measures, but only for a population of objects, not individual ones. As noted earlier, the problem with projected outlines is that valleys are covered up by hills, producing a larger and smoother profile than may exist at any single location. The size results also do not correspond to a common method for measuring particulates by passing them through a series of sieves of different sizes and weighing the fractions on each sieve. The shaking of the sieves allows the grains to find an orientation with the smallest cross-section and to wiggle through the holes.

However, because of the widespread use of relatively easily obtained silhouette images of grains to characterize size and shape, it is interesting to see whether there is enough information in the resulting biased set of images for practical use. To test the applicability of shape measurement of the two-dimensional projected outlines, light microscope images of a dozen samples of sand grains, comprising nearly 900 particles, were recorded at a resolution of 1.32 micrometers per pixel. The sands were selected from the author's personal collection and as shown in Table 5.2 come from locations around the world.

Grains were scattered randomly on a glass slide and the profiles viewed with transmitted light. Visually, the grains are too diverse in shape to be readily recognized by origin. Unlike the tree leaves or arrow points shown in Chapter 4, which are readily identified by an observer, the sand grains have only a few obvious distinguishing features that human vision can extract and use for classification. Figure 5.17 shows the grains from several of the sand samples, especially those with extreme differences; the individual binary profiles have been manually arranged in a single image after acquisition for convenient display and measurement.

TABLE 5.2

Sand Samples

Rub-Al-Khali, Saudi Arabia (burning sands of the Empty Quarter)
Ocean Isle, North Carolina (north of Sunset Beach)
Pisgah National Forest, Brevard, North Carolina (Davidson River)
Dead Sea Valley (Moshav Hatza, Israel)
Wrightsville Beach, North Carolina
Hawaii (Puna Coast, Big Island, Waha'ula Visitors' Center, Volcanoes National Park)
Zion National Park, Utah (West Wall Kayenta Formation)
Grand Bahama Island (Northwest Providence Channel)
Cape Cod (Marconi Station, Wellfleet, Massachusetts)
Iceland (East Coast Beach, Dyrholoey)
White Sands, New Mexico
Mauritania (Akchar Eng, Relict dune)

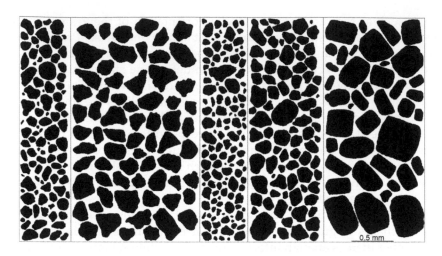

FIGURE 5.17
Projected (silhouette) images of sand samples (from left): Rub-al-Khali, Wrightsville Beach, Zion National Park, Mauritania, White Sands.

Sands that have been rolled back and forth by wave action on a beach and have been graded by this process so that they contain a fairly uniform set of particles (in terms of size and hardness) should be distinguishable from ones formed in a stream bed where all motion is in one direction and there are large as well as small particles. Erosion by wind, which causes sand grains to jump and roll on dunes (saltation), produces surface pitting. Young sands (one of the black sands in the collection from Hawaii contains volcanic glass particles only a few months old when collected) are angular and have sharp corners, distinct from old sands that have become smoothly rounded. As indicated by the comparison chart in Figure 5.16, these variations in shape may be useful for identifying the history of a sand.

Visual inspection of the images in Figure 5.17 indicates that some sands have much larger grains than are present in others, but even these have some small grains similar in size to the finer sands. The goal is to find shape descriptors that can identify the sands. Both dimensionless ratios and Fourier harmonic coefficients were measured and subjected to linear discriminant analysis. As shown by the overlaps of the points and ellipses in the plots in Figure 5.18, it is not possible to separate the various classes of sand grains by this method. Approximately 66% of the individual grains are misclassified using the dimensionless ratios and 78% are misclassified using the Fourier coefficients.

However, it is still possible to distinguish between each pair of samples using this method, if size measures are included as variables. Measurements of a small number of grains separate them as groups, and many of the individual grains are successfully classified. Figure 5.19 shows the separation of the aeolian (wind-blown) Rub-al-Khali sand and the beach sand from Ocean

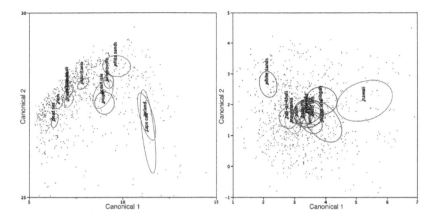

FIGURE 5.18
Inability to distinguish the sands using linear discriminant analysis with measurements of dimensionless ratios (left) or harmonic coefficients (right); the two most significant canonical axes are shown.

Isle using the perimeter, convex perimeter, and solidity; the separation of the Hawaiian (extremely young) and Mauritanian (extremely old) sands using the convex area, equivalent circular diameter, and fractal dimension; and the separation of the White Sands (chemical deposition) and Pisgah (river sand) specimens using roundness, solidity, and radius ratio. Comparable results are obtained for each pair, recognizing that in each case different measurement parameters are used to separate the groups.

Similarly, using Fourier shape descriptors it is possible to distinguish one group from another when they are considered in pairs. With a few dozen particles, the groups are separated, and many of the individual grains are correctly classified. Figure 5.19 also shows these results for the same pairs of sand samples: the Rub-al-Khali and Ocean Isle sands are distinguished by coefficients 1, 5, 10, 11, 14; the Hawaiian and Mauritanian sands by coefficients 1, 8, 13; and the White Sands and Pisgah samples by coefficients 8, 9, 12. As for the dimensionless ratio measurements, different combinations of terms are used for each pairwise comparison, which makes it difficult to generalize about any possible interpretation of the results.

Spherical Harmonics

Harmonic analysis can be applied to three-dimensional shapes as well as to two-dimensional outlines. To introduce the representation of a shape by spherical harmonics, the diagram in Figure 5.20 shows radius r as a function of two angles, ϕ and θ. The radius is a scalar value which can be modeled as an expansion of orthogonal or independent functions. This approach is used, for example, in describing pressure in the Earth's atmosphere, the magnetic field surrounding Mercury, and density in the Sun.

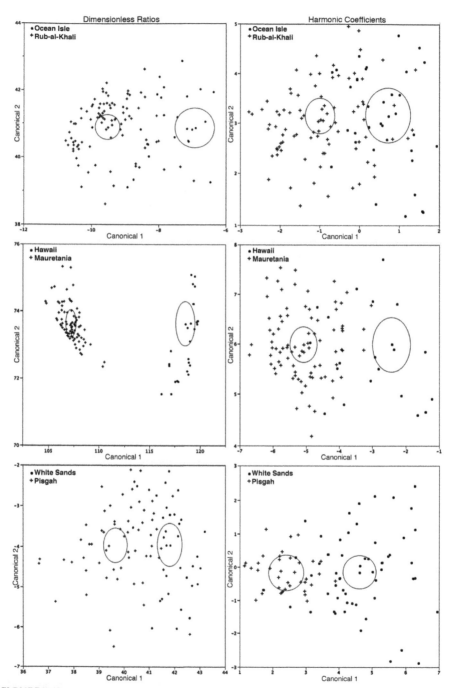

FIGURE 5.19
Separating pairs of sand samples as described in the text: left column, dimensionless ratios; right column, harmonic Fourier coefficients (the two most significant canonical axes are shown). Ellipses represent one standard deviation.

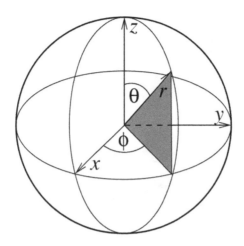

FIGURE 5.20
Coordinate system for radius r as a function of angles ϕ and θ.

Spherical harmonic analysis requires a double summation of Legendre polynomials over a wavenumber space where m is the wavenumber for the angle ϕ (the longitude) and n can be thought of as a "total" wavenumber, from which $(n - m)$ gives the number of zeroes between the north and south poles of the spherical harmonic function. Figure 5.21 shows the regions on the spherical surface where the resulting functions have positive and negative values.

Spherical harmonics form a complete basis set for describing shape (Brechbühler et al., 1995) and can be used to construct an analogous frequency-domain representation of a given 3D object. As for the Fourier series expansions described in Chapter 3, summing these functions with appropriate weighting factors can fit object shapes, provided that the surface is everywhere single valued (i.e., no holes, bridges, loops, or undercuts). Figure 5.22 illustrates the application of the technique to protein–ligand binding sites (Cai et al., 2002; Gutteridge & Thornton, 2004).

The analysis of the coefficients to classify or identify shapes is not simple. In some cases statistical procedures, including stepwise regression, discriminant analysis, or principal components analysis, can be used to find the most important terms or combinations of terms (Dette et al., 2005).

The limitation of the scalar harmonic expansion to shapes that are everywhere single valued in terms of the radius from a central point (usually the centroid, sometimes the center of a bounding or inscribed sphere) is a severe restriction, just as using radius as a function of angle is limiting in two dimensions. Elliptical Fourier functions based on $(x + iy)$ representation for 2D coordinates cannot be extended for 3D shapes, which complicates the possibilities for analysis of complex shapes that may have multiple values for radial

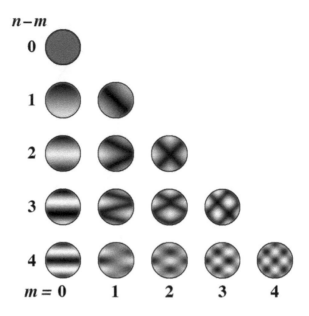

FIGURE 5.21
Regions on the surface of a sphere where the harmonic functions are positive or negative, as a function of the wavenumbers m and n.

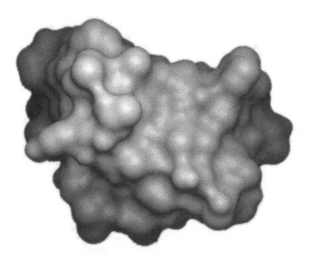

FIGURE 5.22
Spherical harmonics provides a method for mathematically describing the shape of ligands and binding sites; the protein crambin is shown, reconstructed using 40 terms.

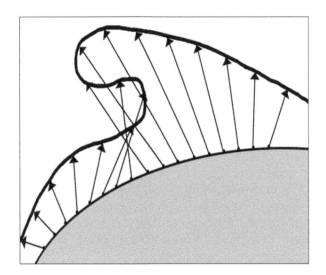

FIGURE 5.23
Diagram of a vector field mapping each point on a surface to a corresponding point on an underlying sphere.

distance from the centroid. A more general approach uses a vector at each point on the inscribed unit sphere that describes the distance and direction to a corresponding point on the object surface. There is a one-to-one mapping of each point on the surface to a point on the sphere, with equal spacing. As indicated schematically in Figure 5.23, this allows for folds, caves, and undercuts (but not holes, bridges, or multiple connectedness) on the surface.

Applying harmonic analysis to the vectors allows for fitting of much more complex shapes. As shown in Figure 5.24, Chung et al. (2007) have modeled the cortical surface of human brains. They report that from 40 to 50 terms in the reconstruction are satisfactory in terms of the appearance of the surface. The terms from such a spherical harmonic analysis have been shown to be useful for classification of anatomical structures, such as differentiating between the hippocampi of normal and schizophrenic patients (Shen et al., 2004). Statistical classifiers based on the spherical harmonic terms achieved 93% accuracy in differentiating between the two populations.

10   20   30   40   50   60   70

FIGURE 5.24
Reconstruction of the human cortical surface as a function of the number of spherical harmonic vector terms. (M. K. Chung et al., 2007, Special Issue of *IEEE Transactions on Medical Imaging on Computational Neuroanatomy* 26:566–581.)

In addition to its use for characterizing the shape of the brain (Gerig et al., 2010) and other medical structures, scalar spherical harmonic analysis has been applied to other classification tasks, using various imaging technologies. Garboczi (2002) classified the shape of aggregates from different sources for use in concrete, and compared the shape descriptions to the performance properties of the final composite. Imaging was accomplished using X-ray tomography.

At an entirely different scale and based on an entirely different imaging method, Tadros et al. (1999) used images from the Infrared Astronomical Satellite (IRAS) to measure the shape of galaxies. Red shift values were used to measure the distance values to obtain the galaxy pattern, and spherical harmonics were used to correlate with the galaxy flux. The earth has also been a subject of analysis. Balmino et al. (1973) used spherical harmonics to analyze the Earth's topography, whereas Whaler and Gubbins (1981) applied the technique to the geomagnetic field.

Spherical Wavelets

Modeling of the human brain's cortical surface has also been carried out using wavelets. Yu et al. (2007) used a model in which the wavelets are scaled copies of an icosahedron. At each level, a new set of vertices is created at the midpoints of edges, and the wavelet function at each level is obtained by removing or "lifting" the functions at the next level scaled to have a zero sum. This is the same procedure illustrated in one dimension in Figure 4.28.

At the seventh level of iteration, the icosahedral mesh has nearly 164,000 vertices (an eighth level would have over 655,000, but the resolution of the imaging device did not justify that step). The resulting functions have local support in both frequency and space, and are consequently somewhat better than Fourier harmonics for localizing morphological variations. Principal components analysis of the wavelet coefficients shows that 98% of the variations can be accounted for by no more than 20 terms. Tracking the most significant terms allows modeling the process of cortical folding with age, as illustrated in Figure 5.25.

Because of their local support, spherical wavelets generally have the ability to detect and pinpoint differences between shapes and locations of high local curvature (Schroder & Sweldens, 1995; Vranic & Saupe, 2002). Nain et al. (2005) compared shapes of similar objects, also from medical imaging, by subtracting a mean shape, calculating wavelet spectra for the residuals, and statistically detecting clusters or bands of wavelets that represented localized shape variations. Laga et al. (2006) applied spherical wavelets to locate similar shapes in a database.

Wavelets, like Fourier coefficients, are capable of fully reconstructing the original shape. But in all of these examples of practical use, the emphasis is on locating significant positions for comparison or matching between

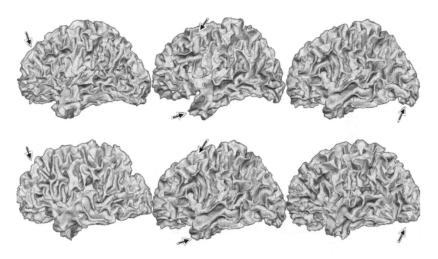

FIGURE 5.25
Models of cortical surfaces with arrows marking locations where the largest variations are detected between the top and bottom images. (P. Yu et al., 2007, *IEEE Transactions on Medical Imaging* 26(4):582–597.)

objects, or localizing changes over time, rather than deriving a compact and statistically useful numeric description of the entire object shape.

Imaging Surfaces

The techniques described are usually applied to 3D voxel arrays. The familiar way that people examine most three-dimensional objects does not provide a volumetric view of the interior but is based on viewing the exterior surface. From the variations in surface brightness and some knowledge about the illumination and the nature of the surface, interpretations of shape and roughness are made visually. There are several ways that this can be done algorithmically to produce a geometric model of the object surface, or at least the portion of it that is directly visible from a single location.

Stereoscopy

Stereoscopy is the basis for some human interpretation of object shape, at least at close range (Marr, 1982). By rotating our eyes in their sockets to bring the same feature in a scene to the central fovea (which has the highest density of cones), the brain receives information from the muscles and nerves telling whether that point is closer or farther away than another one. By examining

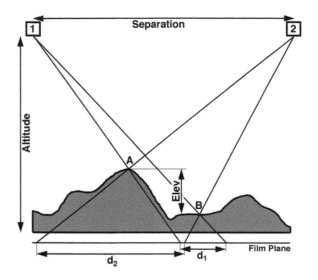

FIGURE 5.26
Geometry for stereoscopic measurement in aerial photography. Two images are taken with a known separation and at a known altitude. By measuring the distances d_1 and d_2 between the images of two points A and B, their elevation difference can be calculated using Equation 5.12.

a series of locations, a crude mental map of relative distances is established. Generally this is only a part of the information available, since shadows, surface brightness, perspective, and other clues are also available. Renaissance painters even included atmospheric haze as a way to show that some objects were farther away from the viewer.

Stereoscopy can be used to measure distances directly, as shown in Figure 5.26. The diagram corresponds to the procedure used in aerial or satellite photography, in which the camera points straight down and records images that overlap. Knowing the altitude and the distance between the two camera locations (given by the time interval and speed) the disparity in distance (lateral shift in the images) between two points in the images allows calculating the elevation difference between them as shown in Equation 5.12 (with the simplifying assumption that the altitude is much greater than the elevation difference).

$$Elevation = \frac{Altitude}{Separation} \cdot (d_1 - d_2) \qquad (5.12)$$

Figure 5.27 shows an example of stereoscopy. Two overlapping images of Gale Crater on Mars (where the Mars rover, *Curiosity*, is headed) were acquired with the Mars Reconnaissance Orbiter, and are printed using the red and cyan (both blue and green) channels. Viewing this through standard stereo-viewing glasses (red filter on the left eye) results in the visual

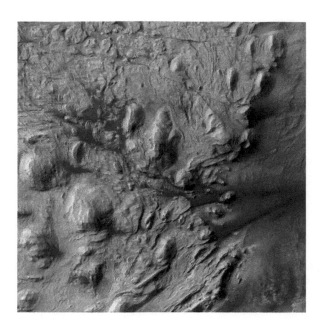

FIGURE 5.27
(See color insert.) Stereo view of the corner of Gale Crater on Mars.

matching of points in the two scenes such that the horizontal offsets are interpreted as elevation differences, which are exaggerated in the figure.

The geometry is slightly different for the case of viewing with two eyes. In that situation, the camera views are pointed inward toward a common point. The angle between the two views is known and the disparity in location of corresponding points allows calculating the height according to Equation 5.13, in which α is the angle between the two views. This is also the geometry that applies when a single viewpoint is used and the object is rotated to obtain two images (Boyde, 1973; Kayaalp et al., 1990; Beil & Carlsen, 1991). The scanning electron microscope typically uses that procedure to capture images like the one shown in Figure 5.28. The only change required in Equation 5.13 is the requirement to divide the disparity $(d_1 - d_2)$ by the image magnification.

$$Height = \frac{(d_1 - d_2)}{2 \cdot \sin\left(\dfrac{\alpha}{2}\right)} \tag{5.13}$$

The technique can be applied to an entire surface by matching many points in the image (usually by cross-correlation, with the added constraint that the points must be in the same order) and generating a range image in which the disparity is shown as a grayscale value (Figure 5.29). The automatic matching

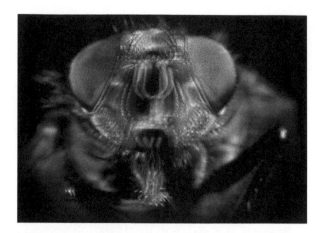

FIGURE 5.28
(See color insert.) Stereo view of a fly's head. The scanning electron microscope pictures were recorded by rotating the specimen by 7 degrees.

of points works best for images of surfaces with rich local detail, and poorest for smooth surfaces that produce many erroneous matches.

Of course, the method only works when the two view directions can see all or at least most of the same points, so extremely rough surfaces require a small angle between the two view directions. Since the error in elevation measurement is proportional to the cosecant of that angle and to the resolution of the images, large angles give the best precision (still usually an order of magnitude worse than the lateral precision with which the point locations can be determined in the images). The range image can be used to create a geometric rendering of the surface, as shown in Figure 5.30, as well as for measurements. The Mex software package (www.Alicona.com) is an example of practical tools that combine stereo images to obtain measurements of surface geometry.

FIGURE 5.29
Combining two stereo images to generate a surface elevation map. Cross-correlation point-by-point on the two images at left determines the horizontal disparity shown as a brightness value in the image at right, which is proportional to the elevation.

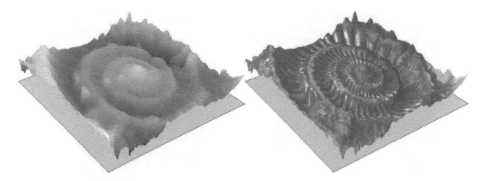

FIGURE 5.30
Using the data from Figure 5.29 to produce a rendered geometric model. At left, the elevation data by itself, and at right the surface appearance from one of the original images superimposed on the geometric model.

Shape from Shading

Another method with a confusingly similar name is photometric stereo, also known as shape from shading, mentioned in Chapter 2. In this case, a single viewpoint is used to collect a sequence of images in which the light source is moved to different known locations. From the change in brightness recorded at each point, it is possible to calculate the local surface orientation angle. The shape-from-shading calculations (Horn & Brooks, 1989) may either be based on a measured brightness versus orientation function or calculated (typically as a cosine function of angle). The method can be confused by local changes in surface albedo, for example, due to compositional variation or chemical discoloration.

An implementation of this technique developed by Tom Malzbender at Hewlett Packard Labs utilizes a hemisphere in which the camera is mounted at the North Pole with multiple lights that can be controlled automatically by the computer to capture separate images. These are then combined to generate a polynomial texture map (PTM) that represents the surface geometry of the object (Malzbender et al., 2001; Mudge & Malzbender, 2006).

With this model, it is possible to show what the surface would look like if it were metallic (giving specular reflections) as a source of illumination is freely positioned at any orientation (the software and example data sets are downloadable from www.hpl.hp.com/research/ptm/downloads/download. html). This technique is particularly well suited to the recording of unique objects such as art and archaeological treasures, forensic evidence, and other similar objects and surfaces (Padfield et al., 2005)

Figure 5.31 shows an example, an ushabti or funerary figurine from the Middle Kingdom in Egypt. Notice that with the surface specular relighting it is possible to see details that are difficult to detect in the original photographs (even the texture of the surface on which the figurine is resting is revealed). The geometric model of the surface slopes can be integrated to obtain elevation data, which is shown in Figure 5.32 as a surface rendering.

FIGURE 5.31
Two images of a funerary figurine. On the left, a camera recording of the actual object, and on the right the computed image using a specular surface reflection model with a light source located at the top right. (Courtesy of Tom Malzbender, Hewlett Packard Labs.)

Other Methods

There are other surface measurement methods as well, which have specialized uses. Shape from focus captures a series of images with a wide aperture, shallow depth of field lens. By detecting the image in which each point is best focused, it is possible to construct a range image (Nayar & Nakagawa, 1990; Subbarao & Choi, 1995). A modification of the method uses a single image and estimates the defocus at each point.

Structured light (Boyer & Kak, 1987; Valkenburg & McIvor, 1998; Chen et al., 2000; Scharstein & Szeliski, 2003; Jia et al., 2010) uses a light source that sweeps a line or a series of points across the image at an angle, whose lateral displacement in a series of images gives the local elevation as diagrammed in Figure 5.33. Figure 2.61 in Chapter 2 shows an example of the results. This

FIGURE 5.32
Surface rendering of the ushabti in Figure 5.31 using the geometric model of the surface.

method is used for industrial applications such as machining, as well as for medical imaging such as measuring the shape of the eye before and after laser surgery.

For crime scene documentation or for recording the positions of artifacts in an archaeological site, a scanning laser rangefinder may be used. The scanner rotates on a tripod, recording the distances to many points. At the same time, a digital camera captures photographs, and these data are combined with the distance values to construct a model of the surfaces in the entire scene. Figure 5.34 shows the panoramic image along with the cloud of points

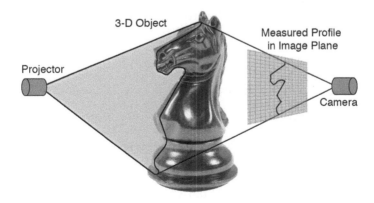

FIGURE 5.33
The principle of structured light imaging. The lateral displacement of points along the projected line as viewed from the camera position reveals the shape of the object. Either rotation of the object or sweeping of the line (or an array of points) may be used to cover the entire surface.

FIGURE 5.34
(See color insert.) Scanning laser rangefinder measurement of the interior of a room: (top) the panoramic image recorded by a digital camera as the unit rotates; (bottom left) the cloud of points whose distances from the unit are measured by the rangefinder; (bottom right) the resulting 3D model of the surfaces combining coordinate positions determined by the rangefinder data with image data from the camera. (Courtesy of Doug Schiff, 3rdTech Inc., Durham, North Carolina.)

recorded from wall, floor, and interior objects in a room, and the resulting 3D reconstruction of the scene, which is freely rotatable and with which dimensional measurements can be made.

Surface Metrology

Surfaces play many important roles. They are responsible for the appearance of objects, and many aspects of mechanical and environmental behavior including adhesion, friction, wear, and chemical activity. They are the principal mediators of the interactions between one object and another, as well as between people and the world around us. The measurement of surfaces, especially ones produced by various machining, molding, and other manufacturing processes, is consequently a key part of quality control operations, as well as being important in research involving surface behavior.

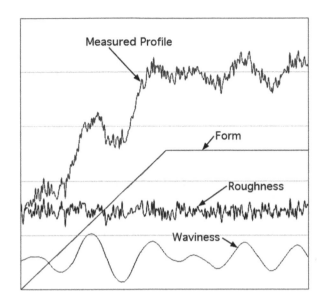

FIGURE 5.35
The measured profile can be considered as the sum of the form, waviness, and roughness.

In engineering and manufacturing, a distinction is often made between shape information representing different relative spatial scales, as shown in Figure 5.35. *Form* (sometimes called *figure*) is the large-scale geometric shape that corresponds to the design, usually specified in engineering drawings, and measured primarily as dimensions between identifiable points such as corners and edges. Although the shape may encompass various kinds of complex curves, such as splines, these are fully defined by a Euclidean geometry. In modern manufacturing the geometric specification may be used directly to control automated lathes, milling machines, and so on.

Waviness represents departures from the specified smooth or flat surfaces and straight or curved edges, typically at spatial scales that are much smaller than the dimensions involved in the form. Waviness is assumed to arise from vibrations or deflections in the machine (or in the workpiece) that occur over time or through indirect connections, allowing relative motion between the workpiece and the tool that may be significant in extent but is relatively low in frequency. Principal components analysis of the waviness profile of tool marks is reported to be useful for quantitative comparisons in forensic applications (Gambino et al., 2011).

Roughness covers the finest-scale irregularities, and for machined parts arises from local interactions between the microstructure of the material in the workpiece and the tool (which includes particles used for grinding or polishing, fixed tools or bits, etc.). This causes vibrations that are generally local, and smaller in magnitude and higher in frequency than waviness. The division between these different spatial scales or frequencies is somewhat arbitrary and may differ according to the size of the part and the nature of the fabrication technique. The cutoffs are often specified as a fraction of the characteristic dimensions or of the length of the scanned profile. Separation between waviness and roughness may be accomplished by applying frequency filters, either in the instrumentation or subsequently to the data.

The accurate measurement of dimensions associated with form is well established and uses devices as simple as calipers or as complex as coordinate measuring machines (CMM) employing laser interferometers to obtain nanometer accuracy over meters of distance. In the case of the Laser Interferometer Gravitational-Wave Observatory (LIGO), the alignment of the interferometer components was performed using a combination of lasers and differential GPS signals over distances of several kilometers. All these methods typically determine the dimensions or coordinates of selected points, which correspond to the design specifications.

For waviness or roughness measurements on surfaces it is necessary to measure a great many points spaced along a line (as in Figure 5.1) or over an area, so that the deviations from ideal form can be analyzed. The data may be presented in summary statistical fashion (e.g., the root-mean-square value of the deviation for points measured every N μm over a total distance of M cm) or in frequency terms (e.g., the power spectrum showing the amplitude of the deviations at various spatial frequencies). There are internationally accepted standards that define the protocols for these tests as used in various industries. Although very important for controlling manufacturing operations, these representations are not readily interpreted in terms of shape as most people understand the term.

When data are obtained for an area of a surface, rather than simply for a single profile across it, more comprehensive measurements of form and departures from form are possible, as shown in Figure 5.36. As mentioned at the beginning of this chapter, these measurement techniques cannot see undercuts in the surface. Because the elevation or height data is both qualitatively and quantitatively different from the lateral dimensions, the results are sometimes called "two-and-a-half-D." The analysis of topography and roughness, the "shape" of the surface, is performed using this data.

The operation of a few widely used metrology instruments for measuring elevation profiles is summarized here as a general background. Techniques for measuring entire surfaces are dealt with later. Much more comprehensive descriptions of a broad range of technologies used for metrology can be found in Whitehouse (1994).

Roundness

The performance of many machined components, such a wheels, gears, automobile engine cylinders, pneumatic cylinders, gun bores, and so forth, requires that they be round within tight tolerances. Standards organizations such as the United States National Institute of Standards

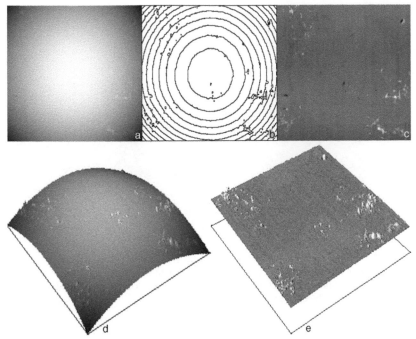

FIGURE 5.36
Scanning profilometer measurements on a ball bearing: (a) presented as a range image in which brightness represents elevation; (b) contour lines drawn at equal increments of height, showing the roughness of the surface and the ellipticity of the shape; (c) after form removal, the surface pits and dirt are more evident; (d) rendered plot of the original surface, showing both form and dirt; (e) after form removal, rendered plot shows the pits and dirt.

and Technology (NIST, formerly the National Bureau of Standards), the British National Physical Laboratories (NPL), the German Physikalisch-Technische Bundesanstalt (PTB), the Japanese Institute of Standards (JIS), and many others who cooperate through the International Organization for Standardization (ISO) in Geneva, have published standard methods for performing and analyzing roundness measurements, and a number of instrument manufacturers offer instruments capable of performing the measurements.

The point was made in Chapter 1 that measuring equal diameters of an object in many directions does not offer proof that the object is circular. That requires measuring the value of the radius from the center. The problem with that geometric definition is that determining the center of the object is also required, which is often not trivial.

Measurements of roundness are typically performed by acquiring 360° traces of the object placed on a turntable, with a stylus connected to a high precision measurement device such as a capacitance or interference gauge. The diagram in Figure 5.37 shows tracing the outer circumference of an object, but

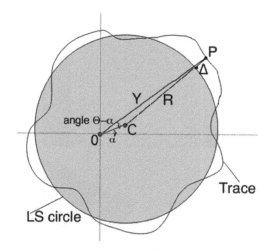

FIGURE 5.37
Tracing of an object as a function of rotation. Y is the distance from the center of the turntable (0) to the stylus (P). R is the radius of the best fit circle from its center (C). Delta (Δ) is the deviation of the trace of the object from the best-fit circle.

many instruments can also be configured to trace the inner circumference of a bored hole.

The least-squares calculation with the measured points is used to determine the best-fit circle and hence the location of the center of the workpiece. Deviations of the trace from the best-fit circle can be calculated to provide a measure of departure from circularity. The weakness of this method is that the deviations contain both the out-of-roundness of the object and the systematic errors in the roundness of the turntable, and these two errors cannot be separated with the single trace.

For high-precision roundness measurements, multiple traces of the object are performed, with the object being rotated on the turntable between traces. Some instruments perform the rotation automatically, for instance, using a Geneva-type mechanism on top of the turntable. This process is continued until n traces have been recorded. Least-squares analysis of the resulting measurements enables the noncircularity of the turntable spindle and bearing to be separated from the profile of the object (Reeve, 1979).

The number of traces that are made on the object is arbitrary but generally not less than four. The mathematics of analyzing the multiple traces is simplest if the object is rotated in equal increments of $360/n$ degrees for each trace, but the measurement accuracy is greater if increments that are mutually prime are used instead. Once the center and profile of the object are determined, the data can be analyzed for departure from circularity in terms of specific failures of form (e.g., ellipticity), as well as waviness and roughness as defined earlier.

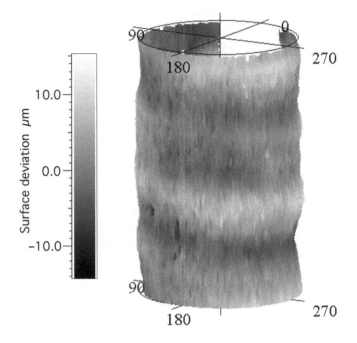

FIGURE 5.38
Shape measurement of a cylinder with 5 μm mean roughness and straightness deviations along the cylinder axis. The method uses interference between a reference beam and light reflected from the surface. (R. Schreiner, 2002, *Optical Engineering* 41:1570.)

Tracing a single circumference around a cylinder does not fully describe its shape. Even collecting several profiles at various positions along the axial length of the part fails to fully represent the roughness or irregularities in that direction. Other techniques, such as optical interferometry, are better able to capture this information as shown in Figure 5.38, with a lateral resolution of the order of μm.

The stylus profilometer procedure works for measuring the roundness of a cylinder, but metrology becomes much more complicated for spheres. For some applications, such as ball bearings, optical lenses, or as gauges to standardize measurements, a variety of specific tests, defined by international standards, are in use. High-precision spheres of steel, glass, and ceramic are available in a variety of sizes and degrees of perfection to meet the needs of industry.

The highest precision sphere ever made is intended to become the new international standard for the kilogram and is made from a perfect crystal of pure silicon, which has a precisely known atomic structure. Project Avogadro, undertaken at the Commonwealth Scientific and Industrial Research Organization (CSIRO) in Australia with international cooperation (the monoisotopic Si 28 was separated in Russia, the single crystal was grown in Germany), aims to assemble Avogadro's number ($6.02214179 \cdot 10^{23}$)

of silicon atoms to become the new standard for the kilogram. The sphere, the roundest object every made, has a diameter of 93.6 mm and has been measured by optical interferometry to have no more than 35 nm deviations from roundness, with a precision of 0.3 nm (approximately a single atomic layer) (Leistner & Giardini, 1991). In performing the measurements, a vacuum environment and tight regulation of temperature are needed, as well as corrections for the presence of surface impurities (a 3–4 nm layer of SiO and SiO_2, which contains some absorbed water, is monitored by ellipsometry). The overall precision is of the order of one part in 10^8.

Straightness

Profilometers are routinely used to perform a scan along a line across a surface to determine its flatness or straightness. Measuring the vertical surface position can be done using a contacting probe, usually a diamond-tipped stylus with a radius of 1–5 μm, and a suitable device to measure its position. This can be a capacitance gauge, an interferometer, or a induction coil, depending on the desired vertical precision and the range of motion that needs to be covered (these same devices are also used to measure profiles having forms that may be much more complicated than flat or nearly flat surfaces, such as optical lenses to determine the surface curvature).

The difficulty with this approach is the requirement to establish a reference datum against which the vertical position can be measured. The mechanism that translates the measuring probe over the part (or, rarely, vice versa) will have its own random and systematic vertical displacement. Separating this motion from the deviations of the surface from straight is not easy.

One practical solution is to produce a reference datum that is as straight and perfect as practically or economically possible, measure it against another more-perfect standard, and store the results. The datum then becomes a secondary standard reference against which the specimen may be compared (as illustrated in Figure 5.39a). The "more perfect standard" may be another physical surface, part of a series of secondary standards that trace their way back to the ultimate reference, or it may be an optical reference such as a laser beam.

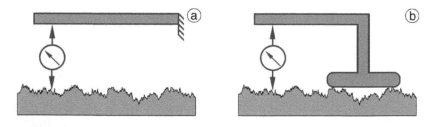

FIGURE 5.39
Principle of waviness and roughness measurement: (a) by comparison to a reference standard; (b) using a skid as a reference.

A configuration that is widely used for practical but less precise measurements uses a "skid" that slides along the highest points on the surface as the reference, while a stylus nearby follows the finer-scale waviness or roughness to generate a profile (Figure 5.39b). Obviously, this method cannot be used to measure large-scale dimensions or form.

Stylus methods encompass both traditional profilometers, with tip radii of a few micrometers, and atomic force microscopes (AFM) with tip radii of a few nanometers (e.g., a single carbon nanotube). The AFM can measure dimensions and roughness down to atomic dimensions, but generally has a much smaller range of measurement, both vertically and laterally. Scanning in the AFM either along a line or in a raster to cover an area may be done using piezo positioners to shift the specimen in x, y, and z while keeping the stylus fixed as a reference point, rather than scanning the stylus across the specimen. Due to problems with creep in piezo devices, some systems incorporate interferometers or capacitance gauges to measure position relative to a fixed reference point.

Noncontacting Measurement

Any contacting probe must exert some force to assure that the tip is in contact with the surface (some AFM instruments use a "tapping mode" in which the tip is repeatedly brought into contact, or close enough to the surface to detect atomic repulsion or attraction forces, but most stylus methods rely on continuous contact). Concern about the effect of a mechanical stylus on the surface being measured and practical issues such as the effect of the scanning speed on the ability to track the profile accurately have encouraged the use of optical techniques such as confocal microscopy or interference microscopy.

Figure 5.40a shows a schematic diagram of an interference microscope. If a monochromatic light source (laser) is used, the interference between the light from the reference and sample arms of the interferometer will produce equal maxima for points on the surface with elevations that differ by an integral number of wavelengths. For very flat surfaces, this is fine. Imaging the entire surface produces a set of fringes or contour lines representing a contour map of the surface. The interference pattern is sometimes called a holographic result, but this should not be confused with fully three-dimensional holograms. Elevations can be determined by analyzing the intensity values in the fringes to provide vertical resolution of better than 1 nm. The horizontal resolution is limited by the light optics to the order of 1 μm.

For rough or irregular surfaces, the fringes are too close together to be distinguished for counting and the alternative approach is to use white light. The multiple wavelengths are all in phase only when the reference and sample arms are the same length. The image of an irregular surface shows only a single contour line marking points at that elevation. Scanning the specimen in the Z-direction brings various points to the

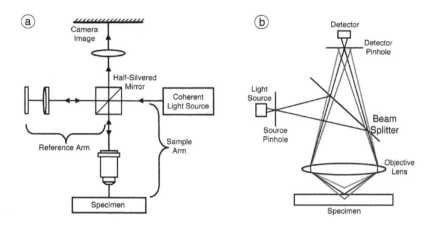

FIGURE 5.40
Schematic diagram of optical surface measuring microscopes: (a) interference optics; (b) confocal optics.

in-phase position and the camera records a series of contours. In practice, the data are generally presented as a range image in which each point is assigned a value (usually shown as a brightness or color) that corresponds to the z-position of the specimen stage when the maximum intensity was recorded at that location.

Confocal optics (Figure 5.40b) can also be used to measure distances. The pinhole apertures at the source and detector, and the in-focus point on the specimen surface, are all optically matched. Light that reflects from any other position on the specimen, either laterally or at different elevations, does not pass through the detector pinhole, so that the measured intensity is very low. Using an objective lens with high numerical aperture (a large incident angle for the light) produces lateral resolution slightly better than 1 μm and depth resolution that is slightly poorer, but interpolation of light intensity with the position as the detector-to-specimen distance is varied can measure distances with a precision of about 0.1 μm.

Both interference or confocal detectors may be used either for a point-by-point measurement along a line, just as a mechanical stylus would be used, with a scanning mechanism and appropriate reference datum. They can also be used to measure the elevation of all points over an area. For a white-light interference microscope this requires a mechanism to raise or lower the specimen relative to the optics. For the confocal microscope, the scanning can be done by moving the specimen laterally but most often it is the light that is scanned using a mirror or prism assembly while the specimen is moved in the Z-direction.

These are the methods that provide the highest resolution, but for practical use in industry a variety of other techniques are also used such as structured light, shape from shading (photometric stereo), stereoscopy, and so on. These methods are usually employed to measure entire areas to produce elevation maps. Light scattered from surfaces contains information about the surface roughness (Chandley, 1976; Welford, 1977) and has been used in industrial instrumentation. A model for Beckmann-Kirchhoff scattering (Ragheb & Hancock, 2006) in the specular direction allows some statistical properties of the surface to be determined, including the mean amplitude of the roughness. Light scattered in other directions depends on the autocorrelation function of the height (Vorburger et al., 1993), which, for example, allows determining the fractal dimension of a surface (Russ, 1994). However, light scattering does not produce a profile or area representation of the actual surface geometry.

Light scattering is also used for particle analysis, particularly in systems built around flow cells. These are able to measure particle sizes from micrometers to millimeters by appropriate design choices, and to process very large numbers of particles in a short time. By monitoring the scatter of the laser light source in multiple directions, the mean diameter and some estimation of shape such as aspect ratio can be obtained (Asano & Yamamoto, 1975; Schuerman et al., 1981; Umhauer & Bottlinger, 1991; Mishchenko, 1993; Jones, 1999). However, for full analysis of particle shape many flow instruments have turned to digital imaging with high-speed strobe illumination to capture and measure images of particle silhouettes, using the techniques shown in previous chapters.

Although the instrumentation descriptions apply to microscope instruments that measure surfaces at resolutions ranging from nanometers to millimeters, the same basic principles apply at any scale. In aerial photogrammetry, elevation profiles (called transects) may be obtained by stereoscopic measurement, with the reference datum being the elevation of the airplane. Most of the land area of the Earth has now been measured stereoscopically using aerial or satellite imagery. For the sea bottom, sonar has been used to measure some areas, but much remains unmapped. Lunar surface elevation data from the Clementine lunar orbiter were obtained one point at a time using a laser rangefinder. The Mars Orbiter Laser Altimeter on the Mars Global Surveyor gathered 27 million individual measurements to construct the map shown in Figure 5.41. For Venus, the Magellan probes used synthetic aperture radar to penetrate the opaque atmosphere.

Image Representation

Devices such as profilometers obtain continuous or analog measurements, with limits to the spatial resolution determined by the hardware (e.g., the radius of the stylus tip, or the size of the light spot in a confocal or interference detector). Both the elevation data and the position values are initially analog, although they are typically digitized by an analog-to-digital

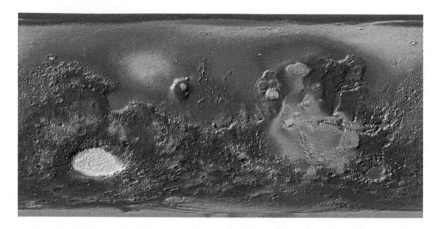

FIGURE 5.41
(See color insert.) Elevation map of Mars. The elevation data have been color coded and shaded to emphasize the relief. (Courtesy of NASA and USGS.)

converter (ADC) so that they can be stored in a computer for analysis. When an image is used for analysis, the camera's array of detectors is inherently analog (the voltage output is proportional to the incident light intensity), but the spacing of the detectors and the conversion of the analog voltage to a digital value (which may take place within the camera electronics) create a digitized set of values for both the position and the brightness.

As shown in Figure 5.42, this ultimately limits the precision with which information can be determined. In a well-designed instrument, the pixel spacing is such that fundamental physical limits (such as the optical resolution) determine the measurement accuracy, and the pixel size is made

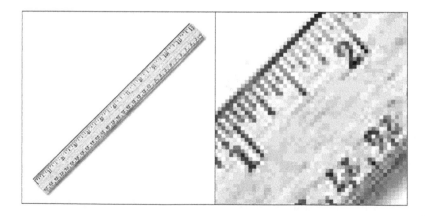

FIGURE 5.42
An image formed with pixels (limited to 16 grayscale values), and an enlarged section showing the effect of the pixel size on determining distances.

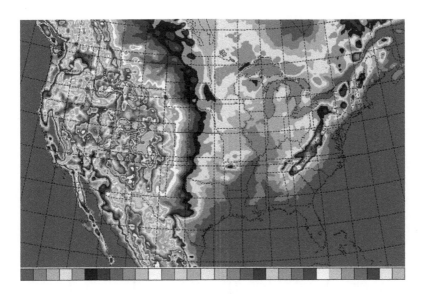

FIGURE 5.43
(See color insert.) Elevation map of the United States, colored in steps of 100 meters.

somewhat smaller than those limits. The same approach is necessary for the range of values that can be assigned to each pixel.

When elevation data are recorded at each pixel location, it is common to display them, as shown in the Mars elevation map in Figure 5.41, either as color or brightness. Since computer displays generally have only 256 shades of brightness available, and since human vision cannot distinguish nearly that many brightness values anyway, some programs use only 256 values to record the intensity (elevation) values. This represents a severe limitation for elevation measurements, which define the shape of a surface.

Consider the Earth: the highest point on the surface is in the Himalayas, at approximately 29,000 feet above sea level; the lowest point is in the Marianas Trench, at approximately 36,000 feet below sea level. If that total range of 65,000 feet is represented in 256 steps (8 bits), each one is about 250 feet. The entire Florida peninsula, much of the Eastern Seaboard and Gulf Coast, and California's Central Valley lie less than 250 feet above sea level and would not be distinguishable from it (Figure 5.43).

A 16-bit (2 bytes per pixel) representation gives an elevation resolution of about 1 foot over this same total range, which is good enough to record the elevation values limited only by the accuracy of most of the measurement devices in use (e.g., on-ground surveying, aerial photogrammetry, or satellite-based measurements). High precision differential GPS has been used to obtain resolution better than 1 cm; representing that level of precision would require an additional byte. Some surface measurement instruments use 4 bytes, giving a potential range-to-resolution greater than $4 \cdot 10^9:1$.

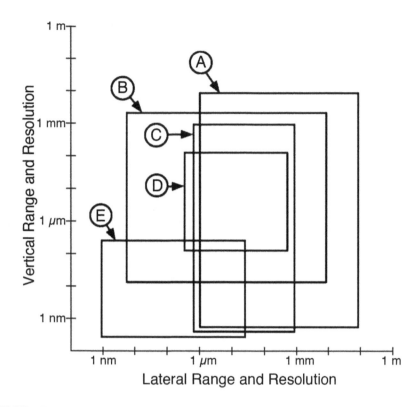

FIGURE 5.44
Stedman diagram comparing the typical range and resolution of several surface measurement technologies: (A) scanned stylus profilometer; (B) scanning electron microscope; (C) interference microscope; (D) confocal microscope; (E) atomic force microscope.

The actual range-to-resolution values for several typical surface measuring tools are summarized in Figure 5.44. This presentation is called a Stedman diagram (named for Margaret Stedman of the British National Physical Laboratories, who introduced it) and is useful for comparing the various instruments. For example, the minimum vertical dimensions detected by the AFM and traditional stylus instruments are about the same, but the AFM has much better lateral resolution and the stylus instruments have a much larger range both laterally and vertically.

Topography

Human vision uses global topographic information to organize information about surfaces (Scott, 1995). The arrangement of hills and dales, ridges,

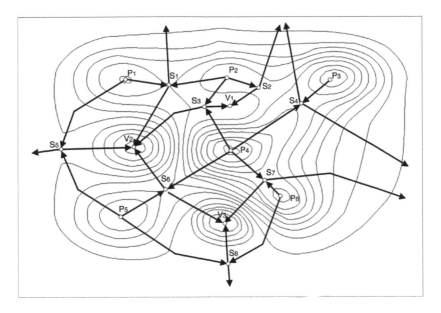

FIGURE 5.45
A contour map of a surface, with the peaks (P), valleys (V), and saddle points (S) marked. The arrows point downhill from peak to saddles to valleys. (After P. J. Scott, 1998, *International Journal of Machine Tools and Manufacture* 38(5–6):559–566.)

courses, and saddle points contains considerable information that describes a surface. A landscape or surface can be divided into regions consisting of hills (points from which all uphill paths lead to one particular peak) and dales (points from which all downhill paths lead to a pit). Figure 5.45 shows a surface with peaks, valleys, and saddle points marked. Boundaries between hills are termed courses and boundaries between dales are called ridge lines.

The Pfalz graph (Pfalz, 1976) and change tree (Figure 5.46) connect the peaks and dales through the respective saddle points where ridge and course lines meet, to summarize the topological structure. These are the syntactical elements that can be used to describe the surface. The change tree shows the height difference and lateral distance between features, which assists in making decisions to ignore features that have either small vertical or lateral extent.

Methods have been proposed (Scott, 1998, 2004; Xiao et al., 2006) for dealing with the finite extent of real images and the corrections necessary for dealing with the intersection of ridges and courses with the boundaries of the image area. It is not yet clear just how this information can be best used for surface characterization and measurement, but parameters such as the volume of connected valleys, the spatial distribution of valleys and peaks across the surface, and orientation of watercourses and ridges are likely to be important.

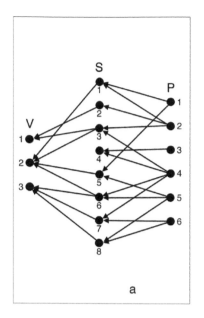
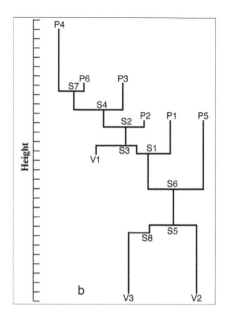

FIGURE 5.46
Surface topology for Figure 5.45: (a) the Pfalz graph showing which peaks (P), saddle points (S), and dales (V) are adjacent; (b) the change tree drawn to show the height difference and lateral distance between them.

Fractal Dimension

At a different and more local scale, many surfaces (but certainly not all) are characterized by a self-similarity or a self-affinity that can be represented by a fractal dimension. Just as for measuring the fractal dimension of outlines in two dimensions, there are several ways to measure this, which do not exactly agree numerically. In addition to the dimension, there must also be a parameter that describes the magnitude of the roughness and perhaps others to characterize the directionality or anisotropy of the surface.

The appeal of the fractal dimension is that it is not dependent on the measurement scale and that it summarizes much of the "roughness" of surfaces in a way that seems to correspond to both the way nature works and the way humans perceive roughness. Given a series of surfaces, the "rougher" the surface as it appears to human interpretation, the higher the fractal dimension. At the same time it must be noted that fractal geometry is comparatively new, and there is a "bandwagon" tendency that may cause it to be applied with more enthusiasm than critical thinking.

In Chapter 3, the fractal dimension of a planar feature is calculated from the boundary using several different algorithms. The dimension lies between a value of 1 (the topological dimension of a line) and 2 (the topological dimension of the plane containing the feature). A perfectly smooth boundary whose observed length does not increase as it is examined at higher and higher resolution has a dimension of 1.0. "Rougher" shapes have increasing dimension values.

Analogous relationships exist for surfaces. The fractal dimension of a surface is a real number greater than 2 (the topological dimension of a surface) and less than 3 (the topological dimension of the space in which the surface exists). A perfectly smooth surface (dimension 2.0) corresponds to Euclidean geometry, and a plot of the measured area as a function of measurement resolution shows no change. But for real surfaces an increase in the magnification or resolution with which examination is carried out reveals more wiggles, nooks, and crannies, and the surface area increases. For a wide variety of natural and man-made surfaces, a plot of the area as a function of resolution is linear on a log–log graph, and the slope of this curve gives the dimension (D).

Measuring the surface area over a range of resolutions is a difficult thing to do; one way is by adsorbing a monolayer of molecules of different sizes and then desorbing them and measuring the amount. This is done for catalyst substrates, for example. But area measurement is not appropriate for many surfaces because they are not ideally self-similar. For most surfaces, the lateral directions and the normal direction are distinct in dimension and physical properties, which means that the scaling or self-similarity that exists in one direction may not be the same as the others. At a sufficiently large scale most surfaces (even the Earth's surface with its mountain ranges and ocean depths) approach an ideal Euclidean form.

Furthermore, for anisotropic surfaces the lateral directions are different. Consequently, many surfaces are mathematically self-affine rather than self-similar. The fact that elevation measurements are single valued and cannot reveal undercuts means that the measured elevation data are self-affine even for a truly self-similar surface (for instance, one produced by diffusion-limited aggregation of particles on a substrate). For self-affine surfaces and data sets there are several correct and useful measurement techniques. A few of the more practical ones are summarized here (and a more complete presentation is available in Russ, 1994).

In most cases, the Fourier power spectrum provides a robust measure of the fractal dimension even for surfaces that are not ideally fractal, nor perfectly isotropic (Russ, 2001). Instead of the usual display mode for the power spectrum, a plot of log (magnitude) versus log (frequency) reveals a fractal surface as a straight line plot whose slope gives the dimension as shown in Equation 5.14.

Notice the similarity to Equation 4.4; the difference is that the relationship in Chapter 4 is applicable to the transform of a line, and the dimension of a line (the boundary of a feature) is equal to or greater than the topological dimension of a line, which is 1.0. Consequently, the expression can yield values between 1.0 and 1.999. For a surface, the fractal dimension is equal to or greater than the topological dimension of a surface, which is 2.0. Equation

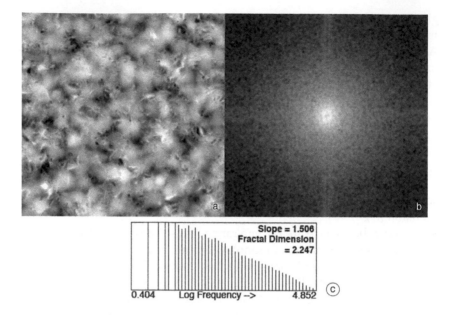

FIGURE 5.47
Fourier analysis of fractal dimension: (a) range image (grayscale proportional to elevation) of a shot-blasted surface; (b) Fourier transform power spectrum of (a); (c) plot of log magnitude versus log frequency averaged over all directions in (b), showing a linear relationship which confirms the fractal behavior, and whose slope gives the dimension using Equation 5.14.

5.14 calculates values between 2.0 and 2.999. The upper limit corresponds to surfaces that extend throughout the space in which they reside.

$$Fractual\ Dimension = \frac{6 - |slope|}{2} \tag{5.14}$$

The principal drawbacks to using the power spectrum plot to measure the dimension are that it tends to overestimate the numerical value of the dimension for relatively smooth surfaces because of finite noise in the data and also that the precision of the measured value is poorer than some of the other methods for images of a given size. Figure 5.47 shows an example of the power spectrum plot for an isotropic fractal surface. The ability to separate the intermediate frequencies from low frequency data that describes the form or figure, and from high frequencies that often reveal instrument limitations or noise, is also useful.

Generating the two-dimensional Fourier transform of the surface range image also reveals any anisotropy. This can be either weak anisotropy in which the fractal dimension is the same in all directions but the magnitude is not, or strong anisotropy in which the dimension also varies with direction. Plotting the slope and intercept of the plot of log (magnitude) versus log (frequency) as a function of orientation provides a quantitative tool to represent the anisotropy.

By itself, the fractal dimension is only a partial description of surface roughness. Stretching the surface in the height direction to increase the magnitude of the roughness does not change the slope of the power spectrum or the fractal dimension. An additional measure, with units of length, is needed to characterize the magnitude. The intercept of the plot of the power spectrum at some selected frequency can be used for this purpose. So can the topothesy, defined as the horizontal distance over which the mean angular change in slope is one radian (this is often an extremely small value and is obtained by extrapolating the measured data).

There are a variety of other measurement approaches that can be used for range images of self-affine fractal surfaces. Two are the covering blanket and the variogram. Both work for isotropic surfaces but do not reveal any anisotropy and do not produce a meaningful average if it is present. The variogram is a plot of the variance in elevation values as a function of the size of the measured region (Dubuc et al., 1989; Hasegawa et al., 1996). Values from areas placed systematically or randomly over the surface are averaged and a mean value obtained. This process is repeated at many different sizes and a plot (Figure 5.48a) made that gives the dimension.

The covering blanket or Minkowski method measures the difference (summed over the entire image) between an upper and lower envelope fitted to the surface, as a function of the size of the neighborhood used. The maximum and minimum values in a neighborhood around each pixel are used to create the envelopes. This is done with a series of different diameter neighborhoods, and the total difference between the minimum and maximum values is summed for each diameter. This is analogous to the Minkowski "sausage" dimension for a profile described in Chapter 3. The covering blanket method produces a plot as shown in Figure 5.48b that gives a dimension.

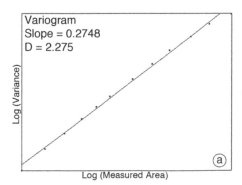
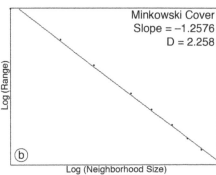

FIGURE 5.48
Measurements of the surface in Figure 5.47: (a) the variogram method; (b) the Minkowski cover.

Notice that these three methods give only approximate agreement as to the numerical value of the dimension. Part of this is the result of limited measurement precision, but part of the difference arises from the fact that all of these techniques measure something that is slightly different. These values are limits to the actual dimension and in general do not agree, so when comparisons are being made between surfaces it is important to use the same technique for all of the measurements.

It is often attractive to perform measurements in a lower dimension, since a smaller number of data points is involved. Historically, much of the work with fractal measurement has been done with boundary lines, whose dimension lies between 1.0 and 1.999.... These procedures can be applied to fractal surfaces by intersecting the surface with a plane and then measuring the dimension of the line that is the intersection. It is important, however, that this plane be parallel to the nominal surface orientation rather than a vertical cut. The vertical cut produces the same profile as that obtained with a profilometer, but because the surface is self-affine and not self-similar the proper measurement and interpretation of this profile are complicated, and the common techniques, including the Richardson plot and the methods shown in Chapter 3, do not apply. Also, of course, a profile is oriented in a particular direction and should not be used with anisotropic surfaces.

The horizontal cut is called a slit-island method (Mandelbrot et al., 1984). The horizontal plane corresponds to sea level and the outlines are the coastlines of the islands produced by the hills that rise above the sea. A plot of the length of these coastlines (Figure 5.49) as a function of measurement scale produces a dimension that is exactly one less than the surface dimension (the difference between the topological dimensions of a surface and a line). The same methods shown in Chapter 3 for measuring the fractal dimension of a boundary can be applied.

Many relationships have been found between fractal dimensions and various aspects of the history and properties of surfaces. Most processes of surface formation that involve brittle fracture or deposit large amounts of energy in small regions tend to produce fractal surfaces, and the numerical value of the dimension is often a signature of the process involved (Mandelbrot et al., 1984; Mecholsky & Passoja, 1985; Mecholsky et al., 1986, 1989; Alexander, 1990; Dauskardt et al., 1990; Fahmy et al., 1991; Lange et al., 1995; Russ, 1997). Mecholsky et al. (1989) show a relationship for fracture toughness as a function of elastic modulus and fractal dimension, which is confirmed by data in Chang et al. (2011).

Also, many surface contact applications (electrical, thermal, etc.) depend upon the relationship between contact area and pressure, and fractal geometry is pertinent to this case. Under some circumstances, friction and wear may also be related to the surface dimension (Zhou et al., 1993). Even the surfaces of fried foods have been found to be fractal (Moreno et al., 2010). Geophysical erosion and deposition processes, cloud formation, and other

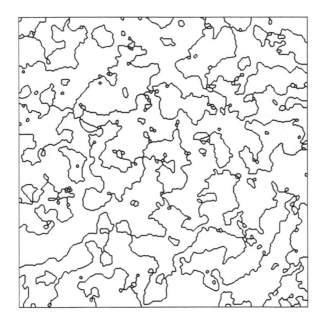

FIGURE 5.49
Coastlines produced by intersecting the elevation data in Figure 5.47 with a horizontal plane.
The Minkowski dimension of the lines (determined using the box-counting method in Chapter
3) is 1.276, corresponding to a surface dimension of 2.276.

large-scale phenomena are frequently characterized by fractal geometry.
Fractal description of surfaces is a relatively new approach to surface characterization. It is applicable to some surfaces but is not appropriate for others
such as ductile deformation.

In addition to describing the roughness of surfaces, fractal geometry
can also be applied to three-dimensional structures. The classic example
is the agglomeration of particulates to form clusters (Stanley & Ostrowsky,
1986; Kaye, 1989; Vicsek, 1992). If the individual particles move in a random walk in three dimensions and stick together whenever they touch, a
very open branching cluster is formed. If the sticking probability is less,
it is possible for particles to migrate down into the openings, valleys, and
crevices, and a denser cluster results. Figure 5.50 shows computer simulations of such 3D clusters formed with 30,000 particles, each corresponding to one voxel.

The fractal dimension of these clusters may be calculated by plotting (on
log–log axes) the mass or number of voxels (or particles) contained in a sphere
of radius m as a function of that radius. For the classic diffusion limited
aggregate (100% sticking probability in Figure 5.50a) the fractal dimension
is 2.5, and it increases toward 2.999... as the cluster becomes more dense and
compact. In addition to the formation of particle clusters, this type of fractal geometry is also applicable to percolation flow through pore networks,

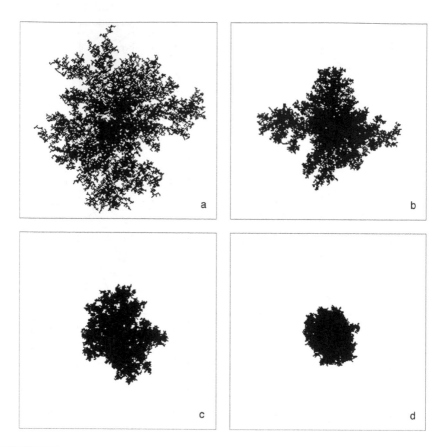

FIGURE 5.50
Projected images of three-dimensional aggregates formed with various sticking probabilities:
(a) 100%; (b) 20%; (c) 5%; (d) 1%.

the spread of forest fires and some illnesses, and other natural phenomena
(Stauffer & Aharony, 1991).

The same method for measurement and the same fractal dimensions pertain to deposition of diffusion-limited aggregates on surfaces
(Figure 5.51). Instead of a radial dimension, the vertical dimension from
the surface is used.

FIGURE 5.51
Diffusion-limited aggregation of frost on the interior surface of a refrigerator.

6

Classification, Comparison, and Correlation

Field Guides

Humans do not identify objects by performing measurements and testing shape factors. Some processes are apparently deeply wired into the brain, such as the visual awareness of faces. The essential elements of the perception process can be described as a semantic or syntactical definition (Sonka et al., 2008):

1. A *face* is an approximately circular part of the human body that contains two eyes, one nose, and one mouth.
2. The *nose* is approximately centered in the face and is vertically elongated.
3. The *eyes* are located above the nose, are approximately circular, and are symmetrically placed to the left and right of center.
4. The *mouth* is centered below the nose and is elongated horizontally.

Figure 6.1 illustrates this knowledge, which is obviously incomplete yet immediately recognizable. Translating semantic definitions like this into computer programs allows digital cameras to locate faces in scenes automatically. Figure 3.5 shows similar syntactical analysis of the shape of a printed letter, and Chapter 4 discusses syntactical analysis in conjunction with the medial axis transform (MAT). It is important to keep in mind that this is a qualitative rather than a quantitative description of the shape. That is why it provides some useful tools for understanding recognition, since human vision is also qualitative (which is in many respects a more powerful attribute than quantitative measurement). However, quantitative measurements are more useful for describing the statistical variation within a group of objects, or for comparing different groups, or providing an estimate of probability for an identification.

In many instances, humans accomplish recognition by a Gestalt comparison of the feature with a catalog, either in hand or in memory. Some recognition tasks require an extensive catalog and organization, as represented by the typical field guides for identifying birds, plants, and so on. These are also

FIGURE 6.1
(Left) The model of a face from the semantic description in the text; (right) a face drawn in the style of Picasso violates the rules.

typically organized as a flow chart, in which a sequence of particular characteristics is used to reach a final conclusion (Figure 1.6 shows a fragment of such a chart).

Figure 6.2 illustrates a simple example of a field guide. During World War II, identifying various planes, both friend and foe, based on their wing shape and engine configuration was an important task for civilians as well as the

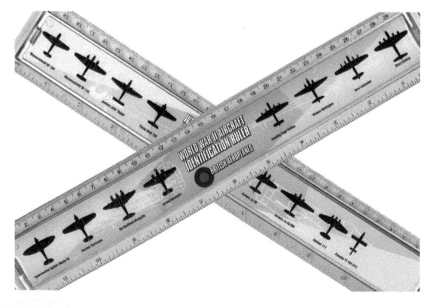

FIGURE 6.2
World War II aircraft identification silhouettes.

military. Portions of the identification guides that were distributed for this purpose are now sold in museum gift shops.

Template Matching and Cross-Correlation

Template matching, described in Chapter 2, can sometimes be used to compare features in images to a set of identification guides. It is easily applied to binary (black-and-white) images and is performed by counting the number or fraction of pixels that do not match. This requires matching the size and orientation of the features to the guide.

Using the sixteen silhouettes from both sides of the ruler shown in Figure 6.2 (eight British and eight German planes, compiled by the Royal Navy Museum), template matching was applied to match each image against each of the others. The score was calculated as the number of matched pixels determined with a Boolean AND minus the number of unmatched pixels determined with a Boolean Exclusive-OR, divided by the average of the number of pixels in the two templates (Figure 6.3). Plotting the score for the matrix of pairwise combinations in Figure 6.4 shows that only the unmanned V-1 "buzz-bomb" has really low scores against all of the other planes. Many of the pairs have scores in excess of 80%, which would make automatic high-confidence recognition difficult if less than ideal viewing conditions were present.

Generally, template matching is less robust than cross-correlation, which is also described in Chapter 2, and which exhibits the strengths and weaknesses of Gestalt comparison. The method can sometimes distinguish those shapes that are more like a target from those that are more different, but interpreting the magnitude of the cross-correlation value or the shape of the cross-correlation peak is rarely possible. For the method to work, the images must be the same approximate size and orientation.

The same pairwise comparisons of the images from the ruler were compared by cross-correlation. As shown in Figure 6.5, the V-1 is again the only

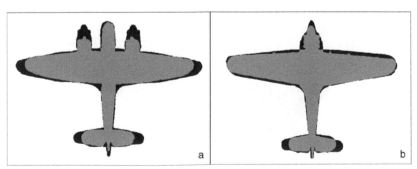

FIGURE 6.3
Template matching. The gray pixels are the overlap (AND) and the black ones the difference (Ex-OR): (a) Blenheim and Junker Ju88; (b) Hurricane and Focke-Wulf 190.

Measuring Shape

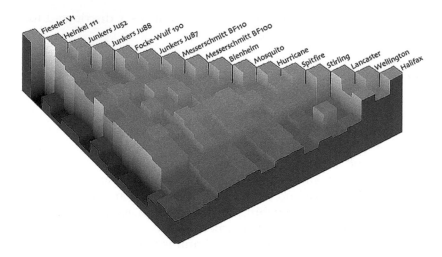

FIGURE 6.4
Template-matching results for the airplane silhouettes. The score for each pairwise combination is plotted as the vertical height of a bar.

shape that is confidently distinguished from the various airplanes. There is not even a consistent trend to match bomber to bomber and fighter to fighter, and some British planes are close matches to some German ones.

Can measurements be employed in this case? Although the actual sizes of the planes vary, the visual extent cannot be used for identification because they may be at different altitudes. But applying the shape measures described in Chapter 3 and Chapter 4 provides a set of data that can be subjected

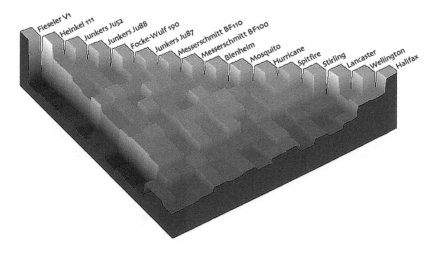

FIGURE 6.5
Cross-correlation results for the airplane silhouettes. Each plane was cross-correlated with each of the others; the peak magnitude values are shown.

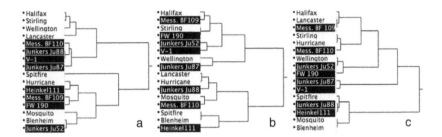

FIGURE 6.6
Dendrograms for the airplane shapes: (a) formfactor; (b) solidity; (c) first invariant moment.
Shading of the names distinguishes British and German aircraft.

to statistical analysis. Three of the more useful parameters turn out to be
solidity, formfactor, and the first invariant moment. If the various aircraft are
ranked according to each of these variables, dendrograms can be drawn for
each variable as shown in Figure 6.6.

The horizontal axis of each plot is the magnitude of the differences in the
parameters for each airplane shape. The lines in the dendrogram are joined
when their differences are less than the value along the horizontal axis. In
this way, it is possible to find clusters of objects that differ by less than a
certain amount or to determine how great a difference corresponds to any
number of clusters. It is immediately clear in this instance that the orders
of the planes are different for each parameter, and that in no case does the
dendrogram resolve into two distinct clusters of British and German aircraft.

This result should not be surprising. The airplanes were designed for different purposes (long- or short-range bombing, interceptor fighters, reconnaissance, etc.) and by different designers. A scatterplot of the values for
formfactor and solidity (Figure 6.7a) shows that the nationalities of the various
planes are not separated. But by applying discriminant analysis (Figure 6.7b)

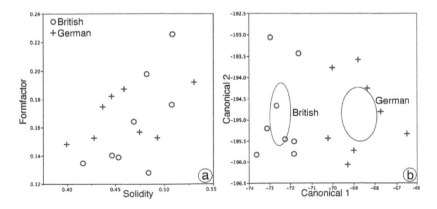

FIGURE 6.7
(a) Scatterplot and (b) linear discriminant results for the airplane shapes.

using all seven invariant moments in addition to the formfactor and solidity, it is possible to separate the nationalities. The canonical variables shown on the axes are linear combinations of the nine measured shape descriptors. This result illustrates both the power and futility of relying blindly on statistical analysis. The separation can be made, but has no practical importance or meaning, and would not be expected to identify the nationality of the next airplane type that might appear in the sky.

One thing that cross-correlation is quite good at is finding a feature in a cluttered or camouflaged scene. In some situations, such as searching reconnaissance images for objects of military interest, the cross-correlation is performed with target images oriented at angles in steps of about 15 to 20 degrees. Figure 6.8 shows the use of a generic face image (which has been heavily smoothed so that only major features remain) as a target to locate all of the faces present. The resulting cross-correlation image has peaks that vary in shape and magnitude depending on the degree to which each face

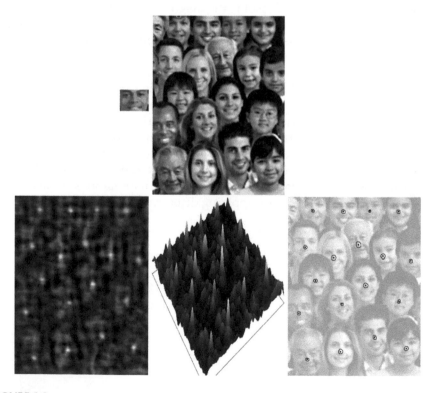

FIGURE 6.8
Cross-correlation. The target face shown at the top left is cross-correlated with the image to locate faces. The amplitude of the result is shown as grayscale brightness and as an amplitude plot, and the location and size of the thresholded peaks are shown superimposed on the original.

matches the template, but which can be thresholded to mark the location of all faces.

However, the cross-correlation image does not provide a useful tool for locating an individual in a crowd or for recognizing an individual, because the variations in the appearance of an individual face can be greater than the differences between persons. That is particularly true when different angles of view may be involved. There are a few applications (Mead & Mahowald, 1988; Mahowald & Mead, 1991) in which images are presented as an array of pixels to train a neural net, which evolves a set of weights that can recognize similar images. Interpreting the weights in the net is difficult. In most applications, identification is based on analysis of measurement data rather than the pixel values.

Cross-correlation is also used to locate landmarks or characteristic features so that images can be aligned by translation and rotation, or by morphing. This is used, for example, in some medical imaging procedures to align images with standardized references or templates for bones or organs, so that local deviations can be observed, or to plan surgery or radiation therapy.

For three-dimensional objects, various approaches have been described and implemented to locate the most similar matching object in a database (Saupe & Vranic, 2001; Funkhouser et al., 2003, 2005; Shilane et al., 2004). Online collections of 3D objects and search engines to explore them use various techniques including spherical harmonics, polygonal models, and (of course) text keywords. Most of the online resources (e.g., the Princeton shape project at http://www.cs.princeton.edu/gfx/proj/shape/) are aimed at finding models that can be described compactly by polygonal surfaces, such as computer-aided design (CAD) drawings, although this has been extended to include approximations of some additional objects such as molecular biology models.

A Simple Identification Example

When shape is used for classification, choosing the appropriate shape parameters may be done by humans based on prior knowledge and experience, or mathematically by techniques such as principal components analysis, stepwise regression, or discriminant analysis. Many parameters are highly correlated, which can make selection of the best reduced set difficult. For example, the equivalent circular diameter is used in calculating a number of dimensionless ratios; if one of these is selected for use, the others may become less important. Statistical methods such as stepwise regression test the significance and effect of adding and removing parameters one at a time to determine the minimal set that gives the best (or adequate) correlation. Prototypical objects selected for training must cover the full range of possibilities to establish limits for each class, but outliers present a problem and it may be necessary to exclude them or accept the possibility that they do not actually belong in the class.

First, consider a case in which human rules can be devised efficiently and effectively. The task is to identify the capital letters A, B, C, D, and E (Figure 6.9a). Because different fonts (both serif and sans serif) are present in different sizes, in various positions and orientations, and with different shades of gray; measurements of size, position, or density by themselves will evidently not be sufficient. By considering how visual recognition of these letters can be accomplished and by making a few measurements on representative letters, it is possible to construct a set of rules and an order for applying them that succeeds based on shape factors.

The thought process that the authors followed, described here, may of course be different from what the reader finds. That is the nature of human decision processes. Some of the measurements that can be performed on the features are indicated in Figure 6.9, panels b through f. They include the maximum and minimum caliper dimensions, and the total skeleton length as well as the length of external or terminal branches, the area and perimeter of each feature (with internal holes filled), as well as the convex hull, and

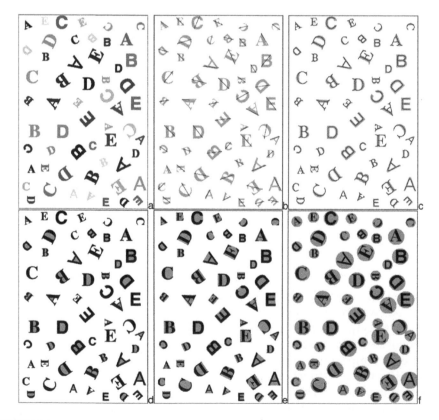

FIGURE 6.9
Measurements of the letters A to E: (a) original image; (b) length (longest chord); (c) skeleton; (d) filled area; (e) convex hull area; (f) circumscribed circle area.

the minimum circumscribed circle. Combining these measures to construct dimensionless ratios is described in Chapter 3.

The first rule to be applied determines the number of holes or loops in each feature. This topological shape property immediately identifies the B as having two loops, while A and D each has one, and the C and E have none (Figure 6.9). To distinguish the A from the D, several possibilities suggest themselves. It is not enough to count the number of end points in the skeleton, because the sans serif A has the same number (2) as the serif D. And because of the different sizes of the letters, it is not easy to discard the short terminal branches of the skeleton, which might remove the legs of a small sans serif A along with the serifs on a larger D. Possibly discarding terminal branches that have lengths less than a fixed proportion of the total skeleton length, as shown in Figure 3.47, would be satisfactory; this more complicated possibility was not tested.

But there are several shape descriptors that can be utilized. One candidate is the ratio of the filled area to the area within the convex hull (the ratio of the filled area to the area within the circumscribed circle also works). For the letter A, the range of values for this dimensionless ratio is from 0.701 to 0.753, whereas for the letter D the range is from 0.866 to 0.994. Since these ranges are distinct, a second rule for features having one loop can be to determine this ratio, test it against 0.80, and thus distinguish the A from the D.

That is not the only possibility. The "roundness" defined in Chapter 3 as $(4 \cdot Area)/(\pi \cdot Length^2)$ can also be used, with a test value of 0.42. And without doubt there are other shape descriptors that can distinguish the A's from the D's.

To distinguish C from E, one possibility is to recognize that the length of the skeleton of the E is greater than that of the C. A dimensionless ratio can be created by dividing the length of the skeleton to the maximum projected length (also called the maximum caliper dimension) of the feature, but this does not completely separate the two letters. The ratio of skeleton length to the square root of the convex area also fails. The problem lies in the large variation of skeleton lengths for letters with and without serifs.

A method that works uses the ratio of the maximum caliper dimension ("length") to the minimum caliper dimension ("breadth"). Examination of the convex hull area in Figure 6.9e shows that the letter E generally occupies an approximately rectangular area, whereas the letter C is much more rounded. The result is that the ratio is greater for E, and a test value of 1.32 can be used to separate them. Again, there are several alternate tests that can be devised. A complete decision tree for a successful identification of the letters is shown in Figure 6.10.

Determining a set of rules for this task is within the capabilities of a human. The variability of the objects to be sorted is relatively small, with example features that can be measured. But most tasks of realistic complexity involve many more rules and much greater variability. Fortunately, there are statistical analysis programs that can assist in the process, discussed later. Humans must still collect representative training populations and

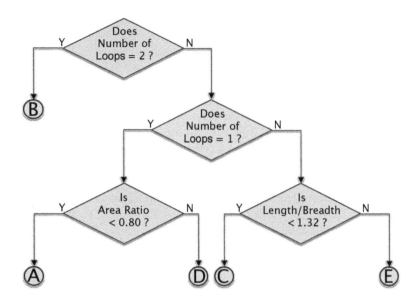

FIGURE 6.10
Flow chart and rules for identifying the letters in Figure 6.9.

provide information about the relative cost of performing various measurements. The software for these "expert systems" decides on the most efficient sequence for applying the rules, sometimes including weighting for the cost of obtaining the inputs (for instance, counting the number of holes or loops is faster than determining the convex hull and the area ratio). But it is the human expert who creates the rules and selects the training examples in the first place.

Defining the Task

There are several different tasks that should be distinguished in the general arena of feature identification, all of which depend on statistical analysis of measurement data, but which emphasize different requirements and have different goals:

1. Identify different classes of objects within a population. This may or may not be done "blind" without knowing the number of such classes or how they may be related to each other. A subset of this task is to search a population of objects to find outliers that may not belong.

2. Identify an individual object as belonging to one of several classes, each of which has previously been characterized by measurements on a representative training set. Success requires great care in establishing and describing the training set. A subset of this task is to find which class an object is most like, if it does not fall within the guidelines for one of them, and to calculate a probability that the object might belong to that class.

3. To identify a collection of objects that is presumed to be of the same type as one of several classes that have been previously characterized. This is a different task than the preceding one, because of the possibility of considering the variability in the objects to be identified.

It is important to keep in mind that the purpose of performing these tasks is not to duplicate or to understand how humans perform recognition or classification tasks. There are no good universal answers to the question "How did you recognize or identify that?" Chapter 1 pointed out the inadequacy of the "grandmother cell" idea (which nonetheless was responsible for developing neural net approaches that do often succeed). Most of the techniques for statistical analysis of data involve complex mathematics that is certainly not what our brains carry out in normal day-to-day life, yet we recognize and classify things routinely without much conscious effort and with generally good success.

The need to successfully recognize food, predators, and possible mates is obviously critical to any organism's evolutionary success, and developing some mechanism for accomplishing these goals is strongly selected by environmental pressure and competition. Some species rely on chemical signatures detected with senses like taste or smell, some utilize sounds or touch, and some use senses that detect information such as electrical or magnetic fields that elude us. Humans are overwhelmingly dependent on vision for information about our environment, and so shape becomes important. The sections that follow describe and illustrate many of the different approaches that are used for these tasks.

Decision Thresholds

For purposes of assigning a new object to one of several established classes, as in the example in Figure 6.10, the measured parameter values are compared against decision thresholds. In some cases this is easy since the different classes are well separated. For the case of the leaves from Figure 3.23, the distributions of measured values shown in Figure 6.11 indicate that convexity is sufficiently different for red oak and white oak that no confusion arises. But for red maple and silver maple, convexity does not adequately distinguish the species. Solidity does provide separation for both the two oaks and the two maples, but cannot distinguish the maples from the oaks.

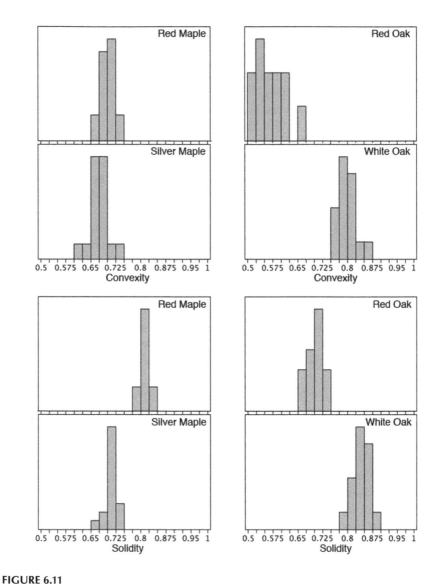

FIGURE 6.11
Histogram distributions for the measurements of solidity and convexity on leaves from different species of maple and oak.

When distributions are slightly overlapped, it may still be possible to establish a decision threshold that produces an acceptably low error rate. Figure 6.12 provides a schematic illustration. In many cases, the decision threshold is placed at the lowest point of the overlap region (0.45 in the illustration), but this is not generally the wisest strategy. The probability of making an error is given by the fraction of the total histogram area that extends beyond the decision point. So for the example shown, the likelihood of

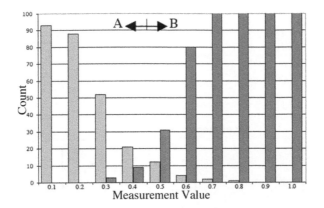

FIGURE 6.12
Example of the overlapped tails of two distributions.

encountering an observation that actually belongs to class A but is identified as belonging to class B is the sum of the counts in the bins above 0.45 divided by the total number of counts in the histogram for A (and vice versa for the opposite error of misidentifying an object that actually belongs to class B).

The decision threshold in Figure 6.13 may be positioned to make the two types of errors equal. On the average, men are taller than women. If height is used in an attempt to classify an individual as a man or woman, a decision threshold at 66.7 inches makes the two error rates equal, but results in approximately 15% of men being classified as women and vice versa. Alternatively, the decision point may be adjusted based on the cost of making a wrong identification. In the example of Figure 6.12, shifting the decision point to 0.25 would increase the probability of misidentifying an object

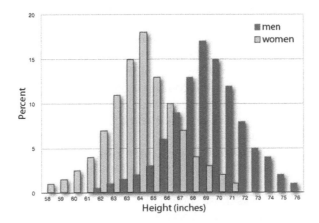

FIGURE 6.13
Height distribution for adult white males and females in the United States. A threshold at 66.7 inches would cause about 15% of men and women to be misidentified based on height.

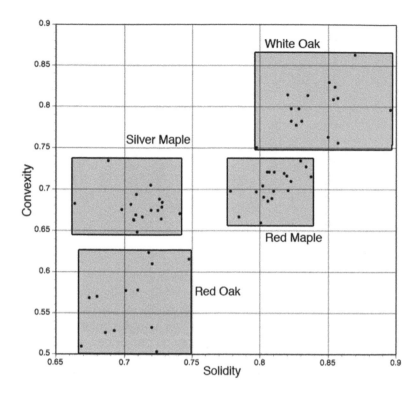

FIGURE 6.14

Two-dimensional scatterplot of solidity and convexity measurements on the oak and maple leaves, showing limit boxes that separate the species.

from class A but would prevent entirely the error of misidentifying an object from class B.

The histograms in Figure 6.11 are one dimensional, based on a single shape measurement. By moving to higher dimensions the separation of classes is generally improved by the ability to combine several measurements. Figure 6.14 shows that the combination of solidity and convexity can completely separate the four species of oak and maple leaves. The technique generalizes directly to N dimensions except that the graphs become more difficult to draw or view.

Hard limits like those shown, corresponding to the boundaries of the boxes, require the training populations to seek and include the extreme values and combinations of values of the various measured parameters. With small populations such as those shown, the absence of extreme examples can create a problem. A different approach seeks to characterize each class using descriptive statistics, usually the mean and variance. Figure 6.15 shows the same data with ellipses that correspond to the two standard deviation (95% confidence) limits for each class.

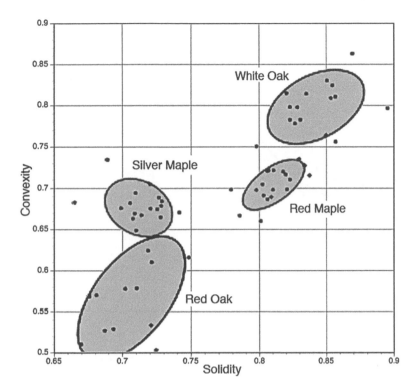

FIGURE 6.15
Solidity and convexity scatterplot with the groups from Figure 6.14 characterized by ellipsoids based on the mean and variance.

This approach has its own set of problems. It presumes that the data in each group are normally distributed, so that the parametric description is valid. While many measures of size (such as the heights of individuals in Figure 6.13) are approximately Gaussian, most shape descriptors, and particularly ratios such as the solidity and convexity used in Figure 6.15, are usually not. It also requires the collection of a training population used to establish the classes that is adequately representative of the population as a whole, and large enough to produce a robust statistical description.

Covariance

Adding more measurements, and moving to a higher dimension space to separate the classes, is a good strategy but raises the issue of how to describe the regions, whether they are boxes, ellipsoids, or other arbitrary shapes. Figure 6.16 illustrates the case in two dimensions. Men are not only taller (on the average) than women as shown in Figure 6.13, they are also heavier. So perhaps height and weight together produce a better ability to separate the sexes. But there is generally some correlation between an individual's height and

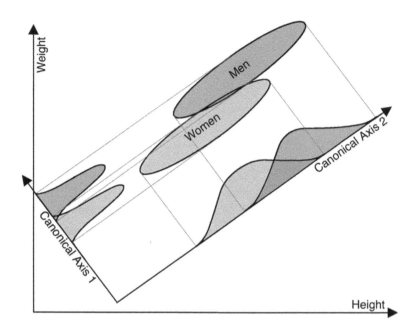

FIGURE 6.16
Fitting canonical axes by linear discriminant analysis provides the maximum separation of classes. (After K. Fukunaga, 1990, *Statistical Pattern Recognition* (2nd ed.), Academic Press, Boston.)

weight values—taller people are likely to be heavier. The resulting graph consists of regions, shown schematically as ellipsoids, that are tilted at an angle with respect to the height–weight axes. Population classes like those shown in Figure 6.15, which are elongated and oriented at an angle to the axes defined by the measurement parameters, reveal a correlation between the parameters. The linear discriminant method uses combinations of the original measured variables (height and weight) to create a set of new axes, called canonical axes, that fit the data. These axes are not necessarily orthogonal.

Because it is more difficult to interpret the distance of any measurement point from the center (the mean), elliptical regions complicate classification. It is desirable to make the population area more equiaxed and preferably circular. The covariance of the data set is defined as

$$C_{i,j} = \frac{1}{(n-1)} \sum_{k=1}^{n} (x_{k,i} - \mu_i) \cdot (x_{k,j} - \mu_j) \tag{6.1}$$

where x is the measurement value, i and j identify the two parameters, μ is the mean value for each parameter, and k runs through the n features in the population. The covariance can vary between $+\sigma_i\sigma_j$ and $-\sigma_i\sigma_j$ where σ is the standard

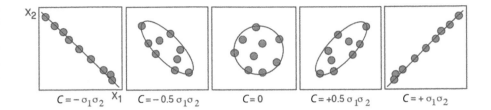

FIGURE 6.17
Examples of covariance. The values vary between perfect positive and perfect negative correlation; a zero value represents uncorrelated values.

deviation value for each parameter (normalized covariance values divided by the product of the two standard deviations are also used; the values then vary between +1.0 and −1.0). A value of zero for $C_{i,j}$ indicates no correlation and the minimum or maximum value indicates perfect correlation as illustrated in Figure 6.17.

The previous examples show only two measured or combined parameters, but in general there may be N axes for a high dimensionality data space. In that case, all of the covariances $C_{i,j}$ for the various parameters measured can be collected into a covariance matrix C. This matrix can be used to transform an irregular population cluster to an equiaxed, circular one. For each measurement vector x (defined by all of the n measurement parameters), the quantity r calculated as

$$r = \sqrt{(x-\mu)' \cdot C^{-1} \cdot (x-\mu)} \tag{6.2}$$

is the Mahalanobis (M-distance) distance from the point representing the feature measurement parameters to the mean of the population, μ, and the prime symbol indicates a transpose. This is a generalization of the usual concept of distance, appropriate for measurements in which the axes have different meanings and scales, or are not independent. In the case of an equiaxed population cluster (zero covariance), the M-distance is the same as the usual Euclidean or Pythagorean distance.

The M-distance is used when distributions are not spherically symmetric. The same value for M-distance implies that two points lie at the same probability density in the distribution. This does not require the distribution to be normal or multinormal, but in the case of a distribution that is, the Mahalanobis transformation based on the covariance converts it to one that is spherically symmetric.

Calculating new canonical variables as linear combinations of the original measurements can remove the covariance as shown in Figure 6.18. Adding a third measurement parameter (roundness) for the same four species of leaves provides even greater separation between the classes as shown in Figure 6.19.

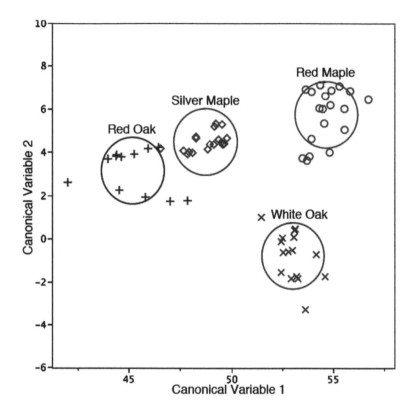

FIGURE 6.18

Two-dimensional scatterplot of the convexity and solidity data from Figure 6.15 using canonical variables to remove covariance between the parameters. The circles are drawn at the two standard deviation (95% confidence) limits.

The canonical variables in Equation 6.3 are linear combinations of the measurements, as described in Chapter 3. Unlike the principal components axes, the canonical axes are not constrained to be perpendicular in the space of the original variables. It is their angular relationship to each other that removes the correlation between the variables so that the ellipses seen in Figure 6.15 become the circles seen in Figure 6.18 and the spheres in Figure 6.19.

The numerical weights shown in Equation 6.3 that are applied to each measured variable are not easily interpreted, because the magnitudes of the values for each variable are different, but combined with the mean values for the variables, they may be qualitatively understood as the angle cosines for the canonical axes with respect to the original measurement axes.

$$Canonical\ 1 = +18.544 \cdot Roundness +35.053 \cdot Solidity +20.063 \cdot Convexity$$
$$Canonical\ 2 = +24.294 \cdot Roundness -11.868 \cdot Solidity -0.8228 \cdot Convexity \qquad (6.3)$$
$$Canonical\ 3 = +3.9342 \cdot Roundness -40.188 \cdot Solidity +33.510 \cdot Convexity$$

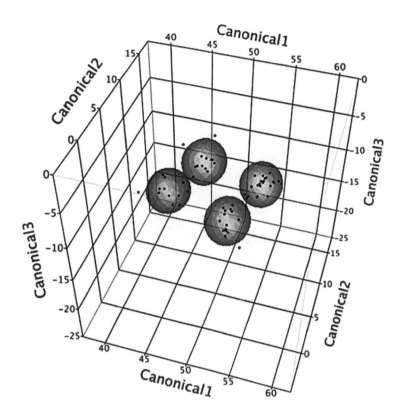

FIGURE 6.19
Adding a third measurement parameter (roundness) for the same four species of leaves. The canonical axes are shown in Equation 6.3.

Example: Mixed Nuts

Of course, not all classification tasks can be successfully handled using just shape parameters. To illustrate this, 300 nuts from a purchased container of mixed nuts were imaged using a flatbed scanner (Figure 6.20) and measured. Shape alone could not distinguish all of the nuts. But by adding a size measurement (equivalent circular diameter: the diameter of a circle with the same area as the feature) and a brightness measurement (the mean brightness value averaged over all pixels within the feature) to three shape parameters (formfactor, solidity, and radius ratio), good results were achieved as shown in Figure 6.21 and Table 6.1. Only 7 out of 300 individual nuts were misidentified (in all cases, ones that were partially broken).

It is useful to compare the identification of the shelled nuts based on dimensionless ratios (plus size and brightness) with the use of invariant moments. As shown in Figure 6.22, the results with moments are poorer, with 12.2% of the individual nuts misclassified. It is also interesting to note that the walnuts, which are perhaps more different visually from the other nuts than

FIGURE 6.20
A few of the mixed nuts that were measured.

any other type, occupy a position in the diagrams between (and partially overlapping) two of the other types. Obviously, in this case moments do not "measure" the same thing as human understanding of shape.

Equipped with these statistical tools, it is interesting to revisit the problem shown in Figure 6.9. The letters A through E shown in Figure 6.23 come from 18 different fonts, some with heavy serifs, some with none, and cover a much greater range of variation than those in the previous example. The simple logic in Figure 6.10 is not adequate to identify each of the letters (although the rule "if there are two loops it is a B" does still apply). All of the various dimensionless ratios and topological shape factors introduced in Chapter 3 were measured for the letters.

A scatterplot of the measured values for the three variables that have the greatest significance in a principal components analysis of the data is shown in Figure 6.24. The various letter classes are not well separated. The complexity of determining a set of rules to accomplish identification is great enough in this case (which is still relatively simple compared to many real-world problems) that using software to apply linear discriminant analysis is essential.

The result of the analysis is shown in Equation 6.4. Notice that the stepwise addition and removal of terms selects as optimum a different combination of measured parameters than those used in Figure 6.10. Plotting the individual letter values on these axes provides robust separation of the classes and the ability to identify the letters, as shown in Figure 6.25.

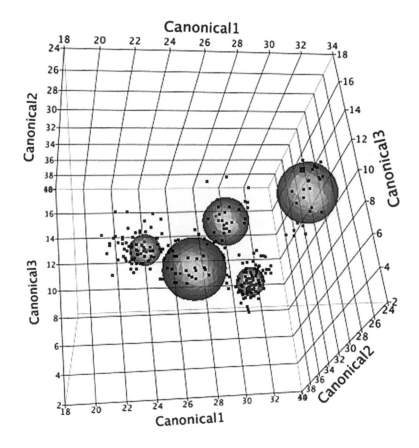

FIGURE 6.21

Linear discriminant plot showing the separation of five classes of shelled nuts (only the most significant canonical axes are shown).

TABLE 6.1

Classification Results for Nuts

Predicted: Actual:	Almond	Brazil	Cashew	Hazelnut	Walnut
Almond	104	0	0	0	0
Brazil	1	28	0	0	1
Cashew	2	1	95	0	1
Hazelnut	0	0	0	31	0
Walnut	1	0	0	0	35

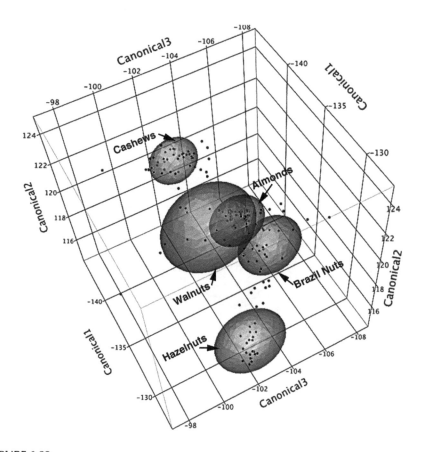

FIGURE 6.22
Discriminant analysis results for the nuts using invariant moments (only the most significant canonical axes are shown).

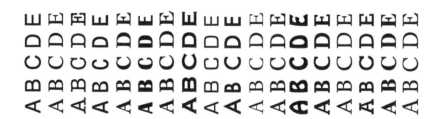

FIGURE 6.23
The letters A through E in 18 different fonts. They are rotated to emphasize that the letters are treated as shapes.

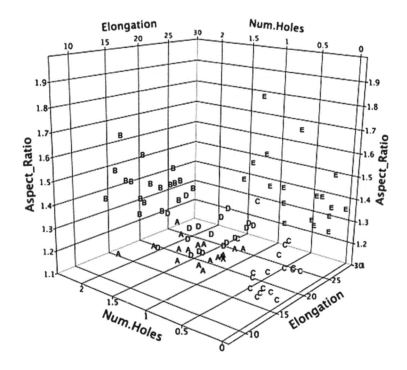

FIGURE 6.24
Scatterplot showing the measured values of the three most significant shape factors for the 90 letters in Figure 6.23.

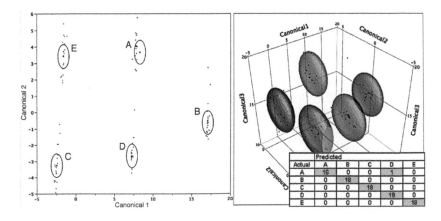

FIGURE 6.25
Canonical plots combining the measured values of shape factors, which identify the letters.

Canonical 1 = +9.4293 · *Num. Holes* −6.1340 · *Roundness* +5.5256 ·
    Solidity −4.3520 · *Convexity* +0.0206 · *Skel. Ends*
Canonical 2 = +0.0031 · *Num. Holes* −44.296 · *Roundness* +35.747 ·
    Solidity −8.0725 · *Convexity* +0.7611 · *Skel. Ends*   (6.4)
Canonical 3 = +1.1953 · *Num. Holes* +2.1537 · *Roundness* −8.8032 ·
    Solidity +30.538 · *Convexity* +0.2162 · *Skel. Ends*

Cluster Analysis

The preceding examples are all based on establishing a description of known
classes using a training population, which is called "supervised training." A
different problem arises when a collection of measurements is obtained but
the actual identity of the individual objects is not known a priori. In this case,
it is often necessary to decide from the data themselves how to group the
objects together. Sometimes the number of classes is known, and sometimes
it is not. Cluster analysis with nonsupervised training is a rich topic that
extends beyond the scope of this text and the measurement of shape (see, for
instance, Pentland, 1986; James, 1988; Fukunaga, 1990; Bow, 1992).

As an example based on shape, consider snowflakes. The fascination with
their beauty and symmetry, and the even deeper fascination with the underlying physics that is responsible for their formation, has captured many
imaginations. Two of the most famous people to photograph snowflakes
and prepare books containing the images are Wilson "Snowflake" Bentley
and Kenneth Libbrecht. Figure 6.26 shows images of a few snowflakes, courtesy of Dr. Libbrecht. No two are alike, but some are more alike than others. Those with "pointier" ends and many branches are typically described
as dendritic, whereas others are often described as platelike. But there is a
visual continuum of shapes, with no obvious or abrupt dividing line between
them. Libbrecht (2006) explains that at a temperature of about 5°C, ice crys-
tals form as thin plates. Depending on the humidity, these vary from solid
hexagons to dendrites. Since each snowflake experiences a different history
of motion through the cloud and encounters different local conditions during the growth process, the flakes that develop vary widely in final shape.

Three shape measurement parameters introduced in Chapter 3 are
selected for illustration: formfactor, which is influenced strongly by the
perimeter; solidity, which is primarily area based; and the number of end
points in the skeleton, which depends on both the interior and periphery
of the feature. The histograms of the measurements shown in Figure 6.27
do not suggest any simple way to subdivide the total population into two
groups. Nor does the 3D scatterplot in Figure 6.28 show any obvious separation into two groups.

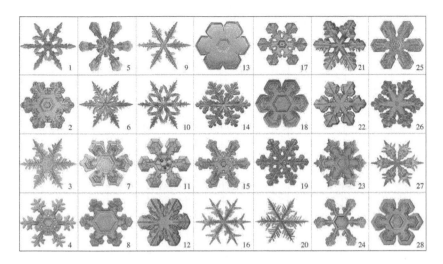

FIGURE 6.26
Twenty-eight snowflake pictures. Measurements on these images are used in the discussion and graphs that follow. (Images courtesy of Dr. Kenneth Libbrecht, California Institute of Technology, Pasadena CA.)

Principal components are linear combinations of the original variables that provide the best fit to the data. This amounts to creating a new set of orthogonal axes that are rotated with respect to the original ones. The first principal component maximizes the dispersion of the data points along the axis, and this captures as much of the variability in the original parameters as possible. The axes are then rotated about this first direction so that each in turn maximizes the remaining scatter of the data points. Equation 6.5 shows the equations for the principal component axes in Figure 6.29; the coefficients for each original variable, scaled by their mean values, are the direction

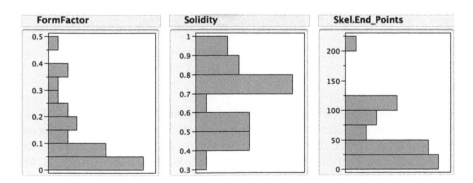

FIGURE 6.27
Distribution of values for formfactor, solidity, and number of skeleton end points measured on the snowflake images.

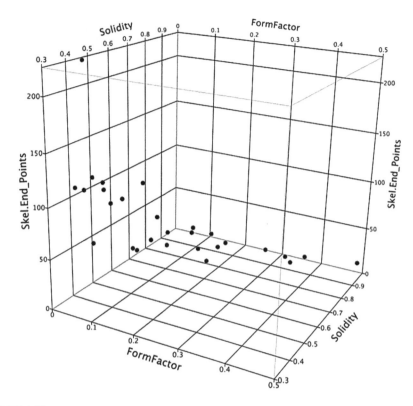

FIGURE 6.28
Three-dimensional scatterplot for formfactor, solidity, and number of skeleton end points in the snowflake images.

cosines for the axes. The three axes account for 83.4%, 12.0%, and 4.6% of the variation in the data, respectively.

Prin 1 = +4.483 · *Formfactor* +3.529 · *Solidity* −0.0116 · *Skel. Ends* −2.346
Prin 2 = +4.472 · *Formfactor* +1.051 · *Solidity* +0.0167 · *Skel. Ends* −2.270   (6.5)
Prin 3 = −4.491 · *Formfactor* +4.569 · *Solidity* −0.0051 · *Skel. Ends* −2.743

Dendrograms

Using the values of the first principal component (from Equation 6.5) as a combination of the three original shape measurements, the dendrogram cluster analysis technique (illustrated in Figure 6.6), also known as hierarchical clustering, is applied to generate the results shown in Figure 6.30. The horizontal axis is the difference in value for the first principal component. The individual flakes are arranged so that ones with a small difference in value are close together. The horizontal lines for each flake are joined at their difference value to become a growing cluster. The clusters then also merge

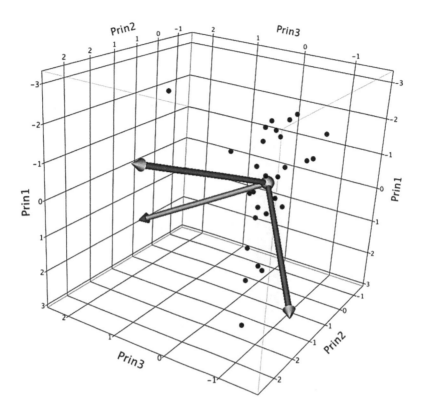

FIGURE 6.29
The data from Figure 6.28 replotted in principal components space. The arrows show the original axes for formfactor, solidity, and number of skeleton end points.

as shown, until finally there is only one single group. But for any value of difference, the identities of the flakes in the various clusters can be identified.

In Figure 6.30, a threshold difference value that produces two clusters is marked. Visually, it does appear that the more "pointy" flakes generally belong to one group and the more "filled-in" flakes to the other. But there is more than one way to accomplish this clustering. The method shown in the figure uses Ward's method, which minimizes the variance in each growing cluster. This tends to join clusters with small numbers of observations and also presupposes that the measurement values in each group are normally distributed so that the variance is a proper parametric description of the values.

Ward's method is the default method in the SAS JMP 9 program used, but there are other possibilities, which lead to different orderings and groupings. The "average linkage" method defines the distance between two clusters as the average distance between all pairs of observations. The "single linkage" method is the minimum distance between an observation in one cluster and the other; this has many desirable theoretical qualities but does not always

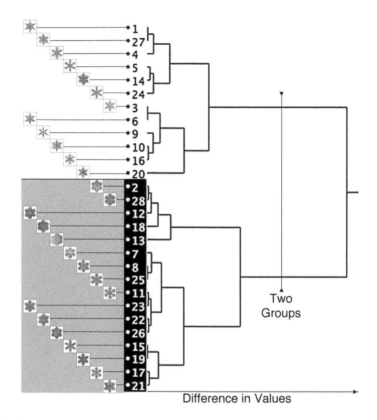

FIGURE 6.30
Dendrogram based on the first principal component value derived by PCA from formfactor, solidity, and number of ends in the skeleton as described in the text, showing the two major groups selected by the cluster analysis.

perform well in practice. As shown in Figure 6.31, these methods produce very different orderings and do not produce two final groups of similar size.

Using Rank Order Instead of Value

An additional concern in the use of the first principal component for the cluster analysis shown is that it accounts for only 83.4% of the scatter in the data. Examination of plots of one measured variable against another (Figure 6.32) shows that they have very nonlinear relationships. The usual method for dealing with this is to replace a plot of the actual values of the measurements with the rank positions of the measurements, as shown in the figure. This Spearman plot has a much higher correlation coefficient (R^2), whose meaning is discussed later.

Using the rank order for each of the same three measurements produces the 3D scatterplot and principal component plot shown in Figure 6.33. The first principal component now accounts for 93.6% of the scatter in the data,

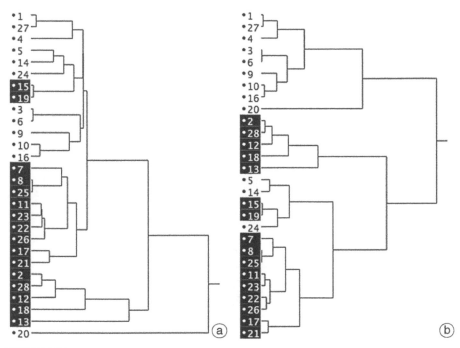

FIGURE 6.31
Dendrograms identical to that in Figure 6.30 except that the clustering methods use (a) average and (b) single linkage criteria. The shading of the numbers indicates the groups from Figure 6.30.

a significant increase, and the axes are calculated as shown in Equation 6.6 (compare to Equation 6.5).

$$Prin\ 1 = +0.0711 \cdot Rank\ Formfactor +0.0697 \cdot Rank\ Solidity -0.0704 \cdot Rank\ Skel.\ Ends -1.028$$

$$Prin\ 2 = -0.0031 \cdot Rank\ Formfactor +0.0876 \cdot Rank\ Solidity +0.0850 \cdot Rank\ Skel.\ Ends -2.352 \tag{6.6}$$

$$Prin\ 3 = +0.0986 \cdot Rank\ Formfactor -0.0475 \cdot Rank\ Solidity +0.0535 \cdot Rank\ Skel.\ Ends -1.562$$

Using these values for each snowflake to perform the same clustering analysis as shown in Figure 6.30 produces the result in Figure 6.34. The ordering of the individual flakes is again different, and two of the flakes (numbers 15 and 19) have changed groups, but the group sizes are equal and the individual identifications are visually plausible.

K-Means and K-Neighbors

The dendrogram method shown in the preceding examples depends on a single value for each observation. Cluster analysis using more than one variable can be performed using K-means clustering. The data are grouped into a specified number of classes such that each point is nearer to the mean or centroid of its class than to the mean of any others. There is an implicit

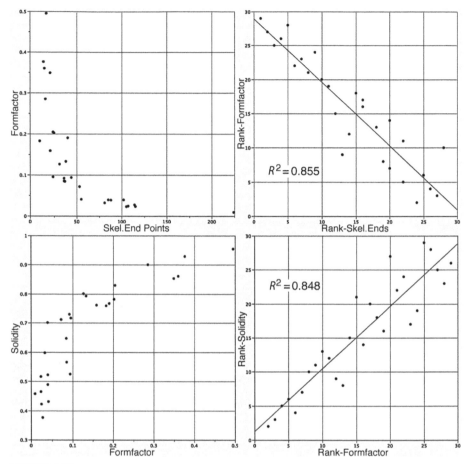

FIGURE 6.32
Nonlinear relationship between formfactor and number of skeleton end points, and between solidity and formfactor, with plots of rank order for each variable. The R^2 values shown are the Spearman correlation coefficients.

assumption that the mean is a proper parametric descriptor of the class, whereas the dendrogram method is nonparametric. The clustering process is iterative, since the inclusion of different points in each group changes the location of the mean or centroid. The iteration ends when no further reassignments are made, and the summation of the distances of all data points from their group centroids has been minimized. An important concern with this approach is whether the distances, calculated in a Pythagorean sense, are meaningful if the various axes have very different scales and meanings.

Applying this procedure to the snowflake data to group the measurements into two classes, using the same three variables used to generate the principal components results in Figure 6.34, produces the clusters marked in Figure 6.35.

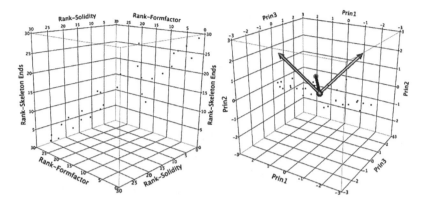

FIGURE 6.33
Three-dimensional scatterplots using the rank orders of the variables and the principal components plot with the original axes shown as arrows.

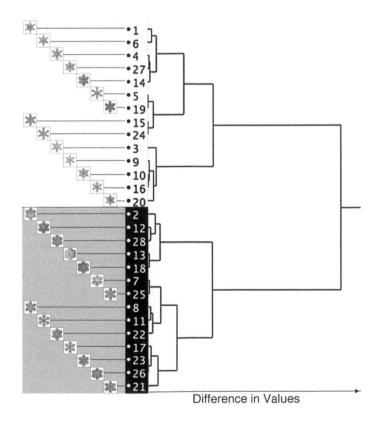

Difference in Values

FIGURE 6.34
Hierarchical clustering (dendrogram) based on principal component of ranked variables (compare to Figure 6.30).

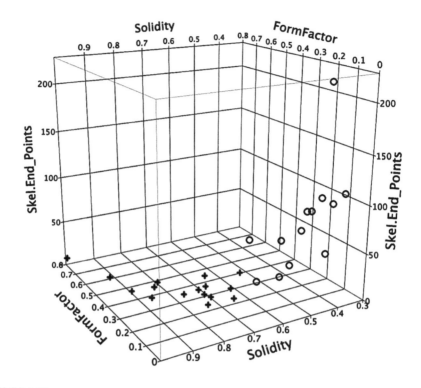

FIGURE 6.35
Three-dimensional scatterplot of the measured values on the snowflakes, showing the two groups identified by *K*-means cluster analysis.

There is no right or wrong result for these snowflake images. It can be left to the eye of the reader to decide which grouping seems most reasonable or indeed if any of them do. The point is that cluster analysis has many possible approaches (of which only a few of the more frequently used are shown here), and that there is often no objective knowledge to test the results against.

The situation is not always made simpler by dealing with groups in two or more dimensions. The extension of dendrograms to more than one dimension is shown next. One difficulty with higher dimensions is that the shapes of the clusters are rarely simple and compact. Figure 6.36 illustrates cases in which one class surrounds another, or is concave, or consists of separated parts.

There are nonparametric methods that do not require a class to have any regular shape (prism, ellipsoid, etc.) or distribution. The most popular is *k*-nearest-neighbor or *k*NN. This classification method works by searching a database containing all previously identified features for those which are "most like" the current measurement as shown schematically for the two-class, two-parameter case in Figure 6.37. This requires finding the distance between the coordinates of the current feature's parameters and those for other features, which means that the time required grows as the database

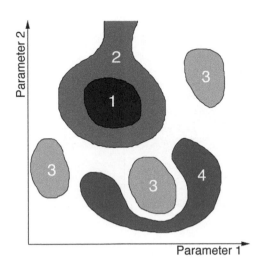

FIGURE 6.36
Examples of nonconvex, disjoint, and surrounded classes that are difficult to represent by simple geometric shapes.

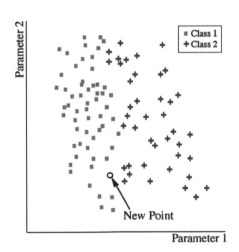

FIGURE 6.37
Example of k-nearest neighbor classification in two dimensions: three of the five nearest neighbors to the new point lie in Class 2, so it is assigned to that class.

grows. The method can be extended to higher-dimension parameter spaces but often becomes significantly more complex.

The distance may be calculated in a Pythagorean sense (square root of the sum of squares of differences), but this overlooks the important fact that the different axes in parameter space have very different units and metrics. Often the distance along each axis is expressed as a fraction of the total

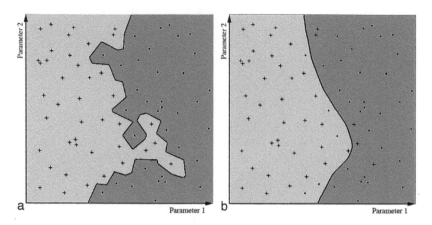

FIGURE 6.38
Nearest-neighbor grouping: (a) the irregular boundary that forms between two classes (whose individual members have parameter values indicated by the points) using single-nearest-neighbor classification; (b) the smoother boundary with five-nearest-neighbor classification (which results in some individual misclassifications).

range of values for that parameter, but there is no real justification for such an assumption. It is sometimes practical to use a normalizing scale such as the standard deviation of values within each class, if this is available, or the Mahalanobis distance. Of course, this resort to parametric values is somewhat contradictory to the fact that kNN methods are used because they are inherently nonparametric.

In the simplest form of k-nearest-neighbor comparison, k is equal to one, and the search is just for the one single prior measurement that is most similar, and which is then assumed to identify the class of the new feature. For features near the boundary between classes, single-nearest-neighbor matching proves to be quite noisy and produces irregular boundaries that do not promote robust identification, as shown in Figure 6.38. Using the identification of the majority of 5, 9, or even more nearest neighbors produces a smoother boundary between classes but is still highly sensitive to the relative size of the populations. If one population has many more members than the other, it effectively shrinks the class limits for the minor group by making it more likely that the majority of matches are with members of the more numerous population. Minor populations representing only a few percent of the total number of observations are strongly discriminated against.

The kNN method can also give a crude estimate of the probability that an observation belongs to a group. If, for example, the nine nearest neighbors are tested and seven of them belong to group A, then the probability that the observation belongs to that group is estimated to be $\frac{7}{9} = 78\%$.

Spanning Trees and Clusters

Human learning seems closer to nearest-neighbor identification than to methods that require mathematical calculation. Children do not initially learn to recognize things in their environment by supervised learning (that happens later, in formal education). Rather, learning proceeds by encountering a wide variety of objects, seeing their similarities and differences, and spontaneously grouping them into categories. Assigning a name to a category is an important step in the process, because we retain information about named classes of things better, and because the names are an important tool for communicating the object classification to someone else.

Unsupervised learning as a computer algorithm is generally treated as a problem of cluster finding and analysis. A multidimensional cluster analysis or dendrogram method that searches for classes in a data set starts by constructing a minimal spanning tree for all of the points present (a significant amount of computation for large data sets). The links in this tree can be weighted either by their Euclidean, normalized, or Mahalanobis length. The lengths of the links are then examined. If the same points are each listed in each other's nearest-neighbor lists, they are probably in the same cluster. Apparently this is pretty much like the spontaneous groupings that develop between "friends" (and friends of friends, etc.) in social networks on the internet.

The spanning tree is then subdivided by cutting the links with the greatest length or the smallest number of matches to leave the identified clusters. Figure 6.39 shows a simplified diagram in two dimensions of a sparse set of measurement values, their minimal spanning tree, and the clusters that result from pruning (after Bow, 1992). The pruning process is halted based on either an independently specified number of clusters or a minimum length of the remaining links. This is a multidimensional generalization of the one-dimensional dendrograms shown earlier.

As an example, Figure 6.40 shows silhouettes of cats and dogs of various types and in various poses but visually recognizable. Can cluster analysis separate them? As shown in Figure 6.41, the measured shape factors can provide a distinction between the classes if the identification of the points is known. However, the individual points do not naturally separate based on cluster analysis. Several of the dog points are closer to a cat point than to any other dog point (marked in Figure 6.41a), and if the pruning method from Figure 6.39 is applied, the points are grouped into two combinations (shown in Figure 6.41b), each of which includes both cats and dogs, plus one single outlier point (which, interestingly, is the poodle clipped for a dog show). Unsupervised learning based on these measured data points and in this parameter space would not naturally determine that there are two groups nor would it correctly identify the members of each group.

If the attributes of objects are not continuous measurements, but the presence or absence of syntactical elements, a different criterion of similarity may be used. Tanimoto and Pavlidis (1975) suggest the ratio of the

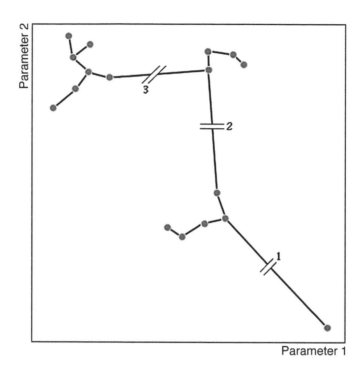

FIGURE 6.39
Schematic diagram showing parameter values plotted as points and the minimal spanning tree connecting them. Cutting the branches with the greatest length separates the clusters.

number of common attributes (ones present in both objects) to the number of those attributes that are present in one or the other object but not in both. This is similar in intent to comparing the number of common pixels (selected with a Boolean AND) to the number of different ones (determined with a Boolean Ex-OR) used for the template matching in Figure 6.3. The ratio converts the yes–no information about the presence of syntactical characteristics into a number that can be used as a measure

FIGURE 6.40
Silhouettes of cats and dogs.

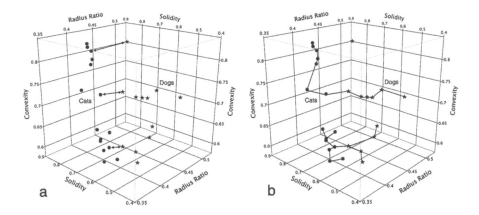

FIGURE 6.41
Measured data from Figure 6.40, with circles marking cats and stars marking dogs, and analysis as discussed in the text.

of distance between points. The clustering analysis then proceeds in the same way as shown in Figure 6.39. This offers an interesting approach to converting syntactical criteria to ones that can be used for statistical or quantitative methods.

The spanning tree method is an example of a divisive or top-down approach to identifying clusters. It is also possible to use an agglomerative or bottom-up method. This begins with the nearest-neighbor points and adds nearby points to that cluster until the next point's distance (defined either as the distance to the nearest point in the cluster or to the centroid of the cluster) is greater than the distance between two as yet unused points. Those then become the center of a new cluster. The process continues until all points have been merged into one or another of the growing clusters.

With both methods the goal is to define a cluster as those objects that are more like each other than they are similar to those in any other cluster. Mathematically, the interset distance between two different clusters is the average squared distance between the individual points in the two sets:

$$D_{12} = \frac{1}{n_1 \cdot n_2} \sum_{i=1}^{n_1} \sum_{j=1}^{n_2} D^2 \left(x_{1i}, x_{2j} \right)$$ (6.7)

The intraset distance within one cluster is similarly the average squared distance between the individual points:

$$D^2 \left(x_i, x_j \right) = \frac{1}{n} \sum \left(x_i - x_j \right)^2$$ (6.8)

so that averaged over all of n points in the m clusters the mean intraset distance becomes:

$$D = \frac{2n}{n-1} \sum_{k=1}^{m} \sigma_k^2 \qquad (6.9)$$

where σ^2 is the variance of the differences within each cluster. Note that the process is nonparametric; this use of the variance follows from the definition of its calculation and does not imply that the clusters have a normal distribution. Some clustering methods do use the variance or standard deviation of points within a cluster to test the distance to a candidate point, and those procedures consequently are parametric. However the distance values are used, the goal of cluster analysis is to make each of the interset distances greater than the mean intraset distance, or in some cases greater than the largest intraset distance.

As for the other analysis methods, the fact that the "distance" values are usually multidimensional vectors, with different scales and meanings for each axis, adds major complications. For both top-down and bottom-up approaches (and for combinations of the two) there are several heuristic algorithms for executing the procedures and no clear choice of an optimum method. Bow (1992) presents a good summary of the mathematics involved and the underlying assumptions in the various methods.

Applying cluster analysis to the data from the arrow points (Figure 4.52) using the size and dimensionless ratio measurements that were found to be most significant for supervised classification (as shown in Figure 4.53) produces the results shown in Figure 6.42. The labels on the points indicate the type for each point. In a static two-dimensional view, it is not readily

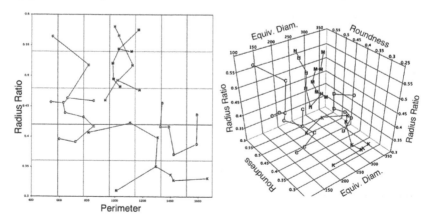

FIGURE 6.42
Cluster plots for the five types of arrow points from Figure 4.52. C = Caraway, H = Hardaway, K = Kirk Serrated, M = Morrow Mountain, and D = Dalton.

apparent that the clusters separate the types completely. Interactively viewing and rotating a plot with higher dimensions reveals the separated clusters.

Populations

For examples using a set of known objects having identities supplied by a human, it is the computer's task to construct a model that uses measurements and combinations of measurements of shape (and sometimes also including size, color, position, etc.) to differentiate the various categories. In this supervised learning, the most difficult part is the selection of the training population. If it is not representative of the larger population that will subsequently be encountered, then the model will fail in unknown and unexpected ways.

If the model consists of absolute limits (the "box" model) for the parameters, then it is vital that the most extreme objects be included. Often this means that objects with extreme values and extreme combinations of various measurement values must be found or estimated. When the model is the more typical one using regression, the mean and standard deviation of the training set are assumed to provide a meaningful and representative description of the population. Each of these situations creates several issues for selecting the training set.

First, the human tendency to select objects based on personal, often unrecognized, aesthetic criteria must be avoided by a randomized selection process. If, for example, some randomizing protocol such as selecting every Nth object that comes down the conveyor belt is used, there may also be bias from the beginning to the end of a sequence (for example, if larger objects tend to rise to the top of the bucket). So it is also necessary to make the selection process uniform over the entire available set. Finally, if two-dimensional images are used to measure three-dimensional objects, their presentation must either be controlled (i.e., all lying on the same side) or randomized. Anything in between will again bias the results.

Testing a model also requires known specimens, but they should not be the same ones used to construct it. The same criteria of random, uniform sampling of the population apply. When the total number of objects available is small, this creates a potential difficulty. One approach that can be used is to use all but one or a few of the known objects to build the model and then test the performance on the holdouts. Repeating this excluding each of the specimens, and then using the average of the models, may provide a satisfactory model as well as an estimate of its performance.

Another issue regarding the characterization of the training set is the choice of mean and standard deviation (or variance) as a compact parametric description of the population. This is an assumption that should be tested. Many observed measurements, such as the height of individuals shown in Figure 6.13, do have a Gaussian distribution. This is the consequence of the central limit theorem in statistics, which states that whenever the number of independent variables (genetics, nutrition, health, etc.) that control an outcome is large, even if they individually have non-Gaussian distributions, they will tend to combine additively in such a way that the outcome is Gaussian. But many of the shape measurements described in preceding chapters are not simple (especially the ratios) and are not controlled by many independent processes. The result is measurements whose distributions are often not Gaussian.

For example, Rovner and Gyulai (2007) measured seeds from more than 1000 plant species to establish a database of seed size and shape. Among their findings:

1. Histograms of size and especially shape measurements are generally not Gaussian in shape and so cannot be adequately described or compared by simple statistical parameters such as mean and standard deviation; the use of nonparametric descriptors and tests is essential.

2. The usual database size of as few as 10 seeds per taxa is too small to show the distributions of size and shape parameters (the authors made a total of 3 million measurements on 150,000 individual seeds, an average of 150 seeds per species).

3. Shape factors based on dimensionless ratios were successful for distinguishing wild from cultivated populations in many plants (e.g., wheat, barley, squash; see Figure 6.43).

In addition to seeds, the shapes of the microscopic silica bodies (phytoliths) that occur between cells in grasses are species specific but do not exhibit a Gaussian distribution. Russ and Rovner (1989) showed that the phytoliths

FIGURE 6.43
Three easily distinguished varieties of squash seeds.

in domestic maize (corn) are statistically distinguishable from those of wild teosinte, the presumed precursor grass from which maize was domesticated. Since the phytoliths survive in archaeological sites, this is important for studying the process and timeline of domestication. Stepwise regression determined a combination of the dimensionless ratios presented in Chapter 3 (formfactor, aspect ratio, and convexity) that distinguished teosinte from maize phytoliths at the 99% confidence level, and furthermore was able to distinguish between two teosinte strains and between three domestic maize strains at the 95% confidence level.

Determining whether a set of measured data can be distinguished from a Gaussian or normal distribution is important because the statistical tests that are appropriate for normally distributed data are well known, relatively easy to apply, available in many programs, and generally easy to understand. Nonparametric analysis is less efficient (more data points are needed to reach the same confidence limits) and generally more difficult to apply.

It is tempting to assume that data can be treated as normal. Figure 6.44 shows distributions of two of the Fourier harmonic coefficients (the first and the fifth) for two of the sand samples shown in Chapter 5. For the Dead Sea sand the first harmonic coefficient (marked f1) cannot be distinguished from the Gaussian distribution shown by the superimposed line, and the fifth coefficient (marked f5) cannot be distinguished from a log-normal distribution shown by the superimposed line, using tests described later. It is not obvious, nor is it suggested, that these fitted distributions would have any particular physical significance. But when the same types of fits are made to the first and fifth harmonic coefficients measured on the Grand Bahama sand, the data do not correspond to the functions and are distinguished from fitted curves. Any analysis applied to the data must not assume normality (or log-normality, or any other simplifying parametric shape) unless that assumption is tested and preferably has some physical justification.

Even for objects whose sizes are normally distributed, the shape measures are often not. Figure 6.45 shows the measurements for the almonds (in their shells) illustrated in Figure 6.53. The lengths of the shells have a distribution that is well fit by a Gaussian curve. The aspect ratio values (the ratio of length to breadth) do not. Several methods are described next for determining whether a distribution can be satisfactorily represented by a Gaussian function.

For non-Gaussian (also called non-normal) data sets, several approaches are possible. One is to find a transformation that can produce a normal set. For example, many biological measurements turn out to be log-normal (Limpert et al., 2001), meaning that the logarithms of the individual values, when plotted as a distribution, have a Gaussian shape and can be analyzed with parametric tests. Figure 3.52 in Chapter 3 shows an example. Another approach is to use the rank order of the measurements rather than the values

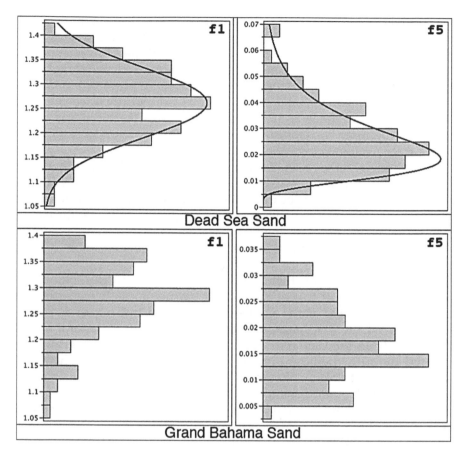

FIGURE 6.44
Distributions of the first and fifth harmonic amplitudes from Fourier analysis of the Dead Sea and Grand Bahama sands, with fitted continuous distribution curves as described in the text.

themselves. This is the basis, for example, of Spearman correlation, and analysis based on the rank orders is illustrated in the snowflake data (see Figures 6.32 to 6.34).

Gaussian (Normal) Distributions

One approach to deciding whether a distribution can be treated as Gaussian or normal is to examine the skew and kurtosis. These are the third and fourth moments, respectively, of the data (the mean and variance are the first and second moments, respectively). For a data set with n values and a mean value μ, the moments are calculated as shown in Equation 6.10.

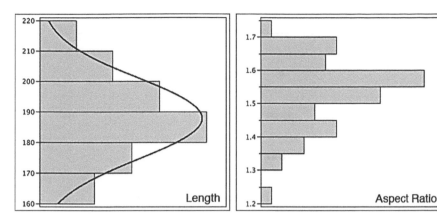

FIGURE 6.45
Measurements of the lengths and aspect ratios of almonds.

$$Skew = \frac{m_3}{m_2^{\frac{3}{2}}}$$

$$Kurtosis = \frac{m_4}{m_2^2} \qquad (6.10)$$

$$m_k = \frac{1}{n}\sum_{i=1}^{n}(x_i - \mu)^k$$

Skew is a measure of how symmetrical the distribution is. A perfectly symmetrical distribution has a skew of zero, while positive and negative values indicate distributions with tails extending to the right (larger values) or left (smaller values), respectively. Kurtosis is a measure of the shape of the distribution. A perfectly Gaussian distribution has a kurtosis value of 3.0; smaller values indicate that the distribution is flatter-topped than the Gaussian, whereas larger values result from a distribution that has a high central peak.

Figure 6.46 shows an example of four sets of 20 data points each that have the same mean and standard deviation but are very different in the way the data are actually distributed. One set (#3 in the figure) does contain values sampled from a normal population. The others do not: one is uniformly spaced, one is bimodal (with two groups), and one has a tight cluster of values with two outliers. The skew and kurtosis values in Table 6.2 reveal this. The outliers in data set #1 produce a positive skew and the clustering produces a large kurtosis. The kurtosis values of the bimodal and uniformly distributed data sets (#2 and #4) are smaller than 3. For the sample of data taken from a normal distribution the values are close to the ideal 0 (skewness) and 3 (kurtosis).

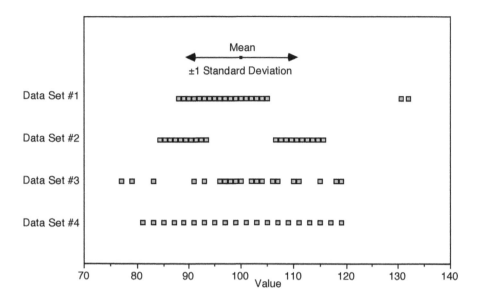

FIGURE 6.46
Four data sets with the same mean and standard deviation.

TABLE 6.2

The Skew and Kurtosis for the Data Sets in Figure 6.46

Data Set #	Mean = μ	Variance = σ^2	Skew	Kurtosis
1	100	140	1.68	5.06
2	100	140	0	1.12
3	100	140	−0.352	2.93
4	100	140	0	1.62

For a given value of skewness or kurtosis calculated for an actual data set, the probability that it could have resulted due to the finite sampling of a population that is actually normal is shown in Figure 6.47, as a function of the number of data values. As the data set grows larger, the constraints on skew and kurtosis narrow. The figure shows the upper limit for the absolute value of skew that a random sample drawn from a Gaussian distribution will exceed, with probability 5% and 1%. The figure also shows the upper and lower limits that the kurtosis for a random sample drawn from a Gaussian distribution will exceed, with probability 5% and 1%. For the small sample size of the data in set #3 of Table 6.2, the skew and kurtosis values are not distinguishable from those for a normal distribution.

There are other techniques for determining the probability that a given distribution can be distinguished from a normal or Gaussian distribution, or from a log-normal distribution, or other specific distributions. Those most

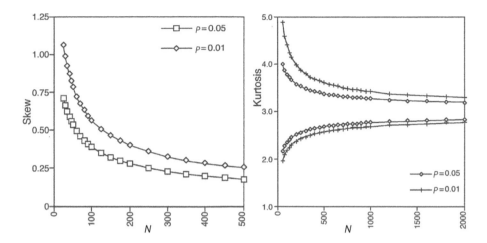

FIGURE 6.47
Probabilities that the value of skew or kurtosis will exceed the limits shown for random samples from a Gaussian distribution, as a function of the number of observations.

commonly encountered are the chi-squared test (Snedecor & Cochran, 1989), the Shapiro–Wilk test (Shapiro & Wilk, 1965), and the Kolmogorov–Smirnov test (Chakravarti et al., 1967). All are well described in advanced statistical analysis texts and included in many statistical analysis programs.

The Shapiro–Wilk test, for example, compares a measured distribution to a Gaussian with the same mean and variance. The test statistic is calculated by dividing the square of a linear combination of sample order statistics by the calculated variance. The resulting W value is then compared to table values based on the number of observations in the data set to determine a probability, p, that the result could be produced by random sampling from a Gaussian distribution.

To illustrate this, 233 pumpkin seeds were imaged by placing them on a flatbed scanner (with the lid open to produce good contrast, scanned at 600 ppi). Measurements of the radius ratio produce a distribution (Figure 6.48a) that can be described by a Gaussian function. The Shapiro–Wilk statistic calculates a p value indicating that there is a 44% likelihood that the distribution cannot be distinguished from normal.

But for the aspect ratio and the equivalent circular diameter the test concludes that distributions are not normal (Figure 6.48b, c). The p value for the aspect ratio is 0.072 (approximately 93% likely that the data are not Gaussian). For the equivalent circular diameter the p value is less than 0.0001 (more than 99.99% likely that the data are not Gaussian). So even though one particular set of measured data may be described by a simple parametric model, the objects represented are not well described by this approach.

Because the Shapiro–Wilk test requires sorting the values into order, it is most appropriate for small data sets. For large sets, the Kolmogorov–Smirnov

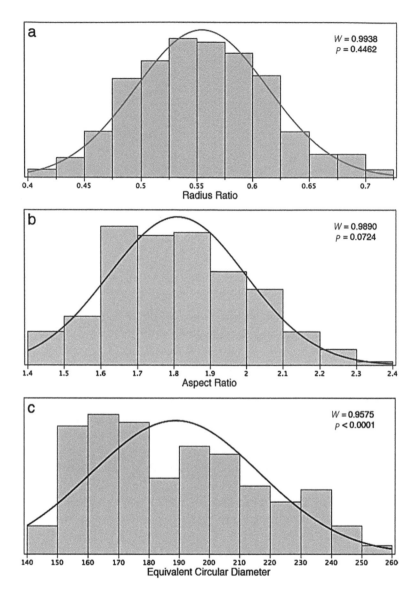

FIGURE 6.48
Distributions of size and shape data for pumpkin seeds, with superimposed best-fit Gaussian curves.

test described later (Equation 6.16, Figure 6.54) may be used. Rather than comparing one data set to another, the comparison is performed between the measured data set and a theoretical Gaussian function with the mean and standard deviation calculated from the actual data. The interpretation of the probability of significant difference is the same.

Comparing Normal and Non-Normal Data Sets

For data sets with a Gaussian distribution, the mean and variance can be used to efficiently characterize the data and to compare one group to another to determine the probability that they really are distinct. Student's t-test calculates a t value as

$$t = \frac{|\mu_1 - \mu_2|}{\sqrt{\dfrac{\sigma_1^2}{n_1} - \dfrac{\sigma_2^2}{n_2}}}$$

$$v = \frac{\left(\dfrac{\sigma_1^2}{n_1} + \dfrac{\sigma_2^2}{n_2}\right)^2}{\dfrac{\left(\dfrac{\sigma_1^2}{n_1}\right)^2}{n_1 - 1} + \dfrac{\left(\dfrac{\sigma_2^2}{n_2}\right)^2}{n_2 - 1}} \tag{6.11}$$

where n is the number of objects in each group, μ is the mean, σ is the standard deviation, and v is the number of degrees of freedom. This t value is then compared to a table or graph that shows the probability, p, that the t value will be exceeded when the two mean and standard deviation values come from populations that are not actually different due to finite sampling. The p values

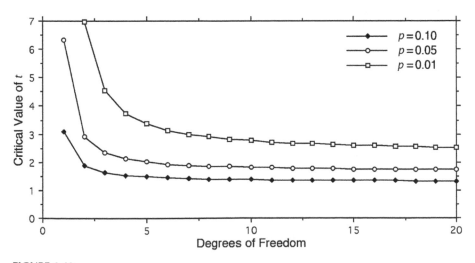

FIGURE 6.49
Critical values of Student's t for confidence limits of 90%, 95%, and 99% as a function of the number of degrees of freedom.

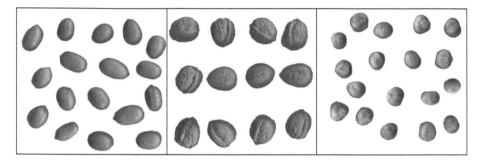

FIGURE 6.50
A few of the pecans, walnuts, and hazelnuts imaged with a flatbed scanner.

shown in Figure 6.49 are for a two-sided test, meaning the probability that the two means are "different." In a one-sided test (deciding whether one mean is "greater" or "less" than the other) the probabilities are halved.

As shown in Figure 6.49, this curve flattens out very quickly as the number of degrees of freedom increases, approaching asymptotic values of $t = 1.282$ ($p = 0.10$), $t = 1.645$ ($p = 0.05$), and $t = 1.960$ ($p = 0.01$), so that for practical use with moderate to large sample sizes it is common engineering practice to regard t values greater than 2 as indicating that two groups are distinct at the 99% confidence level.

As an example, measurements of *roundness* were made on a series of pecan and walnut shells (Figure 6.50), producing the mean values and standard deviations shown in Table 6.3. The t statistic calculated from them is 5.45, indicating that the two groups are distinct and not samples taken at random from the same population (of course, there are other ways to distinguish walnuts from pecans, including size, color, and texture).

TABLE 6.3

Student's t-Test and ANOVA Calculations for Roundness Measurements on Nuts

	Mean	Std. Dev.	n
Walnuts	0.7598	0.0504	46
Pecans	0.6726	0.0769	19
Hazelnuts	0.8461	0.0384	24

t-Test Calculation		ANOVA Calculation	
$\mu_1 - \mu_2$	0.0873	SS_{within}	0.2547
σ_1/n_1	0.000055	SS_{total}	0.5761
σ_2/n_2	0.000312	$SS_{between}$	0.3214
denom.	0.01602	n_1	2
		n_2	86
t	5.448	F	54.26

For more than two groups, the *t*-test generalizes to analysis of variance (ANOVA). This compares the differences between the means of several classes, in terms of their number of observations and variances. To perform the ANOVA test, an *F* value is calculated from the sums-of-squares terms using the observations y_{ij} (i = group, j = observation number). In Equation 6.12, y_s^* is the mean of observations in class i, and y_{mean} is the global average. There are n_i observations in each class, k total classes, and n total observations.

$$SS_{Total} = \sum_i \sum_j \left(y_{ij} - y_{mean} \right)^2$$

$$SS_{Within} = \sum_i \sum_j \left(y_{ij} - y_i^* \right)^2$$

$$SS_{Between} = n \cdot \sum_i \left(y_i^* - y_{mean} \right)^2 = SS_{Total} - SS_{Within} \tag{6.12}$$

$$F = \frac{\dfrac{SS_{Between}}{(k-1)}}{\dfrac{SS_{Within}}{(n-k)}}$$

The degrees of freedom are $(k - 1)$ and $(n - k)$. The value of *F* is then used to determine the probability that the observations in the *k* classes could have been selected randomly from a single parent population. If the value is less than the critical values shown in Table 6.4, then the difference between the groups is not significant at the corresponding level of probability. The table shows critical values for *F* for the case of $p = 0.05$ (95% confidence that not all of the data sets come from the same parent population).

If hazelnuts (which are smaller and rounder) are added to the data set consisting of walnuts and pecans in the preceding *t*-test example, the resulting calculation is shown in Table 6.3. The large *F* value (54.3) indicates that these are distinct groups.

TABLE 6.4

Critical Values of F for $p = 0.05$ (95% Confidence)

$n - k$	$(k - 1) = 3$	$(k - 1) = 5$	$(k - 1) = 10$	$(k - 1) = 40$	$(k - 1) = \infty$
3	9.28	9.01	8.79	8.59	8.53
5	5.41	5.05	4.74	4.46	4.36
10	3.71	3.33	2.98	2.66	2.54
40	2.84	2.45	2.08	1.69	1.51
∞	2.60	2.21	1.83	1.39	1.00

Nonparametric Comparison

There are several nonparametric tests available for comparing sets of data that do not rely on the assumptions of a Gaussian distribution characterized by mean and variance. One approach is based on using the rank order of the measured data rather than their actual numerical values. For small populations, Student's *t*-test can be replaced by the Mann–Whitney test (also called the Wilcoxon test) for two groups. Just as the *t*-test generalizes to the ANOVA, the Mann–Whitney generalizes to the Kruskal–Wallis test for more than two groups. For large populations, the Kolmogorov–Smirnov test shown later is a more convenient procedure.

The Mann–Whitney test compares two data sets by ranking them together into order, and then examining the results (Figure 6.51). The values are used only to perform the ranking, after which only the rank positions are needed. Consider shuffling a deck of cards: if all or most of the red cards end up at one end of the pile, it is an unlikely and even suspicious event. It is possible to calculate (using the binomial theorem) the probability of occurrence of any particular sequence of red and black cards. For the cards that have been ranked into order based on a measurement value, if the number of ways that the sequence could be achieved by random shuffling is high, the conclusion is that the two groups of measurements are not distinguishable and vice versa.

To perform the procedure, the average of the rank positions of the two groups is tallied and whichever is smaller is compared to the average score that would result if the two groups were equal in average rank. In the case of exact ties in value, average rank positions are assigned. The *U* statistic is calculated as

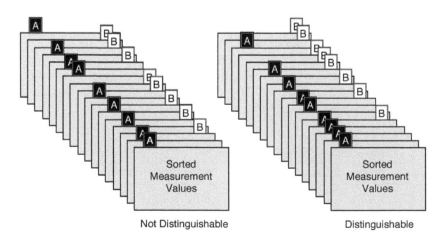

Not Distinguishable   Distinguishable

FIGURE 6.51
Principle of the Wilcoxon or Mann–Whitney test based on rank order.

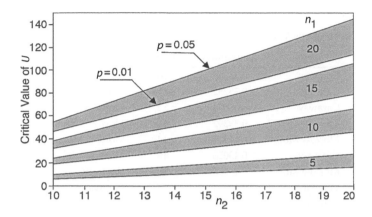

FIGURE 6.52
Critical values of U for 95% and 99% confidence limits for different group sizes.

$$U = \left| \frac{n_1 \cdot (n_1 + 1)}{2} - W_1 \right| \tag{6.13}$$

where W_1 is the sum of rank values and n_i the number of observations in the groups. If the value of U is less than the critical value shown in Figure 6.52, then the two groups are considered to be distinct with the corresponding degree of confidence. For large samples, the significance of z calculated with Equation 6.14 can be taken from tables of the normal distribution:

$$z = \frac{\sqrt{12} \cdot \left(U - \dfrac{n_1 \cdot n_2}{2} \right)}{\sqrt{n_1 \cdot n_2 \cdot (n_1 + n_2 + 1)}} \tag{6.14}$$

To illustrate this procedure, values for the dimensionless shape factor *roundness* measured on two of the types of arrow points shown in Figure 4.52 are used. Table 6.5 shows the ranked values. The test value of U is 89, which from the graph indicates that the two groups cannot be distinguished by this measurement (i.e., the data could have been obtained by randomly sampling from a single population).

In the Kruskal–Wallis test, for more than two groups, the data sets are sorted into a single list and their rank positions are used to calculate the parameter H based on the number of groups k, the total number of objects n, and the number in each group n_i (Equation 6.15). The rank orders for each group R_i are summed. The test value H is compared to critical values (which come from the chi-squared distribution) for the number of degrees of freedom $(k - 1)$ and the probability p that the magnitude of H could have

TABLE 6.5

Ranked Roundness Values for Two Types of Arrow Points

Rank	Type	Roundness	Rank	Type	Roundness
1	Caraway	0.33943	10	Morrow Mtn.	0.45742
2	Caraway	0.35252	11	Caraway	0.46345
3	Caraway	0.36070	12	Caraway	0.46457
4	Caraway	0.39392	13	Morrow Mtn.	0.47669
5	Morrow Mtn.	0.41539	14	Morrow Mtn.	0.48127
6	Caraway	0.43023	15	Morrow Mtn.	0.51002
7	Morrow Mtn.	0.43930	16	Caraway	0.52154
8	Caraway	0.44114	17	Caraway	0.58402
9	Caraway	0.44604			

occurred purely by chance by random sampling from a single parent group (Table 6.6):

$$H = \frac{12}{n \cdot (n+1)} \cdot \sum_{i=1}^{k} \left(n_i \cdot \bar{R}_i^2\right) - 3 \cdot (n+1) \tag{6.15}$$

Applying this procedure to the *aspect ratio* measurement for four classes of the arrow points generates the ranking shown in Table 6.7. The calculated H value is 92.5, which the table indicates is above the critical value. This indicates that the four sets of data are not all samples from the same population, but it does not indicate whether some of them could be indistinguishable. That would require testing them individually.

The ranking operation is unwieldy and slow for large numbers of objects, so the Kolmogorov–Smirnov test may be used instead. This works by constructing cumulative distributions of the number of objects as a function of

TABLE 6.6

Critical Values of H for the Kruskal–Wallis Test

$(k-1)$	$p = 0.05$	$p = 0.025$	$p = 0.01$
1	3.841	5.024	6.635
2	5.991	7.378	9.210
3	7.815	9.348	11.345
4	9.488	11.143	13.277
5	11.070	12.832	15.086
6	12.592	14.449	16.812
7	14.067	16.013	18.475
8	15.507	17.535	20.090
9	16.919	19.023	21.666
10	18.307	20.483	23.209

TABLE 6.7

Ranked Aspect Ratio Values for Four Kinds of Arrow Points

Rank	Type	Aspect Ratio	Rank	Type	Aspect Ratio	Rank	Type	Aspect Ratio
1	Caraway	1.229	11	Dalton	1.508	21	Dalton	1.685
2	Hardaway	1.324	12	Dalton	1.511	22	Caraway	1.701
3	Caraway	1.351	13	Morrow	1.546	23	Caraway	1.726
4	Caraway	1.360	14	Dalton	1.546	24	Dalton	1.796
5	Caraway	1.365	15	Morrow	1.565	25	Morrow	1.798
6	Hardaway	1.436	16	Caraway	1.591	26	Dalton	1.807
7	Hardaway	1.445	17	Morrow	1.593	27	Morrow	1.873
8	Hardaway	1.449	18	Morrow	1.620	28	Caraway	1.935
9	Caraway	1.477	19	Hardaway	1.631	29	Hardaway	1.978
10	Caraway	1.492	20	Caraway	1.666			

TABLE 6.8

Values of A for the Kolmogorov–Smirnov Test

p	0.10	0.05	0.025	0.01
A	1.07	1.22	1.36	1.52

the measured variable, and then finding the greatest difference between the two curves. The test value S is then calculated as shown in Equation 6.16, using the value of A from Table 6.8 corresponding to the desired confidence limit:

$$S = A\sqrt{\frac{n_1 + n_2}{n_1 \cdot n_2}} \qquad (6.16)$$

Based on the number of objects in each group, the probability that the observed maximum difference could occur by chance is used to decide whether the two groups are distinguishable or not. If the maximum difference between the two cumulative distributions exceeds the value of S, the two groups are considered to be distinct with the corresponding probability. Since the cumulative plots have percentage or fraction as the vertical axis, the two groups do not need to be equal in size.

As an example, the *roundness* values measured for Brazil nuts and almonds are compared (Figure 6.53 shows a few of the nuts examined). A total of 159 Brazil nuts and 197 almonds were measured, and the cumulative plots of number versus measured roundness value are shown in Figure 6.54. The greatest difference occurs at a roundness value of 0.6, but only the magnitude of the difference matters, which is 0.259 (25.9%). Using an A value of 1.52, the calculated S from Equation 6.16 is 0.162. Since the observed greatest difference exceeds the test value, the conclusion is that the two sets of data

FIGURE 6.53
A few of the Brazil nuts and almonds imaged with a flatbed scanner.

are different at better than 99% probability (there is less than a 1% chance that the difference is due to random sampling of a single population). Of course, in this case the nuts can also be distinguished in many other ways, including size and color.

As mentioned earlier, the Kolmogorov–Smirnov test may also be used to compare a measured distribution to an ideal one, such as a Gaussian. The comparison and interpretation are performed in the same way shown, using a cumulative fraction calculated from the equation for the ideal distribution for the test distribution.

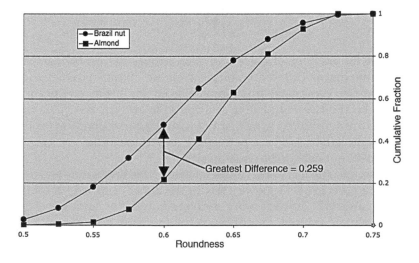

FIGURE 6.54
Cumulative plots of roundness for Brazil nuts and almonds.

Nonparametric Estimation of Covariance

In cases where the measurements are not normally distributed, it is desirable to be able to estimate covariance using nonparametric means. Estimation of the covariance independently of the computation of the linear discriminant axes (demonstrated in Figure 6.15 and Figure 6.18) is useful in quickly troubleshooting situations where the measurements chosen for statistical analysis are highly covariant. In multivariate systems with a large number of measured variables, where a linear discriminant analysis may not show all the independent axes, an analysis of covariance can be used to determine which variables contain nonredundant information.

For normally distributed populations, the degree of association between two measurements is given by the covariance as defined in Equation 6.1. For non-normal distributions, this degree of association is estimated using Kendall's τ statistic, which is computed from a rank ordering of the two measurements. With a pair of independent measurements x and y of a set of objects, the rank order of each measurement is computed so that each object is represented by a pair of ranks, (X_i, Y_i). Then any two pairs of ranked measurements (X_i, Y_i) and (X_j, Y_j) are concordant if $X_j > X_i$ and $Y_j > Y_i$, or if $X_j < X_i$ and $Y_j < Y_i$. Two pairs are discordant if $Y_j > Y_i$ and $X_j < X_i$, or if $Y_j < Y_i$ and $X_j > X_i$.

The Kendall τ statistic is given as the difference of the probabilities of concordance p_c and of discordance p_d over all pairs in the system:

$$\tau = p_c - p_d \tag{6.17}$$

This is computed by counting all the unique pairs of objects in the system, adding 1 when the pairs are concordant, subtracting 1 when the pairs are discordant, and adding 0 if there is a tie in the rank. For a system of n objects, this sum can be calculated with Equation 6.18:

$$\tau = 2 \cdot \sum_{i=1}^{j} \sum_{j>i}^{n} \frac{A_{ij}}{n \cdot (n-1)} \tag{6.18}$$

where $A_{ij} = +1$ if the pairs are concordant, -1 if they are discordant, and 0 otherwise. A full derivation of this parameter is given in Gibbons and Chakraborti (2011). The values of the parameter τ lie in the range -1 to $+1$, with $+1$ indicating that measurements x and y are perfectly correlated, -1 indicating perfect anticorrelation, and 0 indicating true independence. This is analogous to the normalized correlation coefficients shown in Figure 6.17.

As an example, consider some of the dimensionless ratio shape factors used in the analysis of the arrow points (Figure 4.52). The factors chosen do not share the same measurements as their basis:

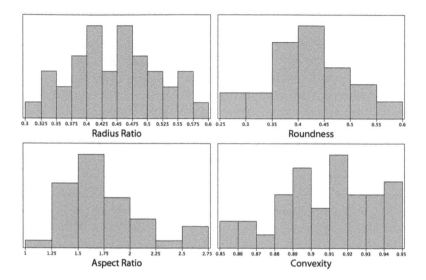

FIGURE 6.55
Distributions of values for the dimensionless ratio shape measurements on the arrow points.

- *Radius ratio* is the ratio of the radii of the inscribed and circumscribed circles.
- *Roundness* is a function of area and maximum projected dimension.
- *Aspect ratio* is the ratio of length to breadth.
- *Convexity* is the ratio of perimeter to the perimeter of the convex bounding polygon.

Distributions of the measured values are shown in Figure 6.55; none of the distributions is adequately described by a Gaussian function.

Plotting the values of roundness, aspect ratio, and convexity against radius ratio (Figure 6.56) shows that the first two are highly correlated ($p < 0.0001$) and even the third has a correlation that appears to be significant at the 95% level ($p = 0.0413$), in spite of the scatter in the data.

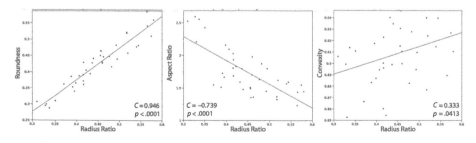

FIGURE 6.56
Plots of roundness, aspect ratio, and convexity versus radius ratio.

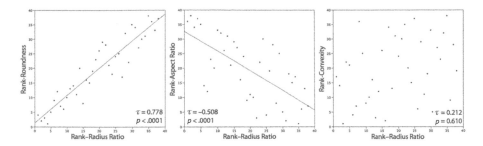

FIGURE 6.57
Plots of the rank order of roundness, aspect ratio, and convexity versus radius ratio.

However, plotting the rank orders of the values and calculating the non-parametric correlation yields a different set of lower results, as shown in Figure 6.57. In computing the τ statistics in this case, the objects i and j are each arrow point in the set, and the measurements x, y are each pairwise combination of radius ratio with roundness, aspect ratio, and convexity.

These are not the only nonparametric tests available, although they are the most commonly used and the ones found in many statistical analysis packages. A comprehensive discussion of these and other nonparametric analysis tools is available in Gibbons and Chakraborti (2011).

Of course, the variable on which the ranking or the cumulative distribution is performed does not have to be a measured parameter. The canonical axes derived in discriminant analysis, or the principal axes from PCA, which calculate a linear combination of multiple measured values, may also be used. The important thing to know about nonparametric tests is that they can be applied to a set of data whether or not the values have a normal distribution. If the data are Gaussian, the penalty for using a nonparametric test is simply one of efficiency. The same confidence level can be reached with fewer data points if the mean and variance properly represent the entire data set. But if it is not certain that the data are normally distributed, the use of a parametric test may be biased and lead to an incorrect conclusion, and the nonparametric approach is preferred.

Outliers

One of the tasks identified at the beginning of this chapter is the need to decide whether a particular observation is really different from the rest—in statistical terms an "outlier." If data are distributed normally, then a probability of significance can be calculated from the Gaussian distribution based on the distance from the mean. The usual procedure is to eliminate a suspect point and calculate the mean and standard deviation using the rest of the data. Then use the ratio of the difference between the mean and the value being tested to the standard deviation to calculate a probability that the difference could have arisen by random sampling.

As shown in Equation 6.19, the probability is calculated by integrating the Gaussian function with a limit given by the normalized distance from the mean (since there is no closed form expression for the integral, values are often taken from tables). In the equation, v is the value, and μ and σ are the mean and standard deviation, respectively, of the distribution as estimated from the remaining measurements:

$$p(x) = 1 - \frac{1}{2\pi} \int_{-\infty}^{x} e^{\frac{-x^2}{2}}$$

(6.19)

$$x = \frac{v - \mu}{\sigma}$$

This is not always a satisfactory solution if the mean and standard deviation are not known precisely. Similarly, if the actual distribution of values is known but not described by parametric equations, the probability of significance can be read directly as the fraction of the distribution that lies beyond the observed value. But in most cases the complete distribution is not available and can only be estimated from a limited number of prior observations.

As an example, Figure 6.58 shows several measured dimensionless ratios for the keys and coins from Figure 2.33. The formfactor values for the coins

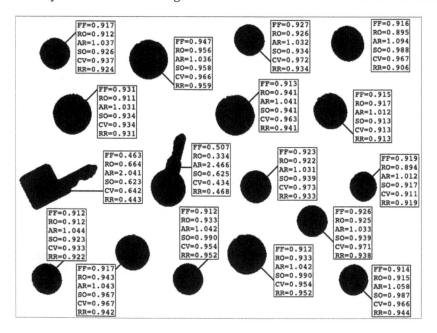

FIGURE 6.58
The keys and coins from Figure 2.33 labeled with the formfactor (FF), roundness (RO), aspect ratio (AR), solidity (SO), convexity (CV), and radius ratio (RR) as defined in Chapter 3.

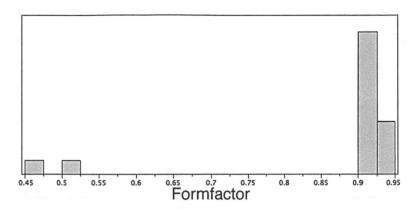

FIGURE 6.59
Distribution of formfactor values for the coins and keys in Figure 6.58.

TABLE 6.9

Effect of Removing Suspect Values One at a Time

	Mean	Std. Dev.	Difference/Std. Dev.	p
All data	0.8689	0.1450		
Remove (.507)	0.8915	0.1146	3.354	0.0004
Remove (.463)	0.8943	0.1037	4.159	0.00002
Remove both	0.9201	0.0096		

range from 0.912 to 0.947, whereas the keys have values of 0.463 and 0.507. The distribution of values is shown in Figure 6.59; the mean value and standard deviation of all 17 measurements are 0.869 ± 0.145.

Eliminating the values measured on the keys one at a time, recalculating the mean and standard deviation of the remaining measurements, and comparing the suspect value to the remaining distribution gives the results shown in Table 6.9. The value p is the probability that the measurement could have been obtained by random sampling from a normally distributed population. In each case, the probability that the measurement is part of the distribution is very small and indicates a probable outlier. Of course, with both values removed the standard deviation shrinks even more. Similar results are obtained with each of the dimensionless ratios shown for these features.

Another example of testing an outlier is shown in Figure 6.60. One of the butterflies (marked at the lower right) is *Heliconius melpomene* (postman); the other 14 are specimens of *Heliconius erato* (red passion flower). The coloration of the two species is similar. Measuring the silhouettes produces the data shown in Figure 6.61 and Table 6.10. The scatterplot suggests that the point marked may be an outlier. Calculating the mean and standard deviation for the other points and testing the possible outlier against them

FIGURE 6.60
Silhouettes of butterflies as discussed in the text. The asterisk marks the different species. (E. Keogh, 2007, *Proceedings of the 13th ACM International Conference on Knowledge Discovery and Data Mining*, Tutorial # 7.)

indicates that for each measurement the probability that the point belongs to the group is very low. Since these measurements are not completely independent (they share area, perimeter, and convex hull in various combinations), the p values cannot be multiplied together to obtain a combined probability.

For determining whether points in a regression analysis may be outliers, Cook's distance (Cook, 1977) measures the effect of deleting a given observation on the remainder of the data. An observation whose omission changes the shape of the distribution of the remaining points is considered more likely to be one that does not belong. Data points with large residuals or high leverage might distort the outcome and accuracy of a regression or other analysis.

The calculation for observation i proceeds by summing for all n observations the square of differences between the prediction Y of the value of the regression model for each observation j and the prediction $Y(i)$ from a refitted regression model from which observation i has been omitted, divided by the product of the number of fitted parameters f and the mean squared error (MSE) of the model.

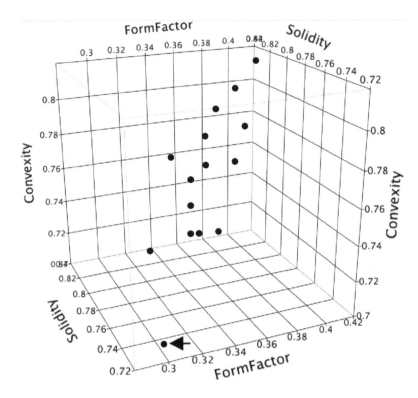

FIGURE 6.61
Scatterplot showing dimensionless ratio shape measurements for the silhouettes in Figure 6.60. The arrow marks the possible outlier as discussed in the text.

TABLE 6.10

Shape Measurements on the Butterflies in Figure 6.60

	Formfactor	Solidity	Convexity
Possible outlier	0.29902	0.72351	0.70818
Mean of the others	0.35579	0.78294	0.77060
Std. Dev. of the others	0.02602	0.02243	0.02415
Difference/Std. Dev.	2.1818	2.6496	2.5847
p	0.0292	0.0080	0.0098

$$D_i = \frac{\sum_{j=1}^{n}\left(\hat{Y}_j - \hat{Y}_j(i)\right)}{f \cdot MSE} \qquad (6.20)$$

Points with a large Cook's distance, often suggested as greater than 1.0, are considered to merit closer examination in the analysis. A more specific

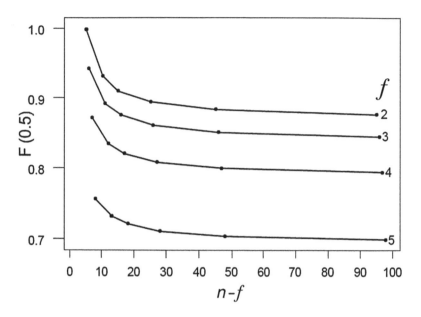

FIGURE 6.62
Median values of the *F*-statistics for various combinations of the number of parameters *f* and number of observations *n*.

test depends upon the fact that the values obey the *F*-distribution, and so the distance can be compared to the median value of the *F*-statistic, which corresponds to a 50% confidence, or placing the point at the edge of the 50% confidence region of the data set. Figure 6.62 shows these median values for various combinations of *f* and *n*. The recommended test value of 1.0 is usually a safe, conservative estimate.

Bayes' Rule

Bayesian inference allows estimating the degree of confidence in a hypothesis by combining multiple conditions. It is well suited to cases in which several different measures of shape (as well as size, color, etc.) are used. Because it is inherently nonparametric, it is directly applicable to measured probability distributions, but can be used with parametric distributions such as a Gaussian when appropriate.

The interpretation of the *p* values (e.g., 0.01, 0.05, etc.) as the probability that an observed result will exceed some limit if the data are actually drawn at random from a normal or other specified mathematical distribution is widely employed in statistical analysis of data. However, it is also possible to use a measured distribution of values determined from a training population to determine probabilities.

Symbolically, the probability of observing some result is written as $P(R|E)$, which is read as the probability (in the range 0...1) of result R given evidence E. Bayes' rule calculates this as shown in Equation 6.21:

$$P(R|E) = \frac{P(E|R) \cdot P(R)}{P(E|R) \cdot P(R) + P(E|\sim R) \cdot P(\sim R)} \tag{6.21}$$

where the symbol $\sim$ indicates "not," so that $P(\sim R)$ is the probability that R is not the result, equal to $1.0 - P(R)$. The equation states that the probability of obtaining the result, R (for example, identification of an object in a particular class), based a measurement, E, can be calculated if the probability, $P(E|R)$, of observing the measurement, E, for objects in that class is known, as well as the overall probability, $P(R)$, of encountering an object in that class. This is particularly helpful in the case of minor classes that are infrequently encountered and in cases in which the measurement values from different classes overlap.

Consider the following example: Figure 6.13 shows the height distributions of adult men and women in the United States. An individual's height is measured as 70 inches. The chart shows that 2% of women and 15% of men have that height. But in the group being measured, it is known that two-thirds are women. What is the probability that the individual is a man? Applying Equation 6.21 with these values gives

$$P(R|E) = \frac{P(70''|\text{man}) \cdot P(\text{man})}{P(70''|\text{man}) \cdot P(\text{man}) + P(70''|\text{woman}) \cdot P(\text{woman})}$$

$$P(\text{man}|70'') = \frac{0.15 \cdot 0.33}{0.15 \cdot 0.33 + 0.02 \cdot 0.67} = 0.787$$

for a result of 78.7%. If instead the opposite question is asked: "What is the probability that the person is a woman?" the denominator does not change, the numerator becomes $0.02 \cdot 0.67$, and the resulting conditional probability that the person is a woman is 21.3%.

In this example the height distributions of men and women are represented by graphs from which the probabilities were obtained. If parametric descriptions of the distributions were applicable, for example, Gaussian distributions each with a known mean and standard deviation, the probabilities could be calculated from that. But whether parametric or nonparametric representation is appropriate, it is necessary to know what the probability distribution is, which means that the populations must be sampled adequately and measured beforehand.

The rules for combining multiple probabilities are handled as shown in Equation 6.22. The probability of obtaining the result, R, based on two independent measurements, E_1 and E_2, is calculated as

$$P\left(R\,|\,E_1 \cap E_2\right) = \frac{\Lambda_1 \cdot \Lambda_2 \cdot P\left(R\right)}{\Lambda_1 \cdot \Lambda_2 \cdot P\left(R\right) + P\left(\sim R\right)} \qquad (6.22)$$

where the symbol $\cap$ denotes AND or the intersection of two conditions, and Λ is the ratio of the probabilities of observing a measurement for objects in the class and for objects not in the class, calculated as shown in Equation 6.23:

$$\Lambda_E = \frac{P\left(E\,|\,R\right)}{P\left(E\,|\sim R\right)} \qquad (6.23)$$

Additional pieces of information, both positive and negative, can be similarly combined. Further information about the use of Bayes' rule can be found in many statistics texts.

Neural Nets

Neural nets provide another approach to both supervised and unsupervised learning. These may be implemented as actual circuits, or simulated by conventional computers, or even treated as a special case of regression, but the principles remain the same. A set of input values (which in some cases are simply the pixel brightness values directly from the camera but more commonly are the various measurements described in previous chapters) is fed to the first layer of the net, and the final output is examined. If the output values are "good" (i.e., if the circuits have succeeded in recognizing an object or differentiating it from others) then the connections that led to the result are strengthened and vice versa. Eventually the circuit "learns" to produce useful results.

The advantage of this approach is that because of the parallel processing it is very fast. The problem is that it is very difficult to understand what the connection values "mean" in terms of the recognition process. The "knowledge" in the neural net is distributed over many individual weights. Also, the systems are very sensitive to the quality and order of the input values used for training, and may zero in on a result that fits the early data and produce conclusions that are unreliable.

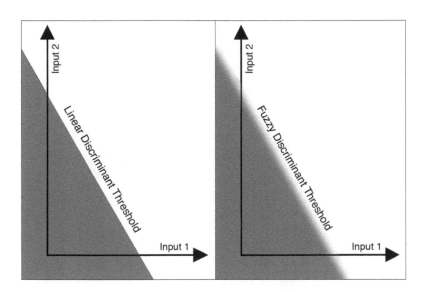

FIGURE 6.63
A linear discriminant threshold is a line in two dimensions, a plane in higher dimensions, separating results above and below the threshold, either sharply or with a fuzzy boundary.

The threshold logic unit (TLU) shown in Figure 1.15 is the basic unit in a neural net. By multiplying each of the input values (e.g., measurements of size, shape, position, and/or color) by appropriate weights (or using the outputs from other logic units, which have in turn responded to those original values), and comparing the result to a predetermined threshold, the TLU makes a decision. The weighting is identical to the linear discriminant process in which a linear combination of input variables is used to form a canonical variable. The threshold value is equivalent to separating the n-dimensional space defined by the input variables into two parts with a hyperplane.

For the case of two input variables, this decision threshold becomes a line as shown in Figure 6.63. On one side of the line is the result in which the linear combination of variables is greater than the threshold; on the other side it is lower. The classic operation of the logic unit uses a discrete threshold that makes an absolute decision, but it is also possible to have a "fuzzy" boundary that produces a probability varying from one result to the other gradually. This may be either linear or, more often, based on a sigmoidal curve or hyperbolic tangent curve, or an actual probability based on a measured distribution as discussed later.

As an example of a single-layer neural net, Figure 6.64 shows the ability to identify each of the arrow point types from Figure 4.52, using only the conventional dimensionless ratio shape factors. The method uses a randomly selected two-thirds of the points to train the net consisting of 3 TLUs, with the remaining points for validation. There are no errors. Each node uses a hyperbolic tangent function to obtain a steep response curve, with weights

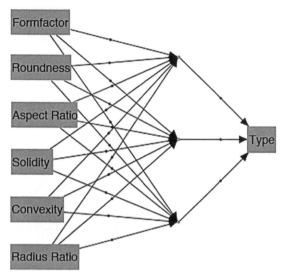

	Node 1	Node 2	Node 3
	TanH	TanH	TanH
Constant	-136.816	-92.09	-15.058
Formfactor	-219.129	3.618	32.384
Roundness	61.893	-15.122	8.196
Aspect Ratio	-9.476	-3.171	6.084
Solidity	20.415	22.479	-21.669
Convexity	219.874	95.177	-23.785
Radius Ratio	53.856	-9.954	56.631

FIGURE 6.64
Schematic diagram of a one-layer neural net that identifies the arrow point types shown in Figure 4.52 based on dimensionless ratio measurements.

for the input measurements that are shown in the table. There is little that can be learned by visual examination of the weights, except that they are large and thus represent very strong, opposing contributions from the various shape factors. This is not generally considered to be an encouraging sign when attempting to create a robust identification tool. Another neural net using harmonic coefficients 4 through 10 from the Fourier analysis of the profile is also 100% successful in identifying the point types.

There are not many interesting cases that can be handled by a single TLU. In a typical neural net there are many units, each of which can be visualized as a set of linear discriminant thresholds. A single layer of logic units cannot handle some kinds of problems, but with two layers, the second layer receiving inputs from each of the units in the first layer, it is possible to define any convex region in the *n*-dimensional space defined by the input variables (with additional layers, nonconvex regions can be handled). Neural nets are

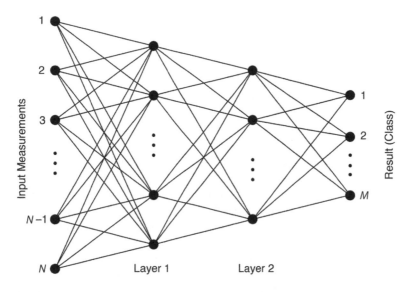

FIGURE 6.65
Schematic diagram of a multilayer neural net. The inputs are measured values for size, shape, position, or color. The result is a classification of the object.

intrinsically nonparametric. Figure 6.65 shows a multilayer neural net schematically, which uses N input values to generate M results (M is often just one: yes or no; this is particularly the case in quality control situations).

With enough units in the network and enough layers, the region corresponding to each class of objects can be adjusted to fit a region in much the same way (and with the same consequence) as the ellipsoids and boxes shown for other approaches but with more complex nonparametric shapes. The problem, of course, is how to determine the weights and thresholds that produce the desired results.

Several procedures have been developed for training a neural net, but the widely used backpropagation algorithm may be taken as illustrative. Initially the weights are set to random values. As each set of inputs corresponding to a member of the training population is presented to the net, the output values are compared to the desired response. The weights are adjusted so that the result moves toward the desired outcome. Then another set of input values is used and the process repeated.

The same training values are presented repeatedly as the weights are adjusted and the outputs approach the desired results. When the results are within some defined mean error or the weights have stabilized, the training is complete. This can be a lengthy procedure, and there are modifications that speed up convergence, primarily by modifying the amounts by which the weights are adjusted in each cycle. In some implementations, the repetitions of the input values are perturbed by adding noise to make the final result more robust.

The advantage of the neural net approach mentioned earlier is the speed with which a decision can be reached once the training is complete, because of the densely parallel arrangement of TLUs. However, in many cases the net is simulated by a conventional computer, which does not yield the speed advantage. One of the drawbacks, also previously mentioned, is that the weights are very difficult to interpret, particularly in the inner layer(s).

The other major difficulty is that while the system may be very good at identifying objects similar to ones it was trained on, the good results may not be due to reasons that correspond to the actual major or important characteristics but to some chance combinations of values present in the training population. This is particularly a problem if the training set is too small, and is called overtraining. Minor deviations from the expected range of objects or encountering a new type of object that differs slightly may produce nonsense results. This is described as an inability to generalize the results.

A modification of the traditional neural net introduces fuzzy logic to the threshold. Instead of being a yes-or-no decision, each unit produces a result that is a probability. For instance, the rule "If X can fly, there is a 70% chance that it is a bird; if X cannot fly there is a 5% chance that it is a bird," accommodates bats, bees, and butterflies, which fly but are not birds, and penguins and ostriches, which are birds but do not fly.

In the traditional neural net, it is only necessary to find some successful path through the network to reach a conclusion. This can be done rather quickly, even it if requires some backtracking. Searching algorithms that work in either direction, from initial data to conclusions, or from possible conclusions back toward the required inputs, or both, are used.

In a fuzzy network, it is necessary to examine all possible paths to see which produces the highest probability outcome (the combination of all the probabilities along the path), which takes longer. Fuzzy neural networks also take longer to train. But they can handle continuous variables such hot versus cold or short versus tall. Would 5 feet 4 inches be considered short or tall? How about 5 feet 10 inches? How about 6 feet 2 inches? Rather than arbitrarily setting each of these to either short or tall, a probability can be assigned that the value would be described as short or tall.

Figure 6.66 illustrates this approach. The specific values are highly context dependent, and would be different for men or women, for adults or children, and for various ethnic groups. The graphs establish probabilities that a particular height might be assigned the label of "short" or "tall." Notice that the two values may not sum to 100%. In other words, an individual 68 inches tall would probably not be called short (10% probability) or tall (30% probability), and would consequently be unlabeled. In some situations this is not appropriate, only two outcomes are possible, and only a single line is needed on the graph.

The graphs shown in Figure 6.66 are arbitrary, intended only as an illustration. In real cases, the probabilities come from external knowledge or judgment. However, in many real instances, measurement of representative

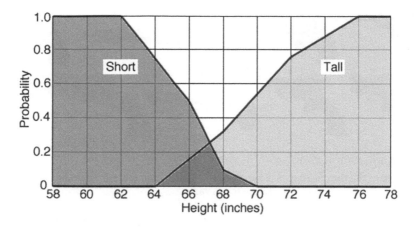

FIGURE 6.66
Illustration of fuzzy probabilities as described in the text.

training populations can establish curves that are meaningful. These are just the distributions of measured values.

This approach is well suited to continuous ranges of shapes, such as deciding whether to characterize a snowflake as "dendritic" or "platelike." Fuzzy logic works well with ideas such as "How much like a circle is it?" that formed the basis of several of the dimensionless ratio measurements in Chapter 3, such as formfactor and roundness. The problem of choosing a measurement value that properly describes the exact way in which the feature shapes depart from being a circle remains, of course.

Comprehensive reviews of fuzzy-logic neural nets can be found in Zadeh, 1965; Negoita & Ralescu, 1975, 1987; Sanchez & Zadeh, 1987; Zimmerman, 1987; Bishop, 1996, 2007. Fuzzy logic can also be described as equivalent to a regression operation such as the linear discriminant classification approach shown previously.

Syntactical Analysis

"Form follows function" applies to manufactured items as well as natural ones. Handguns are a useful example. There are three different modern types: automatic, revolver, and derringer, plus a wide range of antiques. Derringers have a separate barrel for each cartridge (sometimes two, four, or even more) and are among the smallest in size. Revolvers have a separate chamber for each cartridge but share a single barrel. Automatics load cartridges from a clip, usually in the grip, into a single chamber. Antique handguns include flintlocks and percussion designs and are muzzle loaded.

FIGURE 6.67
A selection of the various shapes of handguns (from left): automatics, derringers, antique flint-lock and percussion, and revolvers. The scales vary.

The different design approaches dictate to some extent the physical form of the weapon, but within each class there are major differences because of individual designer's decisions and the somewhat different uses for which each design may be intended. Figure 6.67 illustrates a few of the wide range of shapes that result.

Can the four basic types of guns (automatic, derringer, revolver, and antique) be distinguished by shape alone? A collection of 109 images illustrating the variations in shape was measured to determine dimensionless ratios, moments, and harmonic coefficients. As shown in Figure 6.68 and Table 6.11 the classes can be distinguished by linear discriminant analysis and by a neural net with three hidden nodes, but it requires a minimum of six different measurements in each case, there is still considerable scatter, and some individual errors in identification occur.

Human vision can readily distinguish the different shapes. This is an example of syntactical analysis, described in Figure 6.1. A few characteristic primitives are identified with each type of weapon, although exceptions can be found for each. For example, the shape of the grip is quite different for most revolvers as compared to most automatics, because in most designs the grip of the automatic pistol must house the clip containing the cartridges.

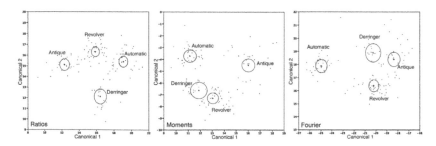

FIGURE 6.68
Linear discriminant results for the handguns using dimensionless ratios, invariant moments, and harmonic Fourier coefficients (only the two most significant canonical axes are shown).

TABLE 6.11

Results of the Linear Discriminant Analysis (LDA) and Neural Net (NN) Classification

Measurements	Variables Used	Number Misidentified
Dimensionless ratios	Formfactor, Roundness, Aspect Ratio, Solidity, Convexity, Radius Ratio	LDA = 3; NN = 9
Moments	Invariant Moments 1, 2, 3, 6, and 7; Affine Invariant Moment 1	LDA = 10; NN = 7
Harmonic analysis	Fourier Coefficients 1, 2, 3, 7, 9, and 14	LDA = 12; NN = 12

Many revolvers have a long, slender barrel and a curved grip. Most derringers do not have a trigger guard and usually have a rounded grip. The overall curved shape and the protruding hammer on an antique handgun is distinctive and so on. Although exceptions mean that no single rule is absolute, isolating a few components of the overall shape produces quick recognition.

The presence of the syntactical primitives can be combined by a threshold logic unit (a "grandmother cell") that decides based on the presence or absence of enough clues. This is part of the "grammar" that ties together the individual syntactical components, along with the spatial relationships between them. It is not clear that computer recognition can be readily adapted to this type of logic because of the difficulty in isolating the primitive elements (slender barrel, curved grip, trigger guard, protruding hammer, etc.). In some cases, cross-correlation can be used with a set of primitive shapes to determine whether they are present in the target. But when the sizes and details of the primitive shapes vary greatly (as for the handgun examples) this becomes impractical.

Medical image analysis almost universally depends upon trained technicians or doctors to examine the images and detect the signs of disease. Some simple applications such as locating metal fillings in dental X-rays could be automated based on the high density of the metal, but this can be done so quickly by a human, who is also likely to be examining the images for other signs of decay or periodontal disease, that there is no strong incentive toward automation. Detecting asbestosis or lung cancer in chest X-rays, or breast cancer in mammograms, or tumors in MRI brain scans may be aided by some of the image processing operations shown in Chapter 2 to improve the visibility of details, but ultimately relies upon human visual examination and diagnosis.

One success for computer-based image measurements is the screening of Pap smears for signs of cervical cancer (Rutenberg & Hall, 2001). Yet even this does not use the computer for diagnosis. Instead, the slide is examined using a digital camera on a microscope to view each cell, and a calculation is made based on measurements of size, shape, density, and internal texture of the nucleus. Stored images of a small number (typically 20) of the cells with the highest calculated probability of being "abnormal" are presented for review by a specialist. The reasoning is that if the most abnormal cells are not cancerous, probably (!) there are no cancerous or precancerous cells present, and no further action is indicated. If the specialist deems it necessary, the stored images of each cell on the entire slide can be reviewed by a pathologist relying on human vision.

The insistence upon only allowing humans to perform diagnosis is understandable because of the high cost and legal liability of errors, as well as the variable cues that must be interpreted. This is also an example of syntactical analysis, in which the syntactical primitives are taught by example or learned by experience. The difficulties faced by this approach include establishing robust criteria that can be communicated effectively and defining the confidence limits of the identification.

The problems of false positives and false negatives are common in medical diagnosis. Boslaugh and Watters (2008) show an example of a test with 95% sensitivity (95% effective in detecting a disease in those who have it) and 99% specificity (only 1% reporting of disease in those who do not). Both of these values are extremely high. But when applied broadly to screen for a disease that affects 1% of the population, Bayesian statistics indicate that the overall false positive rate will be nearly 50%, with obvious negative consequences.

Correlations

Interpreting the meaning and implications of a diagnosis is just one example of the need to correlate the classification or recognition of an object with its further significance.

Many different shape measurements have been used to correlate with a wide range of other factors. While recognizing that correlation does not prove causality, it is still possible to infer relationships between shape and various environmental factors such as temperature or nutrients that might affect shape. Likewise, shape can influence behavior, such as the flow or settling of particulates.

Examining graphs for correlations, and seeking explanations when they are found, is a common exercise. Figure 6.69 shows profiles of a few cars covering the last century. These are not a statistically random representation of the thousands of different models, but those vehicles intended for other uses, such as trucks, minivans, and SUVs are not included. It is obvious that the shapes of cars have changed over time. Reasons for the changes include the vagaries of style as well as efforts to become more "streamlined" and aerodynamically efficient.

The graphs in Figure 6.70 show that aspect ratio follows a generally increasing trend, reaching a climax in the 1980s with long, almost boatlike cars, after which some reduction has taken place. That may be primarily a stylistic change. The increase in convexity may reflect a general departure from

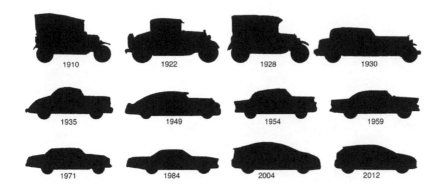

FIGURE 6.69
Silhouettes of several cars (scales vary).

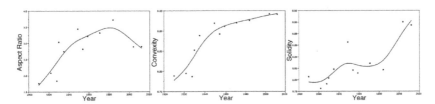

FIGURE 6.70
Plots of the aspect ratio, convexity, and solidity of the car silhouettes in Figure 6.69.

the boxlike shape inherited from the horse-and-carriage toward something more aerodynamic. The graph of solidity does not suggest any particular trend. No interpretation of possible causality can be derived from the plots themselves; finding a correlation does raise a question whose answer must come from independent sources of information.

In many fields of scientific study, correlations of shape measurements of one kind or another have been made with factors that either influence the shape, seem to be the consequence of it, or simply track changes over time. It is an all-too-common error, however, to confuse correlation with causation. Correlations involving shape are not immune to this mistake.

Physical Examples

In some cases, the causality behind an observed correlation or trend in shape can be deduced. Figure 6.71 shows cumin seeds, and the results of grinding them in a mortar and pestle. The ratio of length along the skeleton to the width ("elongation" as defined in Chapter 3) shows a fairly wide range of values but a clear overall trend (Figure 6.72), in which the values initially increase and then fall. The images show that this is due to the breaking up of the seeds into long fibers, which are then progressively broken up into short pieces.

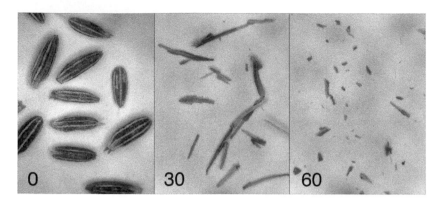

FIGURE 6.71
Cumin seeds, and the results of grinding in a mortar and pestle for the times indicated.

Many other correlations of properties with shape have been reported. Imasogie and Wendt (2004) measured several 3D dimensionless ratios for graphite nodules in cast irons and found that the ratio of the projected area of the particles to the volume of the minimum bounding sphere ($Area^3/36\pi \cdot Volume^2$) correlates with the yield strength, as shown in Figure 6.73.

Cold rolling of steel produces thin sheets (for instance, used in automobiles and appliances). The process results in the elongation of the grains, as shown in Figure 6.74. Measuring the grain shape by placing a grid of lines on the image in various orientations and determining the mean intercept length produces the plots shown. The ratio of the maximum (longitudinal) to minimum (transverse) value correlates with percent reduction in thickness, as shown in Figure 6.75.

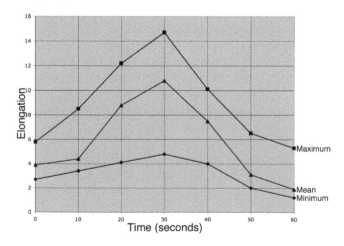

FIGURE 6.72
Elongation values for the cumin seeds and fragments produced by grinding.

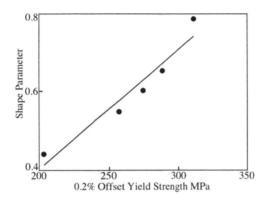

FIGURE 6.73
Correlation between yield strength for cast iron and calculated 3D shape parameter for graphite nodules. (B. I. Imasogie, U. Wendt, 2004, *Journal of Minerals and Materials Characterization and Engineering* 3(1):1–12.)

Mikli et al. (2001) report that for tungsten carbide–cobalt hard metal powders, the circularity of particle cross-sections increases with milling time. Their measured value, called the "spike-parameter quadratic fit," is the cosine of the average apex angle for protrusions outside a circle with equal area and aligned to the particle centroid. Cho et al. (2006) use sphericity and roundness values visually estimated by comparison to the chart in Figure 5.16 to correlate shape with the packing density and mechanical properties of sandy soils. Their data confirm that a decrease in sphericity or roundness produces an increase in void fraction, a decrease in stiffness, and an increase in compressibility.

Hentschel and Page (2003) measure a variety of metal powders (Figure 6.76) produced by methods such as electrolytic deposition, water atomization, and gas atomization, and report that two shape descriptors—aspect ratio and the square root of formfactor—are sufficient to distinguish them. Miyamoto et al. (2002) correlate the linear expansion of particle board construction

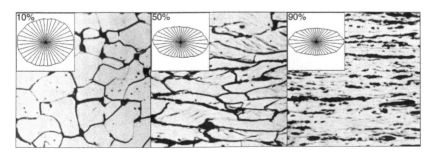

FIGURE 6.74
Microscope images of the grain structure in cold rolled low-carbon steel for reductions in thickness of 10%, 50%, and 90%, with plots of the mean intercept length as a function of direction.

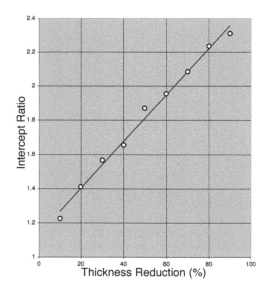

FIGURE 6.75
Correlation between the reduction in thickness of the steel and the mean intercept ratio measured on the grains.

panels with the aspect ratio of the wood particles used in their fabrication (Figure 6.77).

Correlations between fracture properties such as energy absorption and fractal surface geometry have been reported for many different materials, as discussed in Chapter 5. For example, Issa et al. (2003) show a linear correlation between the fractal surface geometry of fracture surfaces in concrete (measured using a modified slit-island method) and the fracture toughness of the material (Figure 6.78). The tougher the material, the higher the fractal dimension.

Hirata (1989) shows that the fractal dimension of the fault system in Japan varies with distance from the central part of the Japan arc (Figure 6.79). Dauskardt et al. (1990) show that in spite of different modes of fracture and different macroscopic appearances, the fractal dimension of fracture surfaces in mild and low-alloy steels are similar and related to microstructural features.

FIGURE 6.76
Copper powders with varying shapes. (M. W. Hentschel, N. W. Page, 2003, *Particle and Particle Systems Characterization* 20:25–38.)

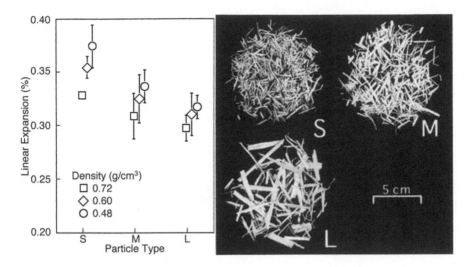

FIGURE 6.77
Correlation of linear expansion with particle shape. (K. Miyamoto et al., 2002, *Journal of Wood Science* 48:185–190.)

Cherepanov et al. (1995) show a correlation for a wide range of ductile materials (Figure 6.80) between the ratio of hardness to yield stress and a parameter that can be interpreted as a hydrodynamic dissipation of energy at the tip of an advancing crack, based on a fractal interpretation of fracture mechanics.

Some papers that discuss a role for particle shape on behavior do not include quantitative shape measurements. Moore and Swanson (1983) report that the effect of abrasive particle shape on wear rate is observed experimentally, but only a qualitative description of the particle shapes is given, and the paper states that no theory incorporating factors such as shape, hardness,

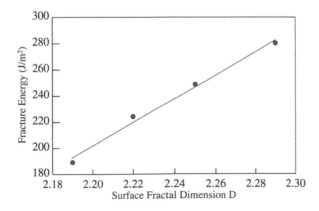

FIGURE 6.78
Correlation between fracture energy and surface fractal dimension in concrete. (M. A. Issa et al., 2003, *Engineering Fracture Mechanics* 70:125–137.)

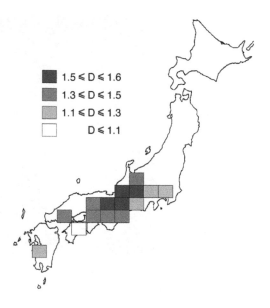

FIGURE 6.79
Variation in fractal dimension of the fault system in Japan with distance from the center of the Japan arc. (T. Hirata, 1989, *Pure and Applied Geophysics* 131(1/2):157–170.)

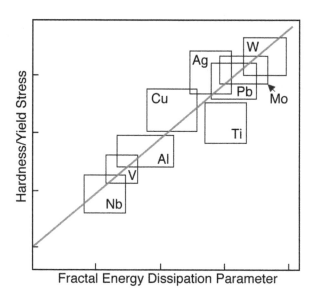

FIGURE 6.80
Correlation between the ratio of hardness (Vickers) to yield stress and an energy dissipation parameter derived from a fractal interpretation of fracture mechanics. (G. P. Cherepanov et al., 1995, *Engineering Fracture Mechanics* 51(6):997–1033.)

debris removal, or load has been developed. Wong and Pilpei (1990) discuss the effect of particle shape on the compression of pharmaceutical powders to form pills but give only qualitative shape information.

However, models relating shape to properties also exist. A comprehensive review of the effect of particle shape on the elastic modulus, thermal expansion coefficient, stress concentration factor, and creep compliance of filled polymers is presented by Chow (1980). Also, Dwyer and Dandy (1990) calculate the effect of particle shape (the axial ratios of ellipsoids) on the Reynolds number in flows, which describes friction drag and heat transfer. Göktürk et al. (1993) use ellipsoids as a model to compare the magnetic permeability of composites containing flake, filament, and equiaxed nickel powders.

In machining, the ductility of the metal and speed of the tool correlate with the shape of the chips produced (Komanduri, 1982; Shaw & Vyas, 1993; Elbastawi et al., 1996; Byrne et al., 2003). Grain shape in metals correlates with the amount, rate, and temperature of deformation (Roberts 1978, 1983; Tanaka, 1981). The hardenability and grain structure of steels can be predicted based on their composition, such as the percentages of elements such as chromium, nickel, manganese, and molybdenum (Kirkaldy & Venugopalan, 1984; Dobrzanski & Sitek, 1998). For some of these correlations, underlying physical reasoning has been established.

Biological Examples

Bertalanffy (1957) wrote, "It is a truism in engineering that any machine requires changes in proportion to remain functional if it is built in different sizes." And, of course, the same thing holds for organic structures, whether bone, feather, or woody stem, and long predates any manufactured object. Correlations involving shape are consequently much studied in the natural sciences, and particularly in studies involving humans. Many relationships between the shapes of bones, brains, muscles, and so forth in ontogeny and phylogeny are shown in Gould (1966, 1977), Alberch et al. (1979), and Cheverud et al. (1983), continuing a tradition that goes back to D'Arcy Thompson (1917).

Similar studies of other animals or families of animals have also been undertaken. The shape of heads has been compared for old world monkeys (Jungers et al., 1995) and the skull shape in frogs has been correlated with their diet (Emerson, 1985). At a much finer scale, the shape of red blood cells has been related to metabolism and adenosine triphosphate (ATP) concentration (Torres et al., 1998), and molecular surface shape has been correlated with drug absorption (Palm et al., 1996). The shape of tumors has been correlated with their metastatic potential (Partin et al., 1989; Delfino et al., 1997).

In an entire book devoted to allometric relationships, McMahon and Bonner (1983) document variations in size and proportions for stringed instruments, a relationship between tree height and cross-section (and a corresponding one for slime molds), between engine horsepower and the

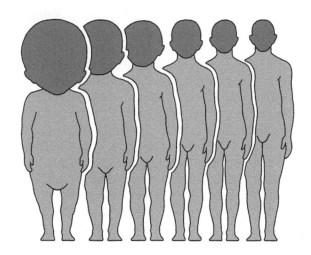

FIGURE 6.81
Changing proportions in the human body with age. (After T. A. McMahon, J. T. Bonner, 1983, *On Size and Life*, W. H. Freeman, New York.)

cylinder displacement and stroke ratio, and so on. They show relationships between bone length and cross-section as well as correlations of brain size and shape to body size, and document the changing proportions of the human body as it ages (Figure 6.81).

A particularly novel use of shape measurements is the study of the changing shape of "idealized" women (e.g., Playboy centerfolds), which has been correlated with dates (Cusumano & Thompson, 1997; Garner et al., 1980).

The nutrient qualities of rice grains (mineral content) have been correlated with aspect ratio (Zhang et al., 2002), and seed longevity in the soil has been correlated with seed shape (Bekker et al., 1998). The shape of fossil leaves has been correlated with paleoclimate factors such as temperature (Royer et al., 2005; Krieger et al., 2007). For flows through porous media, the shape of capillaries has been correlated with resistance to flow (Patzek & Kristensen, 2001) and particle shape has been correlated with packing density (Dixon, 1988). The shape of the sea bottom has been correlated with predator–prey ratios for fish (Rose & Leggett, 1990).

Figure 6.82 shows a plot of data from Sinnot (1936) for bottle gourds, showing a trend of aspect ratio (length/breadth) versus length. The plot of the original values in Figure 6.82a shows evidence of a trend, but it appears to be nonlinear ($R^2 = 0.59$). Rather than trying to fit a polynomial or other nonlinear function to the data, plotting the rank orders of the data produces a linear trend ($R^2 = 0.72$). The value of R^2 is a measure of the degree to which a set of data points is fit by the straight line. Fitting straight lines to data is performed using Equation 6.24, in which m and b are the slope and intercept, x is the independent variable, and y the dependent variable.

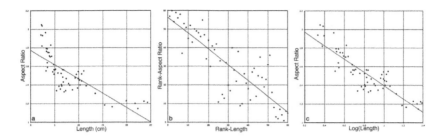

FIGURE 6.82
Correlation plots for aspect ratio versus length of bottle gourds.

The correlation plots in Figure 3.54 and Figure 3.55 provide additional examples.

$$y = m \cdot x + b$$

$$m = \frac{n \cdot \sum_{i=1}^{n} x_i \cdot y_i - \sum_{i=1}^{n} x_i \cdot \sum_{i=1}^{n} y_i}{n \cdot \sum_{i=1}^{n} x_i^2 - \left(\sum_{i=1}^{n} x_i\right)^2}$$

$$b = \frac{\sum_{i=1}^{n} y_i - m \cdot \sum_{i=1}^{n} x_i}{n} \tag{6.24}$$

$$R = \frac{n \cdot \sum_{i=1}^{n} x_i \cdot y_i - \sum_{i=1}^{n} x_i \cdot \sum_{i=1}^{n} y_i}{\sqrt{n \cdot \sum_{i=1}^{n} x_i^2 - \left(\sum_{i=1}^{n} x_i\right)^2} \cdot \sqrt{n \cdot \sum_{i=1}^{n} y_i^2 - \left(\sum_{i=1}^{n} y_i\right)^2}}$$

The graph in Figure 6.83 shows the probability that a given value of R can be exceeded, as a function of the number of data points, when the data are selected from random, uncorrelated populations. For the 60 points in the graph in Figure 6.82b, the R value (0.85) far exceeds the value of 0.4 that corresponds to a 99% confidence limit. Another approach to dealing with nonlinear relationships is to find a functional form that linearizes the data. Figure 6.82c shows a plot of aspect ratio versus the logarithm of length, which improves the R^2 value to 0.83 ($R = 0.91$).

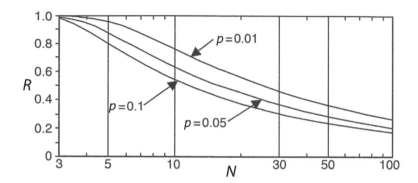

FIGURE 6.83
Probability of exceeding an *R* value for random data.

Example: Animal Cookies

As a summary and demonstration of the use of several of the statistical tests described here and an illustration of the process of deciding which method for measurement and classification to use, a large jar of animal cookies was purchased and the contents sorted. Broken cookies were eliminated, leaving 654 for analysis. The human sorting of the shapes relied on a few syntactical characteristics, such as the elephant's trunk, the camel's hump, and the cat's tail. Some of the shapes were not so easy to distinguish and required some resorting. Since the detail in the cookies was somewhat vague, we used the services of our children and grandchildren to decide on the identities of each of the 13 groups, and those names are used here without regard to what the cookie producer thinks the "correct" names should be for the animals represented.

The cookies were placed on a flatbed scanner and images obtained at a resolution of 600 pixels per inch. Since the scanner lid was open, the background of the images was dark and the cookie shapes were easily thresholded without any need for image processing. Figure 6.84 shows examples of the various cookie shapes (the reader can judge the identifying names), but it is not claimed that these particular cookies are "representative" or "average" as the shapes within each class varied considerably.

The first set of measurements utilizes the dimensionless ratios described in Chapter 3. Examination of the distributions and descriptive statistical parameters for these measurements shows that while each pair of classes can be distinguished using a selected dimensionless ratio, no general separation of the classes of shapes is provided by any single measurement parameter. Figure 6.85 shows the distributions of values for the four parameters that proved to be the most useful: roundness, solidity, convexity, and form-factor. For the readers' convenience, the definitions of these dimensionless

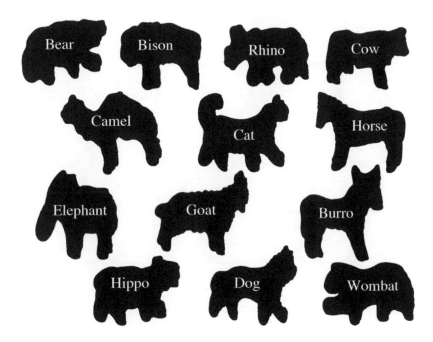

FIGURE 6.84
Thresholded binary images of cookies from each of the groups.

ratios are repeated in Equation 6.25. Statistical summaries for the roundness measurements are collected in Table 6.12.

$$Roundness = \frac{4 \cdot Area}{\pi \cdot Max\ Dimension}$$

$$Solidity = \frac{Area}{Convex\ Hull\ Area}$$

$$Convexity = \frac{Perimeter}{Convex\ Hull\ Perimeter}$$

$$Formfactor = \frac{4\pi \cdot Area}{Perimeter^2}$$

(6.25)

Using the roundness data as an example, a Student's *t*-test on two groups of measurements (bison and wombat) yields a *t*-statistic of 1.442 with 87 degrees of freedom using Equation 6.7. This corresponds to a 92% probability that the two groups are different (from Figure 6.49).

However, comparison of the skew and kurtosis values in Table 6.12 to the limits shown in Figure 6.47 indicates that none of the distributions should

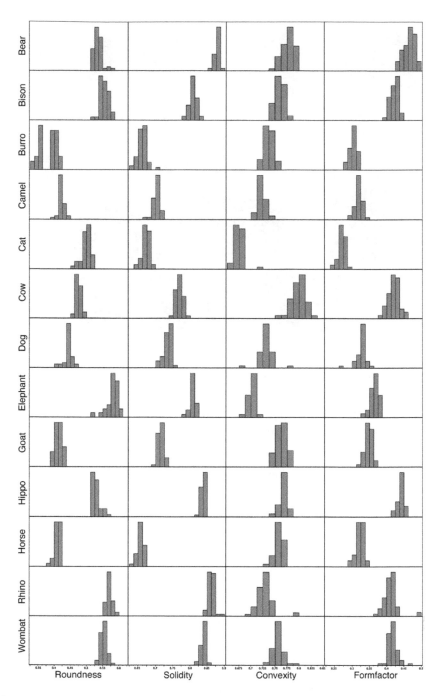

FIGURE 6.85

Distributions of measured dimensionless ratios for each shape class.

TABLE 6.12

Summary Descriptive Statistics for Roundness Measurements

Class	Number	Mean	Std. Dev.	Skew	Kurtosis
Bear	63	0.5351	0.0125	1.0817	2.2141
Bison	48	0.5546	0.0118	−0.2007	−0.1448
Burro	60	0.3822	0.0269	−0.3546	1.5032
Camel	44	0.4216	0.0092	−0.3267	1.8123
Cat	48	0.4979	0.0125	−1.1357	1.3636
Cow	45	0.4749	0.0082	0.4528	−0.1090
Dog	45	0.4435	0.0127	−1.2960	3.2285
Elephant	29	0.5804	0.0179	−1.4786	3.3663
Goat	50	0.4132	0.0102	−0.4656	−0.0425
Hippo	51	0.5292	0.0117	1.4497	1.6909
Horse	38	0.4103	0.0092	−0.9020	0.8872
Rhino	67	0.5695	0.0083	0.6351	−0.1045
Wombat	66	0.5528	0.0093	0.2226	−0.2048

be treated as a Gaussian. Instead of a parametric test such as the *t*-test, the Kolmogorov–Smirnov test can be applied. The cumulative plots of the bison and wombat distributions (Figure 6.86) show that the greatest difference is 0.138. For 95% confidence, the S value from Equation 6.13 is 0.288, and even for 90% confidence it is 0.202, so based on this test using this parameter (roundness) the two classes are not distinguishable.

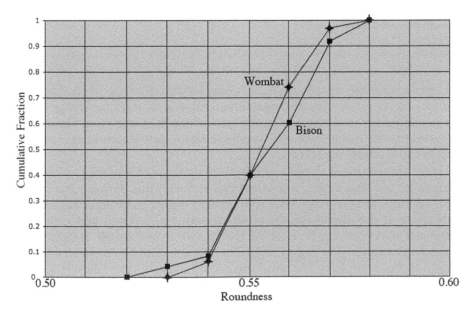

FIGURE 6.86
Cumulative plots of the roundness values for the bison and wombat shapes.

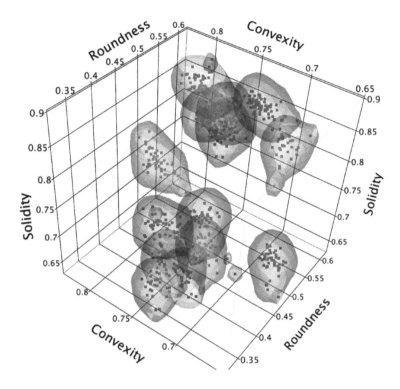

FIGURE 6.87
(See color insert.) Scatterplot of the measured values, color coded by class with nonparametric surfaces surrounding each cluster.

Furthermore, the distributions are considerably overlapped so that individual objects cannot be distinguished, and this is true for many of the classes and measured values. A scatterplot of the data, shown in Figure 6.87, suggests that by using several of the shape factors in combination, it may be possible to separate the classes.

Linear discriminant analysis is based on analysis of variance, which is a parametric test that ideally would not be used with this data set. But it does offer an efficient way to see how well these groups can be separated. Using only three of the dimensionless ratios: roundness, solidity, and convexity, canonical variables can be derived that perform fairly well at separating the groups as shown in Figure 6.88. The relationships for the variables are shown in Equation 6.26. The resulting model correctly identifies 94% of the individual cookies, with 42 errors. Adding formfactor as part of the analysis further improves the error rate to 16 individual errors out of the 654 cookie shapes. If all 12 of the dimensionless ratios shown in Equations 3.2, 3.3, and 3.4 are used, only 5 individual objects are misclassified.

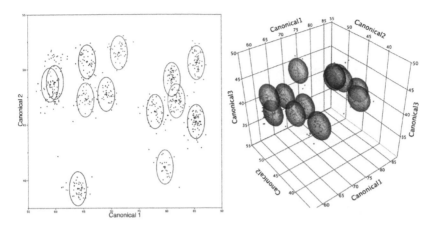

FIGURE 6.88
(See color insert.) Linear discriminant results using three dimensionless ratios.

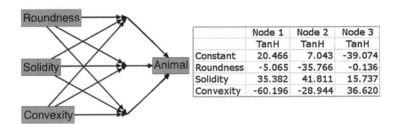

| | Node 1 | Node 2 | Node 3 |
	TanH	TanH	TanH
Constant	20.466	7.043	-39.074
Roundness	-5.065	-35.766	-0.136
Solidity	35.382	41.811	15.737
Convexity	-60.196	-28.944	36.620

FIGURE 6.89
Single-layer neural net with three inputs and the weights for each node.

$$Canonical\ 1 = +9.212 \cdot Roundness +116.98 \cdot Solidity -28.537 \cdot Convexity$$
$$Canonical\ 2 = -51.632 \cdot Roundness +32.871 \cdot Solidity +63.578 \cdot Convexity \quad (6.26)$$
$$Canonical\ 3 = +61.330 \cdot Roundness -54.806 \cdot Solidity +73.345 \cdot Convexity$$

A somewhat different approach, which is nonparametric, uses a neural net to place the decision surfaces between the various groups. A single-layer neural net trained using a randomized two-thirds of the observations and validated using the remaining third, and based on just the roundness, solidity, and convexity, is shown in Figure 6.89 The surfaces in the three-dimensional parameter space are shown as a set of regions in Figure 6.90. The model produces a total of 47 individual misclassifications out of 654 shapes, or a 93% success rate. Adding formfactor reduces the number misclassified to 18, and using all 12 of the dimensionless ratios reduces the error rate to 9 misclassified objects (1.4% error rate).

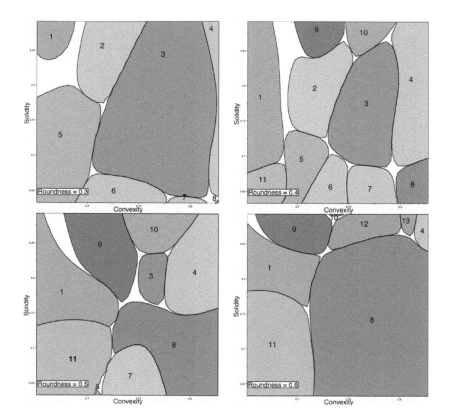

FIGURE 6.90
(See color insert.) Contours for solidity versus convexity at different levels of roundness, showing the decision boundaries generated by the neural net model. Numbers identify the same region in the various slices.

Additional Measurements

All of the analysis results shown used dimensionless ratios as measures of shape. Although this is frequently a first choice because of the ease of measurement and availability of the procedures in many software packages, it is also possible, of course, to perform other measurements. Harmonic analysis and invariant moments both lend themselves to numeric analysis. The first 20 terms in the Fourier analysis of the boundary produce data that, as for the dimensionless ratios, do not have Gaussian distributions. Figure 6.91 shows the values for just the second harmonic amplitude to illustrate this. None of the other terms are any better behaved.

A scatterplot of the terms (Figure 6.92) does show some clusters, which indicates that linear discriminant analysis may succeed. It does, as shown in Figure 6.93. Using the magnitude of harmonic terms 2 through 7, the classes are successfully separated with only five individual misclassified cookies.

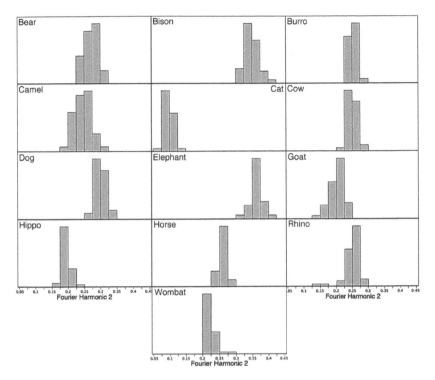

FIGURE 6.91
Comparison of the magnitude values of the second Fourier harmonic term for the 13 classes.

Using these same terms in a neural net with three hidden nodes (using two-thirds of the data for training and the remainder for validation) generates a result with only two misclassified cookies.

Moments also produce values that can be used for classification. Figure 6.94 shows the scatterplot for three of the affine invariant moments, which reveals clustering of the values for each class. Using principal components analysis on the data produces a better separation of the classes as shown. The first three principal components account for 50.8%, 28.4%, and 8.5%, respectively, of the data variation.

Using all of the invariant moments in a linear discriminant model produces the result in Figure 6.95; 13 cookies are misclassified. Using the same input to generate a neural net results in eight misclassifications.

Since it is possible to produce useful classification results with each of these sets of data, which one is best? The neural net generated with six Fourier amplitudes from harmonic analysis has the very best error rate, producing only two individual misclassifications (0.3%). Linear discriminant analysis based on four dimensionless ratios results in 16 misclassified cookies (2.4%) but has the advantages that the input values may be easier to measure, are supported in many software packages, and the model may be easier to

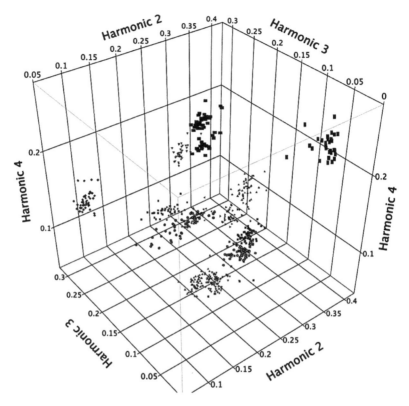

FIGURE 6.92
Scatterplot of the magnitude terms from harmonic analysis.

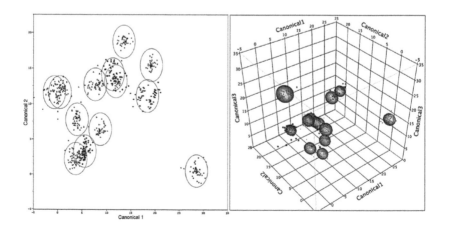

FIGURE 6.93
Linear discriminant classification of the harmonic coefficients for the cookies (only the most significant canonical axes are shown).

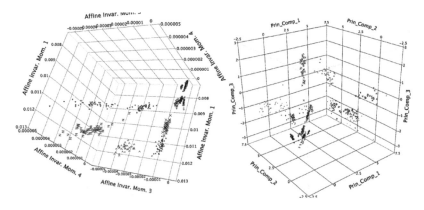

FIGURE 6.94
Scatterplots of three invariant moments and the principal components values calculated for the invariant moments.

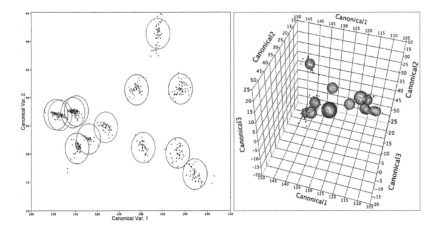

FIGURE 6.95
Linear discriminant results using all of the invariant moments (only the most significant canonical axes are shown).

understand. The invariant moments are the easiest properties to calculate, requiring only sums based on pixel coordinates with no complicated mathematics, and do not use any measurements of the object periphery (which might be important if imaging of the objects was limited in resolution or noisy). The error rates for discriminant analysis and neural net models are 2.0% and 1.2%, respectively. Table 6.13 summarizes the performance of the various methods.

The ultimate choice among these methods depends on the individual user's preferences, the available software, and the purpose to which the model will be put.

TABLE 6.13

Summary of Results for the Cookies in Figure 6.84

Measurements	Calculation Method	Misclassified
Four dimensionless ratios	Discriminant analysis	16 (2.4%)
	Neural net	18 (2.8%)
Twelve dimensionless ratios	Discriminant analysis	5 (0.8%)
	Neural net	9 (1.4%)
Harmonic coefficients 2–7	Discriminant analysis	5 (0.8%)
	Neural net	2 (0.3%)
Invariant moments	Discriminant analysis	13 (2.0%)
	Neural net	8 (1.2%)

Heuristic Classification

Although each of the analytical approaches illustrated can give excellent results, the method by which we acquired names for the different types of animals suggests that a heuristic approach may also be applicable. In cases where the shapes are indistinct using available imaging techniques, or where it is not feasible to remove incomplete, broken, or obscured shapes (for example, when discarding such items could introduce undesirable bias into the data), it is possible to "crowdsource" the classification.

Crowdsourcing is a distributed problem-solving model that broadcasts a problem to an unknown group of solvers in the form of an open call for solutions. Users, known as the crowd, submit solutions and also sort through the solutions, finding the best ones. This technique has been used successfully to add annotations to large databases of images (Deng et al., 2009), with relatively high accuracy as measured by the utility of the annotations in subsequent analysis. Sorokin and Forsyth (2008) show that this technique can be exceptionally cost effective for classifying large numbers of images and provide guidance in constructing effective classification strategies.

It is critical that any crowdsourced technique be robust in light of the tendency to "game" the system and provide low-quality or maliciously erroneous data (Kittur et al., 2008). Strategies exist (Sorokin & Forsyth, 2008) for minimizing the effect of low-quality data, including collecting multiple votes for each object to be classified, using standards to gauge effectiveness, and simplifying the classification task as much as possible such as by providing limited choices for classification. Crowdsourcing strategies can be effective at solving computationally difficult tasks. This is especially true of tasks like image classification that leverage the sophistication of human visual processing relative to the state of the art in computation techniques.

The Leafsnap program (see Chapter 3 and Belhumeur et al., 2008) is another example of the use of crowdsourcing to not only identify and classify objects, but to also build databases of images along with substantial and robust metadata (such as location). Cooper et al. (2010) show that this is also useful for more abstract problem-solving tasks by using a crowdsourcing technique to solve protein-folding processes, another computationally difficult task (Khatib et al., 2011).

The success of crowdsourcing in such cases suggests that there is fertile ground for using human intelligence on the large scale to solve truly intractable problems in imaging, particularly in cases with corrupted or incomplete data. Despite recent efforts such as Leafsnap, the full potential of what is often called the cognitive surplus (Shirky, 2011) for image and object classification remains largely unexplored at this time.

Conclusions

From a philosophical point of view, shape may be considered either as an intrinsic property of objects, or as a relationship to space, or as dependent upon the context in which an object exists or is encountered. For humans, and increasingly for computer programs, the description and measurement of shape is a tool, and this book is concerned with implementing and exploiting that tool.

The shape of objects is the most important key to classification, identification, and recognition both for people and computers. In most cases, unless there are tight limits on viewpoint, lighting, and object presentation, shape is more important than size, color, or position. In some industrial process control situations such constraints are possible but not in most real-world situations. When the apparent size of an object can vary in an image (for instance, due to distance) or color can vary with lighting, these become at best secondary clues. The apparent shape of three-dimensional objects does vary as they are represented by two-dimensional images, but different viewpoints can be accommodated by using different sets of criteria.

People and computers describe shape in different ways. Most computer methods begin with some kind of measurement that produces numerical values. There may be just a few, highly reductive ones such as fractal dimension or formfactor, or an entire sequence of values such as the harmonic Fourier coefficients or moments. In either case, the ability to apply statistical procedures to select those numbers that are most significant for a particular task, whether it is quality control, correlation with history or properties, classification, or recognition, is well suited to dealing with numerical values.

With adequate training, which is not always easy to accomplish, computer methods perform with high reliability, using combinations of different measurements in ways that are fundamentally different from human vision and understanding. The statistical processes and decision paths may be difficult for humans to follow, but the results are more consistent and tolerant of extraneous factors than is usually possible for human vision, especially when multiple observers are involved.

Some methods for image processing do not reduce a shape to numbers. Syntactical analysis tries to break up a complex shape into a few simple forms and qualitatively describe their relationships to each other. The medial axis transform encodes the shape as a continuous set of width values along the midlines. This is an image processing more than an image measurement operation. Although these offer interesting approaches to shape encoding that may have some connections to human visual processing, the approaches do not readily provide tools that assist in the traditional goals of classification or correlation.

Human vision is generally more flexible than computer programs for recognition. Since the visual system is comparative, rather than based on measurement, it depends largely on canonical examples. These may represent an entire object or the key parts (a form of syntactical analysis). In some cases, direct comparison to another object is possible. This is like the child's game of "which of these things is most like the other." It depends on our ability to mentally rotate objects for comparison and to decide how much tolerance for variation is acceptable. Of course, that tolerance varies depending on experience and circumstances.

When the comparison is to an atlas or field guide, or to memory, the error rate increases. But human vision generally tolerates a high error rate (either false positives or negatives—recognizing something incorrectly or failing to recognize something) as a trade-off for speed. The error rates increase dramatically in noisy cases, where there is a lot of unimportant or unrelated variation in shape, in the presence of extraneous clutter or distractions, or when there are intermediate cases that require more careful decision making.

Because computer methods differ from those employed in human vision, they are often more robust in the presence of noise and distraction, but also less able to cope with the new and unexpected. The methods shown in the preceding chapters for obtaining measurements and in this chapter for analyzing them document the variety of approaches used in computer analysis, their strengths and limitations, and the overall success that is possible in a very broad spectrum of applications.

And it is helpful to remember the caution of Werner Heisenberg (1958):

> *Every word or concept*
> *Clear as it may seem to be*
> *Has only a limited range*
> *Of applicability.*

References

M. S. Abdaheer, E. Khan (2009) Shape based classification of breast tumors using fractal analysis, *IEEE Multimedia Signal Processing and Communication Technologies* 272–275.

A. Acquisti, R. Gross, F. Stutzman (2011) Faces of Facebook Privacy in the Age of Augmented Reality, draft document online at http://www.heinz.cmu.edu/~acquisti/face-recognition-study-FAQ/

P. Alberch, S. J. Gould, G. F. Oster, D. B. Wake (1979) Size and shape in ontogeny and phylogeny, *Paleobiology* 5(3):296–317.

A. M. Albert, K. Ricanek, E. Patterson (2007) A review of the literature on the aging adult skull and face: Implications for forensic science research and applications, *Forensic Science International* 172(1):1–9.

D. J. Alexander (1990) Quantitative analysis of fracture surfaces using fractals, pp. 39–51 in *Quantitative Methods in Fractography*, STP 1085, ASTM, Philadelphia.

S. K. Alibhai, Z. C. Jewell (2008) Identifying white rhino (*Ceratotherium simum*) by a footprint identification technique, at the individual and species levels, *Endangered Species Research* 4:219–225.

J. Alirezaie, M. E. Jernigan, C. Nahmias (1997) Neural network-based segmentation of magnetic resonance images of the brain, *IEEE Transactions on Nuclear Science* 44(2):194–198.

G. Amayeh et al. (2006) Hand-based verification and identification using high-order Zernike moments, *IEEE Computer Vision and Pattern Recognition Workshops*, 40–47.

G. Amayeh, G. Bebis, M. Nicolescu (2008) Gender classification from hand shape, *IEEE Computer Vision and Pattern Recognition Workshops*, 1–7.

A. Amini, J. Duncan (1992) Bending and stretching models for LV wall motion analysis from curves and surfaces, *Image and Vision Computing* 10(6):418.

J. A. Anderson (1995) *An Introduction to Neural Networks*, MIT Press, Cambridge, MA.

J.-P. Antoine et al. (1996) Multiscale shape analysis using the continuous wavelet transform, *International Conference on Image Processing* 1:291–294.

S. Asano, G. Yamamoto (1975) Light scattering by a spheroidal particle, *Applied Optics* 14(1):29–49.

D. Attali, A. Montanvert (1997) Computing and simplifying 2D and 3D continuous skeletons, *Computer Vision and Image Understanding* 67(3):161–273.

A. Baddeley, E. B. Vedel Jensen (2005) *Stereology for Statisticians*, Chapman & Hall, Boca Raton, FL.

J. W. Baish, R. K. Jain (2000) Fractals and cancer, *Cancer Research* 61(22):8347–8350.

D. H. Ballard, C. M. Brown (1982) *Computer Vision*, Prentice Hall, Englewood Cliffs, NJ.

G. Balmino et al. (1973) A spherical harmonic analysis of the earth's topography, *Journal of Geophysical Research* 78(2):478–481.

J. R. Banavar et al. (1997) Sculpting of a fractal river basin, *Physical Review Letters* 78(23):4522–4525.

H. G. Barth, S.-T. Sun (1985) Particle size analysis, *Analytical Chemistry* 57:151.

P. J. Barrett (1980) The shape of rock particles, a critical review, *Sedimentology*, 27:291–303.

D. E. Barton, F. N. David (1962) The analysis of chromosome patterns in the normal cell, *Annals of Human Genetics* 25(4):323–329.

W. Bauer, C. D. Mackenzie (1995) Cancer detection via determination of fractal cell dimension, Michigan State University report CL-980.

J. K. Beddow, G. C. Philip, A. F. Vetter (1977) On relating some particle profiles characteristics to the profile Fourier coefficients, *Powder Technology* 18:15–19.

W. Beil, I. C. Carlsen (1991) Surface reconstruction from stereoscopy and "shape from shading" in SEM images, *Machine Vision and Applications* 4:271–285.

R. M. Bekker et al. (1998) Seed size, shape and vertical distribution in the sol: Indicators of seed longevity, *Functional Ecology* 12(5):834–842.

P. N. Belhumeur et al. (2008) Searching the world's herbaria: A system for visual identification of plant species, in D. Forsyth et al. (eds.) ECCV 2008 Part IV, LNCS 5305, pp. 116–119, Springer Verlag, Berlin.

S. O. Belkasim, M. Shridhar, M. Ahmadi (1991) Pattern recognition with moment invariants: A comparative study and new results, *Pattern Recognition* 24(12): 1117–1138.

P. R Belyea, R. C. Thunell (1984) Fourier shape analysis and planktonic foraminifera evolution, *Journal of Paleontology* 58(4):1026–1040.

S. H. Bennett et al. (2000) Origin of fractal branching complexity in the lung, www. stat.rice.edu/riedi/UCDavisHemoglobin/fractal3.pdf

L. von Bertalanffy (1957) Quantitative laws in metabolism and growth, *The Quarterly Review of Biology* 32(3):217.

A. Bertillon (1890) *La Photographie Judiciaire, Avec un Appendice sur la Classification et l'identification Anthropometriques*, Gauthier-Villars, Paris.

P. J. Besl, N. D. McKay (1992) A method for registration of two 3-D shapes, *IEEE Transactions on Pattern Analysis and Machine Intelligence* 14(2):239–256.

J. V. Beusekom, M. Schreyer, T. M. Breuel (2010) Automatic counterfeit protection system code classification, *Proceedings of the SPIE* 7541:75410F.

J. L. Bird, D. T. Eppler, D. M. Checkley, Jr. (1986) Comparisons of herring otoliths using Fourier series shape analysis, *Canadian Journal of Fisheries and Aquatic Science* 43:1228–1234.

H. Bischof, W. Schneider, A. J. Pinz (1992) Multispectral classification of Landsat-images using neural networks, *IEEE Transactions on Geoscience and Remote Sensing* 30(3):482–490.

C. M. Bishop (1996) *Neural Networks for Pattern Recognition*, Oxford University Press, Oxford, UK.

C. M. Bishop (2007) *Pattern Recognition and Machine Learning*, Springer, New York.

H. R. Bittner, P. Wlczek, M. Sernetz (1989) Characterization of fractal biological objects by image analysis, *Acta Stereologica* 8:31–40.

P. M. Blough, L. K. Slavin (1987) Reaction time assessments of gender differences in visual-spatial performance, *Perception & Psychophysics* 41(3):276–281.

H. Blum (1967) A transformation for extracting new descriptors of shape, in *Models for the Perception of Speech and Visual Forms*, W. Walthen-Dunn (ed.), pp. 362–380, MIT Press, Cambridge, MA.

H. Blum (1973) Biological shape and visual science, Part I, *Journal of Theoretical Biology* 38:205–287.

B. Bollabas (2002) *Modern Graph Theory*, Springer, Berlin.

F. L. Bookstein (1991) *Morphometric Tools for Landmark Data*, Cambridge University Press, Cambridge, UK.

S. Boslaugh, P. A. Watters (2008) *Statistics in a Nutshell*, O'Reilly, Beijing.

S.-T. Bow (1992) *Pattern Recognition and Image Preprocessing*, Marcel Dekker, New York.

E. T. Bowman et al. (2001) Particle shape characterisation using Fourier analysis, *Geotechnique* 51(6):545–554.

A. Boyde (1973) Quantitative photogrammetric analysis and qualitative stereoscopic analysis of SEM images, *Journal of Microscopy* 98:452–471.

K. L. Boyer, A. C. Kak (1987) Color-encoded structured light for rapid active ranging, *IEEE Transactions on Pattern Analysis and Machine Intelligence* 9:14–28.

L. Brabant et al. (2010) Three-dimensional analysis of high resolution X-ray computed tomography data with Morpho+, *Microscopy & Microanalysis* 17:252–263.

R. N. Bracewell (1989) The Fourier transform, *Scientific American*, June, 62–69.

C. Brechbühler, G. Gerig, O. Kübler (1995) Parameterization of closed surfaces for 3-D shape description, *Computer Vision and Image Understanding* 61(2):154–170.

J. Bueller (1992) Institute of Archaeology, Hebrew University of Jerusalem, private communication.

D. Burschka et al. (2005) Scale-invariant registration of monocular endoscopic images to CT-scans for sinus surgery, *Medical Image Analysis* 9(5):413–426.

G. Byrne, D. Dornfeld, B. Denkena (2003) Advanced cutting technology, *CIRP Annals–Manufacturing Technology* 52(2):483–500.

W. Cai et al. (2002) Protein–ligand recognition using spherical harmonic molecular surfaces: Towards a fast and efficient filter for large virtual throughput screening, *Journal of Molecular Graphics and Modelling* 20(4):313–328.

O. Campàs et al. (2010) Scaling and shear transformations capture beak shape variation in Darwin's finches, *Proceedings of the National Academy of Sciences* 107(8):3356–3360.

D. J. Canfield, R. L. Anstey (1981) Harmonic analysis of cephalopod suture patterns, *Mathematical Geology* 13(1):23–35.

Y. Capowiez et al. (1998) 3D skeleton reconstructions of natural earthworm burrow systems using CAT scan images of soil cores, *Biology and Fertility of Soils* 27(1):51–59.

R. M. Carter, Y. Yan (2005) Measurement of particle shape using digital imaging techniques, *Journal of Physics: Conference Series* 15:177–182.

I. M. Chakravarti, R. G. Laha, J. Roy (1967) *Handbook of Methods of Applied Statistics* Vol. 1, Wiley, New York.

P. J. Chandley (1976) Surface roughness measurements from coherent light scattering, *Optical and Quantum Electronics* 8:323–327.

Q. Chang et al. (2011) Three-dimensional fractal analysis of fracture surfaces in titanium-iron particular reinforced hydroxyapatite composites: Relationship between fracture toughness and fractal dimension, *Journal of Materials Science* 46:6118–6123.

F. Chen, G. M. Brown, M. Song (2000) Overview of three-dimensional shape measurement using optical methods, *Optical Engineering* 39(1):10–22.

Y. Cheng et al. (2011) Accurate 3D registration of magnetic resonance images for detecting local changes in cartilage thickness, *Journal of Electronic Imaging* 20(2):023002.

G. P. Cherepanov, A. S. Balankin, V. S. Ivanova (1995) Fractal fracture mechanics—A review, *Engineering Fracture Mechanics* 51(6):997–1033.

J. L. Chermant, M. Coster (1978) Fractal object in image analysis, International Symposium on Quantitative Metallography, Florence, Associazione Italiana di Mettalurgica.

J. L. Chermant, M. Coster (1987) Fractal methods in profilometric analysis: Application to rupture of cold-worked brass, *Acta Stereologica* 6(III):845–850.

J. M. Cheverud, J. J. Rutledge, W. R. Atchley (1983) Quantitative genetics of development: Genetic correlations among age-specific trait values and the evolution of ontogeny, *Evolution* 37(5):895–905.

J. L. Chim et al. (2004) Examination of counterfeit banknotes printed by all-in-one color inkjet printers, *Journal of American Society of Questioned Document Examiners* 7(2):69–75.

G.-C. Cho, J. Dodds, J. C. Santamarina (2006) Particle shape effects on packing density, stiffness and strength: Natural and crushed sands, *Journal of Geotechnical and Geoenvironmental Engineering*, May, 591–602.

T. S. Chow (1980) The effect of particle shape on the mechanical properties of filled polymers, *Journal of Materials Science* 15:1873–1888.

M. K. Chung et al. (2007) Weighted Fourier series representation and its application to quantifying the amount of gray matter, Special Issue of *IEEE Transactions on Medical Imaging on Computational Neuroanatomy* 26:566–581.

M. Cieplak et al. (1998) Models of fractal river basins, *Journal of Statistical Physics* 91(1–2):1–15.

N. Clark (1986) Three techniques for implementing digital fractal analysis of particle shape, *Powder Technology* 46(1):45–52.

H. E. Cline, W. E. Lorensen (1987) System and method for the display of surface structures contained within the interior region of a solid body, U.S. Patent 4,710,876.

C. E. Connor (2005) Friends and grandmothers, *Nature*, 1036–1037.

D. R. Cook (1977) Detection of influential observations in linear regression, *Technometrics* 19(1):15–18.

J. W. Cooley, J. W. Tukey (1965) An algorithm for the machine calculation of complex Fourier series, *Mathematics of Computation* 19(90):297–301.

L. A. Cooper, P. Podgorny (1976) Mental transformations and visual comparison processes: Effects of complexity and similarity, *Journal of Experimental Psychology: Human Perception and Performance* 2(4):503–514.

S. Cooper, F. Khatib et al. (2010) Predicting protein structures with a multiplayer online game, *Nature* 466:756–760.

T. Cootes et al. (2005) Modeling facial shape and appearance, in *Handbook of Face Recognition*, S. Z. Li, A. K. Jain (eds.), Springer, New York.

L. F. Costa, R. M. Cesar (2009) *Shape Analysis and Classification*, CRC Press, Boca Raton, FL.

J. S. Crampton (1995) Elliptic Fourier shape analysis of fossil bivalves: Some practical considerations, *Lethaia* 28(2):179–186.

D. L. Cusumano, J. K. Thompson (1997) Body image and body shape ideals in magazines, *Sex Roles* 37(9–10):701–731.

I. Daubechies (1992) *Ten Lectures on Wavelets*, Society for Industrial and Applied Mathematics, Philadelphia, PA.

J. Daugman (2004) How iris recognition works, *IEEE Transactions on Circuits and Systems for Video Technology* 14(1):21–30.

R. H. Dauskardt, F. Haubensak, R. O. Ritchie (1990) On the interpretation of the fractal character of fracture surfaces, *Acta Metallurgica et Materialia* 38(2):143–159.

R. Davies, C. Twining, C. Taylor (2008) *Statistical Models of Shape*, Springer, London.

R. T. DeHoff, E. H. Aigeltinger, K. R. Craig (1972) Experimental determination of the topological properties of three-dimensional microstructures, *Journal of Microscopy* 95(1):69–91.

V. P. Delfino, T. Lettini, E. Vacca (1997) Heuristic adequacy of Fourier descriptors: Methodologic aspects and applications in morphology, in P. E. Lestrel (ed.) *Fourier Descriptors and Their Applications in Biology*, Cambridge University Press, Cambridge, UK.

M. DeMarsicoi et al. (1997) Indexing pictorial documents by their content: A survey of current techniques, *Image and Vision Computing* 15:119–141.

J. Deng, W. Dong et al. (2009) ImageNet: A large-scale hierarchical image database, 2009 *IEEE Conference on Computer Vision and Pattern Recognition*, 248–255.

S. Derrode, F. Ghorbel (2004) Shape analysis and symmetry detection in gray-level objects using the analytical Fourier–Mellin representation, *Signal Processing* 84(1):25–39.

H. Dette, V. Melas, A. Pepleyshev (2005) Optimal designs for three-dimensional shape analysis with spherical harmonic descriptors, *Annals of Statistics* 33(6):2758–2788.

G. Diaz et al. (1989) Elliptic Fourier analysis of cell and nuclear shapes, *Computers and Biomedical Research* 22:405–414.

G. Diaz et al. (1990) Recognition of cell surface modulation by elliptic Fourier analysis, *Computer Methods and Programs in Biomedicine* 31:57–62.

G. Diaz et al. (1997) Elliptical Fourier descriptors of cell and nuclear shapes, in P. E. Lestrel (ed.) *Fourier Descriptors and Their Applications in Biology*, Cambridge University Press, Cambridge, UK.

A. G. Dixon (1988) Correlations for wall and particle shape effects on fixed bed bulk voidage, *Canadian Journal of Chemical Engineering* 66(5):705–708.

L. A. Dobrzanski, W. Sitek (1998) Application of a neural network in modelling of hardenability of constructional steels, *Journal of Materials Processing Technology* 78(1–3):59–66.

J. A. Dowdeswel (1982) Scanning electron micrographs of quartz sand grains from cold environments examined using Fourier shape analysis, *Journal of Sedimentary Petrology* 52:1315–1323.

I. L. Dryden, K. V. Mardia (1998) *Statistical Shape Analysis*, Wiley, New York.

B. Dubuc et al. (1989) Evaluating the fractal dimension of profiles, *Physical Review* A 39:1500–1512.

G. A. Dunn, A. F. Brown (1986) Alignment of fibroblasts on grooved surfaces described by a simple geometric transformation, *Journal of Cell Science* 83:313–340.

H. A. Dwyer, D. S. Dandy (1990) Some influences of particle shape on drag and heat transfer, *Physics of Fluids* A 2(12):2110–2118.

M. Egmont-Petersen, D. de Ridder, H. Handels (2002) Image processing with neural networks—a review, *Pattern Recognition* 35(1):2279–2301.

R. Ehrlich, B. Weinberg (1970) An exact method for characterization of grain shape, *Journal of Sedimentary Petrology* 40:205–212.

R. Ehrlich et al. (1984) Petrographic image analysis: 1. Analysis of reservoir pore complexes, *Journal of Sedimentary Petrology* 54:1365–1378.

M. A. Elbestawi, A. K. Srivastava, T. I. El-Wardany (1996) A model for chip formation during machining of hardened steel, *CIRP Annals–Manufacturing Technology* 45(1):71–76.

S. B. Emerson (1985) Skull shape in frogs: Correlation with diet, *Herpetologica* 41(2):177–188.

T. Eppler, R. Ehrlich (1983) Sources of shape variation in lunar impact craters: Fourier shape analysis, *Geological Society of America Bulletin* 94:274–291.

P. Evison, R. W. V. Bruegge (2010) *Computer-Aided Forensic Facial Comparison*, CRC Press, Boca Raton, FL.

Y. Fahmy et al. (1991) Application of fractal geometry measurements to the evaluation of fracture toughness of brittle intermetallics, *Journal of Materials Research* 6(9):1856–1861.

C. Faloutsos et al. (1994) Efficient and effective querying by image content, *Journal of Intelligent Information Systems*, 3:231–262.

J. Feder (1988) *Fractals*, Plenum Press, New York.

L. Fei-Fei, R. Fergus, P. Perona (2007) Learning generative visual models from few training examples: An incremental Bayesian approach tested on 101 object categories, *Computer Vision and Image Understanding* 106(1):59–70.

S. F. Ferson, F. J. Rohlf, R. K. Koehn (1985) Measuring shape variation of two-dimensional outlines, *Systematic Zoology* 34:59–68.

M. Flickner et al. (1995) Query by image and video content: The QBIC system, *IEEE Computer* 28(9):23–32.

A. G. Flook (1978) Use of dilation logic on the Quantimet to achieve fractal dimension characterization of texture and structured profiles, *Powder Technology* 21:295–298.

A. G. Flook (1982) Fourier analysis of particle shape, in *Particle Size Analysis 1981–2*, N. G. Stanley-Wood, T. Allen (eds.), Wiley Heyden, London.

J. Flusser, T. Suk (1993) Pattern recognition by affine moment invariants, *Pattern Recognition* 26(1):167–174.

J. Flusser, T. Suk (2006) Rotation moment invariants for recognition of symmetric objects, *IEEE Transactions on Image Processing* 15:3784–3790.

J. D. Foley et al. (1996) *Computer Graphics: Principles and Practice*, Addison Wesley, Reading MA.

H. Freeman, L. S. Davis (1977) A corner finding algorithm for chain-code curves, *IEEE Transactions on Computers* 26:297–303.

M. Frigo, S. G. Johnson (2005) The design and implementation of FFTW3, *Proceedings of the IEEE* 93(2):216–231.

J. P. Frisby (1980) *Seeing: Illusion, Brain and Mind*, Oxford University Press, Oxford, UK.

J. P. Frisby, J. V. Stone (2010) *Seeing: The Computational Approach to Biological Vision* (2nd ed.), MIT Press, Cambridge, MA.

K. S. Fu (1974) *Syntactic Methods in Pattern Recognition*, Academic Press, Boston.

K. S. Fu (1982) *Syntactic Pattern Recognition and Applications*, Prentice-Hall, Englewood Cliffs, NJ.

K. Fukunaga (1990) *Statistical Pattern Recognition* (2nd ed.), Academic Press, Boston.

T. Funkhouser et al. (2003) A search engine for 3D models, *ACM Transactions on Graphics* 22(1):83–105.

T. Funkhouser et al. (2005) Shape-based retrieval and analysis of 3D models, *Communications of the ACM* 48(6):58–64.

F. Galton (1892) *Finger Prints*, Macmillan, London.

C. Gambino, P. McLaughlin et al. (2011) Forensic surface metrology: Tool mark evidence, *Scanning* 33:272–278.

E. J. Garboczi (2002) Three-dimensional mathematical analysis of particle shape using X-ray tomography and spherical harmonics: Application to aggregates used in concrete, *Cement and Concrete Research* 32(10):1621–1638.

D. M. Garner et al. (1980) Cultural expectations of thinness in women, *Psychological Reports* 47(2):483–491.

G. Gerig et al. (2010) Shape vs size: Improved understanding of the morphology of brain structures, medical image computing and computer assisted intervention, *Lecture Notes in Computer Science* 2208:24–32.

M. Ghaemi et al. (2002) Differentiating multiple system atrophy from Parkinson's disease: Contribution of striatal and midbrain MRI volumetry and multi-tracer PET imaging, *Journal of Neurology, Neurosurgery and Psychiatry* 73:517–523.

P. K. Ghosh, K. Deguchi (2008) *Mathematics of Shape Description*, Wiley, Singapore.

J. D. Gibbons, S. Chakraborti (2011) *Nonparametric Statistical Inference* (5th ed.), Chapman & Hall/CRC, Boca Raton, FL.

H. S. Göktürk, T. J. Fiske, D. M. Kalyon (1993) Effects of particle shape and size distributions on the electrical and magnetic properties of nickel/polyethylene composites, *Journal of Applied Polymer Science* 50:1891–1901.

C. Goodall (1991) Procrustes methods in the statistical analysis of shape, *Journal of the Royal Statistical Society* (b) 53:285–339.

M. F. Goodchild (1980) Fractals and the accuracy of geographical measures, *Mathematical Geology* 12(2):85–98.

S. J. Gould (1966) Allometry and size in ontogeny and phylogeny, *Biological Reviews* 41(4):58–638.

S. J. Gould (1977) *Ontogeny and Phylogeny*, Harvard University Press, Cambridge, MA.

G. H. Granlund (1972) Fourier preprocessing for hand print character recognition, *IEEE Transactions on Computers* C21(2):195–201.

R. L. Gregory (2009) *Seeing through Illusions*, Oxford University Press, Oxford, UK.

J. Grum, R. Stürm (1995) Computer supported recognition of graphite particle forms in cast iron, *Acta Stereologica* 14(1):91–96.

K. Gurney (2003) *An Introduction to Neural Networks*, CRC Press, Boca Raton, FL.

A. Gutteridge, J. Thornton (2004) Conformational change in substrate binding, catalysis and product release: An open and shut case? *FEBS Letters* 567(1):67–73.

A. J. Haines, J. S. Crampton (2000) Improvements to the method of Fourier shape analysis as applied in morphometric studies, *Paleontology* 43(4):765–783.

J. Hajnal et al. (2001) *Medical Image Registration*, CRC Press, Boca Raton, FL.

F. Harary (1969) *Graph Theory*, Addison Wesley, Reading, MA.

M. Hasegawa et al. (1996) Calculation of fractal dimensions of machined surface profiles, *Wear* 192:40–45.

Z. He, T. Tan et al. (2008) Towards accurate and fast iris segmentation for iris biometrics, *IEEE Transactions on Pattern Analysis and Machine Intelligence* 31(9):1670–1684.

N. Healy-Williams, D. F. Williams (1981) Fourier analysis of test shape of planktonic foraminifera, *Nature* 289:485–487.

N. Healy-Williams, R. Ehrlich, W. Full (1997) Closed-form Fourier analysis: A procedure for extraction of ecological information about foraminiferal test morphology, in P. E. Lestrel (ed.) *Fourier Descriptors and Their Applications in Biology*, Cambridge University Press, Cambridge, UK.

D. O. Hebb (1968) Concerning imagery, *Psychological Review* 75(6):466–477.

W. Heisenberg (1958) *Physics & Philosophy*, Penguin Books, Harlow UK.

M. W. Hentschel, N. W. Page (2003) Selection of descriptors for particle shape characterization, *Particle and Particle Systems Characterization* 20:25–38.

H. Heywood (1954) Particle shape coefficients, *Journal of the Imperial College Chemical Society*, 8:15–33.

J. Hilditch, D. Rutovitz (1969) Chromosome recognition, *Annals of the New York Academy of Sciences* 157:339–364.

T. Hirata (1989) Fractal dimension of fault systems in Japan, *Pure and Applied Geophysics* 131(1/2):157–170.

B. K. P. Horn, M. J. Brooks (1989) *Shape from Shading*, MIT Press, Cambridge, MA.

M. S. Hosseini, B. N. Araabi, H. Soltanian-Zadeh (2010) Pigment melanin: Pattern for iris recognition, *IEEE Transactions on Instruments and Measurements* 59(4):792–804.

C. V. Howard, M. G. Reed (2005) *Unbiased Stereology* (2nd ed.), Garland Science, New York.

C.-B. Hsu, S.-S. Hao, J.-C. Lee (2011) Personal authentication through dorsal hand vein patterns, *Optical Engineering* 50:087201.

M. K. Hu (1962) Visual pattern recognition by moment invariants, *IEEE Transactions on Information Theory* 8:179–187.

D. H. Hubel (1988) *Eye, Brain, and Vision*, W. H. Freeman, New York.

J. M. Hughes, D. J. Graham, D. N. Rockmore (2010) Quantification of artistic style through sparse coding analysis in the drawings of Pieter Breugel the Elder, *Proceedings of the National Academy of Sciences* 1074:1279–1283.

D. J. Hurley, B. Arbab-Zavar, M. S. Nixon (2007) The ear as a biometric, in *Handbook of Biometrics*, pp. 131–150, Springer, New York.

A. Iannarelli (1989) *Ear Identification*, Paramount Publishing, Freemont, CA.

M. I. S. Ibrahim, M. S. Nixon, S. Mahmoodi (2010) Shaped wavelets for curvilinear structures for ear biometrics, *International Symposium on Visual Computing*, Las Vegas, NV.

B. I. Imasogie, U. Wendt (2004) Characterization of graphite particle shape in spheroidal graphite iron using a computer-based image analyzer, *Journal of Minerals and Materials Characterization and Engineering* 3(1):1–12.

S. Ings (2008) *A Natural History of Seeing*, W. W. Norton, New York.

M. A. Issa, M S. Isla, A. Chudnovsky (2003) Fractal dimension—a measure of fracture roughness and toughness of concrete, *Engineering Fracture Mechanics* 70:125–137.

V. K. Ivanov et al. (2007) Fractal analysis of sea ices images, *Sixth International Kharkov Symposium on Physics and Engineering of Microwaves*, June 25–30:989–991.

H. Iwata, Y. Ukai (2002) SHAPE: A computer program package for quantitative evaluation of biological shapes based on elliptic Fourier descriptors, *Journal of Heredity* 93:384–385.

B. Jacobshagen (1997) Craniofacial variability in hominoidea, in P. E. Lestrel (ed.) *Fourier Descriptors and Their Applications in Biology*, Cambridge University Press, Cambridge, UK.

M. James (1988) *Pattern Recognition*, Blackwell Scientific, London.

G. Jayalaitha, R. Uthayakumar (2007) Estimating the skin cancer using fractals, *Proceedings of the International Conference on Computational Intelligence*, Dec. 13–15:306–311.

X. Jia et al. (2010) Model and error analysis for coded structured light measurement system, *Optical Engineering* 49(12):123603.

C. R. Johnson et al. (2008) Image processing for artist identification, *IEEE Transactions on Signal Processing* 25(4):37–48.

D. R. Johnson (1997) Fourier descriptors and shape differences: Studies on the upper vertebral column of the mouse, in P. E. Lestrel (ed.) *Fourier Descriptors and Their Applications in Biology*, Cambridge University Press, Cambridge, UK.

D. R. Johnson et al. (1985) Measurement of biological shape: A general method applied to mouse vertebrae, *Journal of Embryology and Experimental Morphology* 90:363–377.

S. G. Johnson, M. Frigo (2007) A modified split-radix FFT with fewer arithmetic operations, *IEEE Transactions on Signal Processing* 55(1):111–119.

A. R. Jones (1999) Light scattering for particle characterization, *Progress in Energy and Combustion Science*, 25(1):1–53.

W. L. Jungers, A. B. Falsetti, C. E. Wall (1995) Shape, relative size, and size-adjustments in morphometrics, *American Journal of Physical Anthropology* 38(Suppl. 2):137–161.

R. A. Katz, S. M. Pizer (2003) Untangling the Blum medial axis transform, *International Journal of Computer Vision* 55(2/3):139–153.

A. Kayaalp, A. R. Rao, R. Jain (1990) Scanning electron microscope-based stereo analysis, *Machine Vision and Applications* 3:231–246.

B. H. Kaye (1989) *A Random Walk through Fractal Dimensions*, VCH Verlagsgesellschaft, Weinheim.

B. H. Kaye et al. (1983) A study of physical significance of three-dimensional signature waveforms, *Proceeding of the Fineparticle Characterization Conference*, Hawaii.

D. G. Kendall (1989) The statistical theory of shape, *Statistical Science* 4(2):87–120.

E. Keogh (2007) Mining shape and time series databases with symbolic representations, *Proceedings of the 13th ACM International Conference on Knowledge Discovery and Data Mining*, Tutorial # 7.

D. Keren (2003) Recognizing image "style" and activities in video using local features and naive Bayes, *Pattern Recognition Letters* 24:2913–2922.

F. Khatib, F. DiMaio et al. (2011) Crystal structure of a monomeric retroviral protease solved by protein folding fame players, *Nature Structural and Molecular Biology*, doi 10.1038/nsmb.2119

A. Khotanzad, Y. H. Hong (1990) Invariant image recognition by Zernike moments, *IEEE Transactions on Pattern Analysis and Machine Intelligence* 12(5):489–497.

A. Kikuchi et al. (2002) Fractal tumor growth of ovarian cancer: Sonographic evaluation, *Gynecologic Oncology* 87(3):295–302.

D. T. Kincaid, R. B. Schneider (1983) Quantification of lead shape with a microcomputer and Fourier transform, *Canadian Journal of Botany* 61:2333–2342.

J. S. Kirkaldy, D. Venugopalan (1984) Prediction of microstructure and hardenability in low-alloy steels, pp. 125–148 in *Phase Transformations in Ferrous Alloys*, ASTM, Philadelphia.

J. L. Kirsch, R. A. Kirsch (1988) The anatomy of painting style: Description with computer rules, *Leonardo* 21(4):437–444.

A. Kittur, E. H. Chi, B. Suh (2008) Crowdsourcing user studies with Mechanical Turk, *Proceedings of the Twenty-Sixth Annual SIGCHI Conference on Human Factors in Computing Systems*, 453–456.

R. Komanduri (1982) Some clarifications on the mechanics of chip formation when machining titanium alloys, *Wear* 76(1):15–34.

J. D. Krieger, R. P. Guralnick, D. M. Smith (2007) Generating empirically determined, continuous measures of leaf shape for paleoclimate reconstruction, *Palaios* 22(2):212–219.

W. C. Krumbein (1941) Measurement and geological significance of shape and roundness of sedimentary particles, *Journal of Sedimentary Petrology*, 11:64–72.

W. C. Krumbein, L. L. Sloss (1963) *Stratigraphy and Sedimentation*, Freeman, San Francisco, CA.

F. P. Kuhl, C. R. Giardina (1982) Elliptic Fourier features of a closed contour, *Computer Graphics and Image Processing* 18:236–258.

P. Kumar et al. (2008) Grasping molecular structures through publication-integrated 3D models, *Trends in Biochemical Sciences* 33(9):408–412.

H. Kurz, K. Sandau (1997) Modeling of blood vessel development—bifurcation pattern and hemodynamics, optimality and allometry, *Comments on Theoretical Biology* 4/4:261–291.

H. Laga, H. Takahasi, M. Nakajima (2006) Spherical wavelet descriptors for content-based 3D model retrieval, *IEEE SMI* 2006:15–25.

D. A. Lange et al. (1995) Relationship between fracture surface roughness and fracture behavior of cement paste and mortar, *Journal of the American Ceramic Society* 76(3):589–597.

P. R. Law. Quantifying variation in footprints of white rhino (*Ceratotherium simium*) using shape and size, private communication.

A. Leistner, W. Giardini (1991) Fabrication and testing of precision spheres, *Metrologia* 28(6):503.

S. R. Lele, J. T. Richtsmeier (2000) *An Invariant Approach to Statistical Analysis of Shapes*, Chapman & Hall/CRC, Boca Raton, FL.

A. Leonardis, B. Schiele, M. J. Tarr (eds.) (2009) *Object Categorization, Computer and Human Vision Perspectives*, Cambridge University Press, Cambridge, UK.

P. E. Lestrel (ed.) (1997) *Fourier Descriptors and Their Applications in Biology*, Cambridge University Press, Cambridge, UK.

P. E. Lestrel (2000) *Morphometrics for the Life Sciences*, World Scientific, Singapore.

P. E. Lestrel, H D. Brown (1976) Fourier analysis of adolescent grown of the cranial vault: A longitudinal study, *Human Biology* 48:517–528.

P. E. Lestrel, A. F. Roche (1986) Cranial base shape variation with age: A longitudinal study using Fourier analysis, *Human Biology* 58:527–540.

G. Li et al. (2008) Accuracy of 3D volumetric image registration based on CT, MR and PET/CT phantom experiments, *Journal of Applied Clinical Medical Physics* 9(4).

H. Li, R. Hartley (2007) The 3D-3D registration problem revisited, *Proceedings of ICCV* 2007:1–8.

S. X. Liao, M. Pawlak (1996) On image analysis by moments, *IEEE Transactions on Pattern Analysis and Machine Intelligence* 18(3):254–266.

K. Libbrecht (2006) *Field Guide to Snowflakes*, Voyageur Press, St. Paul, MN.

E. Limpert et al. (2001) Log-normal distributions across the sciences: Keys and clues, *BioScience* 51(5):341–352.

T.-W. Lin, Y.-F. Chou (2003) A comparative study of Zernike moments for image retrieval, *16th IPPR Conference on Computer Vision, Graphics and Image Processing*, 621–629.

P. Lippman (1987) An introduction to computing with neural nets, *IEEE ASSP* 3(4):4–22.

D. P. Livingston et al. (2011) 3D volumes constructed for pixel-based images by digitally clearing plant and animal tissue, *Journal of Microscopy* 240(2):122–129.

S. Lobregt, P. W. Verbeek, F. C. A. Groen (1980) Three dimensional skeletonization: Principle and algorithm, *IEEE Transactions on Pattern Analysis and Machine Intelligence* 2:75–77.

S. Loncaric (1998) A survey of shape analysis techniques, *Pattern Recognition* 31(8):983–1001.

W. E. Lorensen, H. E. Cline (1987) Marching cubes: A high resolution 3d surface construction algorithm, *Computer Graphics (Proc. SIGGRAPH 87)* 21(4):163–169.

S. Lovejoy (1982) Area-perimeter relation for rain and cloud areas, *Science* 216:185–187.

R. Magritte (1927) Les mots et les images, *Les Révolution Surréaliste*, December 1927.

R. Magritte (1935) Surrealist Manifesto, *Cahiers d'Art*, 1935.

M. A. Mahowald, C. A. Mead (1991) The silicon retina, *Scientific American* 264:76–82.

J. B. A. Maintz, M. A. Viergever (1998) A survey of medical image registration, *Medical Image Analysis* 2(1):1–36.

C. von der Malsburg (1988) Pattern recognition by labeled graph matching, *Neural Networks* 1:141–148.

T. Malzbender, D. Gelb, H. Wolters (2001) Polynomial texture maps, In *SIGGRAPH '01: Proceedings of the 28th Annual Conference on Computer Graphics and Interactive Techniques*, pp. 519–528, ACM Press, New York.

B. B. Mandelbrot (1967) How long is the coast of Britain? Statistical self-similarity and fractional dimension, *Science* 155:636–638.

B. B. Mandelbrot (1982) *The Fractal Geometry of Nature*, W. H. Freeman, San Francisco.

B. B. Mandelbrot et al. (1984) Fractal character of fracture surfaces of metals, *Nature* 308:721.

D. Marr (1982) *Vision*, W. H. Freeman, San Francisco.

M. Martin-Landrove et al. (2007) Fractal analysis of tumoral lesions in brain, *IEEE Engineering in Medicine and Biology Society*, pp. 1306–1309.

W. S. McCulloch, W. Pitts (1943) A logical calculus of the ideas immanent in nervous activity, *Bulletin of Mathematical Biology* 5:115.

T. A. McMahon, J. T. Bonner (1983) *On Size and Life*, W. H. Freeman, New York.

C. A. Mead, M. A. Mahowald (1988) A silicon model of early visual processing, *Neural Networks* 1(1):91–97.

J. J. Mecholsky, D. E. Passoja (1985) Fractals and brittle fracture, in *Fractal Aspects of Materials* pp. 117–119, Materials Research Society, Pittsburgh, PA.

J. J. Mecholsky et al. (1986) Crack propagation in brittle materials as a fractal process, in *Fractal Aspects of Materials II*, Materials Research Society, Pittsburgh, PA.

J. J. Mecholsky et al. (1989) Quantitative analysis of brittle fracture surfaces using fractal geometry, *Journal of the American Ceramic Society* 72:60.

T. P. Meloy (1977) Fast Fourier transforms applied to shape analysis of particle silhouettes to obtain morphological data, *Powder Technology* 17(1):27–35.

V. Mikli, H. Kaerdi et al. (2001) Characterization of powder particle morphology, *Proceedings of the Estonian Academy of Sciences Engineering* 7(1):22–34.

M. Mishchenko (1993) Light scattering by size-shape distributions of randomly oriented axially symmetric particles of a size comparable to a wavelength, *Applied Optics* 32(24):4652–4666.

K. Miyamoto, S. Nakahara, S. Suzuki (2002) Effect of particle shape on linear expansion of particleboard, *Journal of Wood Science* 48:185–190.

F. Mokhtarian, A. K. Mackworth (1986) Scale-based description and recognition of planar curves and two- dimensional shapes, *IEEE Transactions on Pattern Analysis and Machine Intelligence* 8:34–43.

F. Mokhtarian, S. Abbasi, J. Kittler (1996) Robust and efficient shape indexing through curvature scale space, *British Machine Vision Conference*, 53–62.

P. L. Mokhtarian, S. Abbasi, J. Kittler (1996) Robust and efficient shape indexing through curvature scale space, *British Machine Vision Conference*, 53–62.

M. A. Moore, P. A. Swanson (1983) The effect of particle shape on abrasive wear: A comparison of theory and experiment, *Wear of Materials*, April 11–14, 1–11.

M. C. Moreno, P. Bouchon, C. A. Brown (2010) Evaluating the ability of different characterization parameters to describe the surface of fried foods, *Scanning* 32(4):212–218.

M. Mudge, T. Malzbender (2006) New reflection transformation imaging methods for rock art and multiple-viewpoint display, *IEEE Symposium on Virtual Analytics Science and Technology*, 195–202.

D. Mumford (1991) Mathematical theories of shape: Do they model perception? *Geometric Methods in Computer Vision* 1570:2–10.

D. Nain et al. (2005) Multiscale 3D shape analysis using spherical wavelets, *Proceedings of the MICCAI* 3750:459–467.

S. K. Nayar, Y. Nakagawa (1990) Shape from focus: An effective approach for rough surfaces, *IEEE Proceedings of Robotics and Automation*, 218–225.

C. V. Negoita, D. A. Ralescu (1975) *Applications of Fuzzy Sets to Systems Analysis*, Halsted Press, New York.

C. V. Negoita, D. A. Ralescu (1987) *Simulation, Knowledge-Based Computing, and Fuzzy Statistics*, Van Nostrand Reinhold, New York.

J. C. Neto et al. (2006) Plant species identification using elliptic Fourier leaf shape analysis, *Computers and Electronics in Agriculture* 50:121–134.

W. Niblack (ed.) (1993) *Storage and Retrieval for Image and Video Databases*, SPIE Proceedings Vol. 1908.

R. E. Nisbett (2004) *The Geography of Thought: How Asians and Westerners Think Differently ... and Why*, Free Press, New York.

R. E. Nisbett, T. Masuda (2003) Culture and point of view, *Proceedings of the National Academy of Sciences*, 100(19):11163–11170.

J. Noh, K. Rhee (2005) Palmprint identification algorithm using Hu invariant moments, in *Fuzzy Systems and Knowledge Discovery: Lecture Notes in Computer Science*, 3614:91–94, Springer, Heidelberg.

R. Ogniewicz (1994) Skeleton space: A multiscale shape description combining region and boundary information, *Proceedings of the Conference on Computer Vision and Pattern Recognition*, 746–751.

P. O'Higgins (1997) Methodological issues in the description of forms, in P. E. Lestrel (ed.) *Fourier Descriptors and Their Applications in Biology*, Cambridge University Press, Cambridge, UK.

F. Ohtsuki et al. (1997) Fourier analysis of size and shape changes in the Japanese skull, in P. E. Lestrel (ed.) *Fourier Descriptors and Their Applications in Biology*, Cambridge University Press, Cambridge, UK.

G. O'Keeffe (1976) *Georgia O'Keeffe*, Viking Press, New York.

A. Ono (2003) Face recognition with Zernike moments, *Systems and Computers in Japan* 34(10):26–35.

J. D. Orford, W. H. Whaley (2006) The use of fractal dimension to quantify the morphology of irregular-shaped particles, *Sedimentology* 30(5):655–668.

G. C. Ostermeier et al. (2001) Relationship of bull fertility to sperm nuclear shape, *Journal of Andrology* 22(4):595–603.

M. Ozkan, B. M. Dawant, R. J. Maciunas (1993) Neural-network-based segmentation of multi-modal medical images: A comparative and prospective study, *IEEE Transactions on Medical Imaging* 12(3):534–544.

J. Padfield et al. (2005) Polynomial texture mapping: A new tool for examining the surface of paintings, *ICOM Committee for Conservation* 1:504–510.

K. Palm et al. (1996) Correlation of drug absorption with molecular surface properties, *Journal of Pharmaceutical Science* 85(1):32–39.

F. Pan, M. Keane (1994) A new set of moment invariants for handwritten numeral recognition, *IEEE International Conference on Image Processing ICIP* (1):154–158.

T. E. Parks ed. (2001) *Looking at Looking*, Sage Publications, Thousand Oaks, CA.

A. W. Partin et al. (1989) Fourier analysis of cell motility: Correlation of motility with metastatic potential, *Proceedings of the National Academy of Sciences* 86(4):1254–1258.

T. W. Patzek, J. G. Kristensen (2001) Shape factor correlations of hydraulic conductance in noncircular capillaries, *Journal of Colloid and Interface Science* 236(2):305–317.

T. Pavlidis (1977) *Structural Pattern Recognition*, Springer Verlag, New York.

T. Pavlidis (1978) A review of algorithms for shape analysis, *Computer Graphics and Image Processing* 7:243–258.

P. Pentland, ed. (1986) *From Pixels to Predicates*, Ablex, Norwood, NJ.

E. Persoon, K.-S. Fu (1977) Shape discrimination using Fourier descriptors, *IEEE Transactions on Systems, Man, and Cybernetics* 7:170–179.

J. L. Pfalz (1976) Surface networks, *Geographical Analysis* 8(2):77–93.

A. Pierret et al. (2002) 3D reconstruction and quantification of macropores using X-ray computed tomography and image analysis, *Geoderma* 106(3–4):247–271.

R. Pilgram et al. (2006) Shape discrimination of healthy and diseased cardiac ventricles using medial representation, *International Journal of Computer Assisted Radiology and Surgery* 1(1):33–38.

Z. Pincus, J. A. Theriot (2007) Comparison of quantitative methods for cell-shape analysis, *Journal of Microscopy* 227(2):140–156.

S. M. Pizer et al. (1999) Segmentation, registration and measurement of shape variation via image object shape, *IEEE Transactions on Medical Imaging* 18(10):851–865.

F. Poczeck (1997) A shape factor to assess the shape of particles using image analysis, *Powder Technology* 93:47–53.

M. I. Posner, M. E. Raichle (1994) *Images of Mind*, W. H. Freeman, New York.

L. Pothaud et al. (2000) A new method for three-dimensional skeleton graph analysis of porous media: Application to trabecular bone microarchitecture, *Journal of Microscopy* 199(2):149–161.

P. Prakash, V. D. Mytri, P. S. Hiremath (2011a) Comparative analysis of spectral and spatial features for classification of graphite grains in cast iron, *International Journal of Advanced Science and Technology* 29:31–40.

P. Prakash, V. D. Mytri, P. S. Hiremath (2011b) Fuzzy rule based classification and quantification of graphite grains from microstructure images of cast iron, *Microscopy and Microanalysis* 17(6):896–902.

P. Prusinkiewicz, A Lindenmayer (1990) *The Algorithmic Beauty of Plants*, Springer, New York.

I. Przerada, A. Bochinek (1990) Microfractographical aspects of fracture toughness in microalloyed steel, in *Stereology in Materials Science*, Polish Society for Stereology, Krakow.

S. Psarra, T. Grajewski (2001) Describing shape and shape complexity using local properties, *Proceedings of the Third International Space Syntax Symposium*, Atlanta, #28.

H. A. Qader et al. (2007) Fingerprint recognition using Zernike moments, *International Arab Journal of Information Technology* 4(4):372–376.

H. Ragheb, E. R. Hancock (2006) The modified Beckmann-Kirchhoff scattering theory for rough surface analysis, *Pattern Recognition* 40(7):2004–2020.

S. V. Raj (2010) Microstructural Characterization of Metal Foams: An Examination of the Applicability of the Theoretical Models for Modeling Foams, NASA/TM-2010-216342.

N. Ramanathan, R. Chellappa, S. Biswas (2009) Computational methods for modeling facial aging: A survey, *Journal of Visual Languages and Computing* 20(3):131–144.

K. K. Ray, G. Mandala (1992) Study of correlation between fractal dimension and impact energy in high strength low alloy steel, *Acta Metallurgica et Materialia* 40(3):463.

C. P. Reeve (1979) Calibration designs for roundness standards, National Bureau of Standards NBSIR 79-1758.

W. L. Roberts (1978) *Cold Rolling of Steel*, Marcel Dekker, New York.

W. L. Roberts (1983) *Hot Rolling of Steel*, Marcel Dekker, New York.

I. Rock (1984) *Perception*, W. H. Freeman, New York.

I. Rodriguez-Iturbe, A. Rinaldo (1997) *Fractal River Basins: Change and Self-Organization*, Cambridge University Press, Cambridge, UK.

F. J. Rohlf (1990) Morphometrics, *Annual Review of Ecology and Systematics* 21:299–316.

F. J. Rohlf, J. W. Archie (1984) A comparison of Fourier methods for the description of wing shape in mosquitoes (Diptera: Culicidae), *Systematic Zoology* 33:302–317.

A. Rose, W. C. Leggett (1990) The importance of scale to predator-prey spatial correlations: An example of Atlantic fishes, *Ecology* 71(1):33–43.

R. Rosenblatt (1962) *Principles of Neurodynamics*, Spartan Books, Washington, DC.

I. Rovner, F. Gyulai (2007) Computer-assisted morphometry: A new method for assessing and distinguishing morphological variation in wild and domestic seed populations, *Economic Botany* 61(2):154–172.

D. L. Royer et al. (2005) Correlations of climate and plant ecology to leaf size and shape: Potential proxies for the fossil record, *American Journal of Botany* 92:1141–1151.

J. C. Russ (1994) *Fractal Surfaces*, Plenum Press, New York.

J. C. Russ (1997) Fractal dimension measurement of engineering surfaces, in *7th International Conference on Metrology and Properties of Engineering Surfaces* (B. G. Rosen, R. J. Crafoord, eds.), Chalmers University, Göteborg Sweden, 170–174.

J. C. Russ (2001) Fractal geometry in engineering metrology, pp. 43–82 in E. Mainsah et al. (eds.) *Metrology and Properties of Engineering Surfaces*, Kluwer Academic Publishers, London.

J. C. Russ (2011) *The Image Processing Handbook* (6th ed.), CRC Press, Boca Raton, FL.

J. C. Russ, I. Rovner (1989) Stereological identification of opal phytolith populations from wild and cultivated zea, *American Antiquity* 54(4):784–792.

J. C. Russell et al. (2009) Automatic track recognition of footprints for identifying cryptic species, *Ecology* 90(7):2007–2013.

M. R. Rutenberg, T. L. Hall (2001) Automated Cytological Specimen Classification System and Method, U.S. Patent 6,327,377.

B. Ruthensteiner, N. Baeumler, D. G. Barnes (2010) Interactive 3D volume rendering in biomedical publications, *Micron* 41:886.e1–886.e17.

F. Sachse et al. (1996) Segmentation and tissue-classification of the visible man dataset using computer tomographic scans and thin-section photos, *Proceedings of First Users Conference of the National Library of Medicine's Visible Human Project*, 123–126.

E. Sanchez, L. A. Zadeh (eds.) (1987) *Approximate Reasoning in Intelligent System Decision and Control*, Oxford Press, New York.

K. Sandau, H. Kurz (1994) Modelling of vascular growth processes: A stochastic biophysical approach to embryonic angiogenesis, *Journal of Microscopy* 175:205–213.

L. M. Sander (1986) Fractal growth processes, *Nature* 322(Aug.):789–793.

X. Sang et al. (2011) Applications of digital holography to measurements and optical characterization, *Optical Engineering* 50:091311.

J. C. Santamarina, G. C. Cho (2004) Soil behaviour: The role of particle shape, *Proceedings of the Skempton Conference*, London, 1–14.

B. Sapoval (1991) Fractal electrodes, fractal membranes, and fractal catalysts, in *Fractals and Disordered Systems*, Springer Verlag, Berlin.

D. Saupe, D. Vranic (2001) 3D model retrieval with spherical harmonics and moments, *Pattern Recognition* 2001:89–93.

R. J. Schalkoff (1991) *Pattern Recognition: Statistical, Syntactical and Neural Approaches*, Wiley, New York.

D. Scharstein, R. Szeliski (2003) High-accuracy stereo depth maps using structured light, *IEEE Proceedings on Computer Society Conference on Computer Vision and Pattern Recognition*, 195–202.

R. Schmidt et al. (2008) Longitudinal multimodal imaging in mild to moderate Alzheimer disease: A pilot study with memantine, *Journal of Neurology, Neurosurgery, and Psychiatry* 79:1312–1317.

R. Schreiner (2002) Interferometric shape measurement of rough surfaces as grazing incidence, *Optical Engineering* 41:1570.

P. Schroder, W. Sweldens (1995) Spherical wavelets: Efficiently representing functions on the sphere, *Proceedings of SIGGRAPH 95*:161–172.

D. W. Schuerman et al. (1981) Systematic studies of light scattering. 1: Particle shape, *Applied Optics* 20(23):4039–4050.

D. Schumacher (1992) General filtered image rescaling, in D. Kirk (ed.) *Graphic Gems III*, Academic Press, San Diego, CA.

H. P. Schwartz, K. C. Shane (1969) Measurement of particle shape by Fourier analysis, *Sedimentology* 13:213–231.

P. J. Scott (1995) Recent advances in areal characterization, IX Intern. Oberflächen-kolloq, Technical University Chemnitz-Zwickau, 151–158.

P. J. Scott (1998) Foundations of topological characterization of surface texture, *International Journal of Machine Tools and Manufacture* 38(5–6):559–566.

P. J. Scott (2004) Pattern analysis and metrology: The extraction of stable features from observable measurements, *Proceedings of the Royal Society of London* A 460:2845–2864.

S. Seebacher, W. Osten, W. Jüptner (1998) Measuring shape and deformation of small objects using digital holography, *SPIE Conference on Laser Interferometry* IX:104–115.

D. Shaked, A. Brukstein (1998) Pruning medial axes, *Computer Vision and Image Understanding* 69(2):156–169.

S. S. Shapiro, M. B. Wilk (1965) An analysis of variance test for normality (complete samples), *Biometricka* 52(3–4):591–611.

M. C. Shaw, A. Vyas (1993) Chip formation in the machining of hardened steel, *CIRP Annals-Manufacturing Technology* 42(1):29–33.

D. Shen, H. Ip, K. T. Cheung, E. K. Teoh (1999) Symmetry detection by generalized complex moments: A close-form solution, *IEEE Transactions on Pattern Analysis and Machine Intelligence* 21:466–476.

L. Shen, J. Ford, F. Makedon, A. Saykin (2004) A surface-based approach for classification of 3D neuroanatomic structures, *Intelligent Data Analysis* 8(6):514–542.

P. Shilane et al. (2004) The Princeton shape benchmark, *IEEE SMI* 04:1–12.

C. Shirky (2011) *Cognitive Surplus: How Technology Makes Consumers into Collaborators*, Penguin, New York.

K. Siddiqi, B. B. Kimia (1995) Toward a shock grammar for recognition, Technical Report LEMS-143, Brown University, Providence, RI.

K. Siddiqi, A. Shokoufandeh et al. (1999) Shock graphs and shape matching, *International Journal of Computer Vision* 35:13–32.

W. von Siemens, W. C. Coupland. (1893) *Personal Recollections of Werner von Siemens*. D. Appleton and Company, New York.

D. Sims (1994) Biometric recognition: Our hands, eyes and faces give us away, *IEEE Computer Graphics and Applications* 14(5):14–15.

E. Sinnott (1936) A developmental analysis of inherited shape differences in cucurbit fruits, *American Naturalist* 70:245–254.

C. Sinthanayothin et al. (1999) Automated localisation of the optic disc, fovea, and retinal blood vessels from digital colour fundus images, *British Journal of Ophthalmology* 83:902–910.

S. P. Smith, A. K. Jain (1982) Chord distribution for shape matching, *Computer Graphics and Image Processing* 20:259–265.

T. G. Smith et al. (1989) A fractal analysis of cell images, *Journal of Neuroscience Methods* 37:274–280.

G. W. Snedecor, W. G. Cochran (1989) *Statistical Methods*, Iowa State University Press, Ames, IA.

D. R. Soll, E. Voss et al. (1988) Dynamic morphology system: A method for quantitating changes in shape, pseudopod formatin, and motion in normal and mutant amoeba of *Dictostelium discoideum, Journal of Cellular Biochemistry* 37:177–192.

M. Sonka, V. Hlavac, R. Boyle (2008) *Image Processing, Analysis and Machine Vision* (3rd ed.) Thomson Learning, Toronto.

A. Sorokin, D. Forsyth (2008) Utility data annotation with Amazon Mechanical Turk, *2008 IEEE Computer Society Conference on Computer Vision and Pattern Recognition Workshops*, 1–8.

H. E. Stanley, N. Ostrowsky (1986) *On Growth and Form*, Martinus Nijhoff, Boston.

D. Stauffer, A. Aharony (1991) *Introduction to Percolation Theory*, Taylor & Francis, London.

W. J. Stemp B. E. Childs, S. Vionnet (2010) Laser profilometry and length-scale analysis of stone tools, *Scanning* 32(4):233–243.

M. Subbarao, Y. Choi (1995) Accurate recovery of three-dimensional shape from image focus, *IEEE Transactions on Pattern Analysis and Machine Intelligence* 3:266–274.

W. Sun, X. Yang (2011) Nonrigid image registration based on control point matching and shifting, *Optical Engineering* 50:027006.

J. Suo et al. (2010) A compositional and dynamic model for face aging, *IEEE Transactions on Pattern Analysis and Machine Intelligence* 32(3):385–401.

H. Tadros et al. (1999) Spherical harmonic analysis of the PSCz galaxy catalogue: Redshift distortions and the real-space power spectrum, *Monthly Notices of the Royal Astronomical Society* 305(3):527–546.

R. Tam, W. Heidrich (2002) Feature-preserving medial axis noise removal, *Proceedings of the European Conference on Computer Vision*, 2351:672–686.

R. Tam, W. Heidrich (2003) Shape simplification based on the medial axis transform, *Proceedings of IEEE Visualization*, 481–488.

T. Tanaka (1981) Controlled rolling of steel plate and strip, *International Materials Reviews* 26:185–212.

S. Tanimoto, T. Pavlidis (1975) A hierarchical data structure for picture processing, *Computer Graphics and Image Processing* 4:104–119.

S. Tapley, M. Bryden (1977) An investigation of sex differences in spatial ability: Mental rotation of three-dimensional objects, *Canadian Journal of Psychology*, 31:123–130.

M. R. Teague (1980) Image analysis via the general theory of moments, *Journal of the Optical Society of America* 70(9):920–930.

M. C. Thomas et al. (1995) The use of Fourier descriptors in the classification of particle shape, *Sedimentology* 42(4):635–645.

D. W. Thompson (1917) *On Growth and Form*, Cambridge University Press, Cambridge, UK.

P. Toharia et al. (2007) A study of Zernike invariants for content-based image retrieval, *Proceedings of the 2nd Pacific Rim Conference on Advances in Image and Video Technology*, Springer, Berlin, 944–957.

C. Torrence, G. P. Compo (1998) A practical guide to wavelet analysis, *Bulletin of the American Meteorological Society* 79(1):61–78.

A. M. Torres et al. (1998) Characterization of erythrocyte shapes and sizes by NMR, *Magnetic Resonance Imaging* 16(4):423–434.

J. T. Tou, R. C. Gonzalez (1981) *Pattern Recognition Principles*, Addison Wesley, Reading, MA.

M. J. Tovée (2008) *An Introduction to the Visual System*, Cambridge University Press, Cambridge, UK.

S. Tran, L. Shih (2005) Efficient 3D binary image skeletonization, *IEEE Computational Systems Bioinformatics Conference* 2005:364–372.

M. Twain (1883) *Life on the Mississippi*, Osgood, Boston.

T. Ueda, Y. Kobatake (1983) Quantitative analysis of changes in cell shape of *Amoeba proteus* during locomotion and upon responses to salt stimuli, *Experimental Cell Research* 147(2):466–471.

H. Umhauer, M. Bottlinger (1991) Effect of particle shape and structure on the results of single-particle light-scattering size analysis, *Applied Optics* 30(33):4980–4986.

R. J. Valkenburg, A. M. McIvor (1998) Accurate 3D measurement using a structured light system, *Image and Vision Computing* 16(2):99–110.

D. S. Van Nieuwenhuise et al. (1978) Source of shoaling in Charleston Harbor: Fourier grain shape analyses, *Journal of Sedimentary Petrology* 48(2):373–383.

P. J. Vermeer (1994) Medial axis transform to boundary representation conversion, Ph.D. thesis, Purdue University, West Lafayette, IN.

H. Verschuelen et al. (1993) Methods for computer assisted analysis of lymphoid cell shape and motility, including Fourier analysis of cell outlines, *Journal of Immunological Methods* 163:99–113.

T. Vicsek (1992) *Fractal Growth Phenomena*, World Scientific, Singapore.

T. V. Vorburger, E. Marx, T. R. Lettieri (1993) Regimes of surface roughness measurable with light scattering, *Applied Optics* 32(19):3401–3408.

D. V. Vranic, D. Saupe (2002) Description of 3D-shape using a complex function on the sphere, *IEEE Proceedings on ICME* 2002:177–180.

H. Wadell (1932) Volume, shape and roundness of rock particles, *Journal of Geology* 40:443–451.

J. Wasen, R. Warren (1990) *Catalogue of Stereological Characteristics of Selected Solid Bodies, Volume 1: Polyhedrons*, Chalmers University, Göteborg, Sweden.

J. Wasen et al. (1996a) *Catalogue of Stereological Characteristics of Selected Solid Bodies, Volume 2: Hexagonal Prisms*, Chalmers University, Göteborg, Sweden.

J. Wasen et al. (1996b) *Catalogue of Stereological Characteristics of Selected Solid Bodies, Volume 3: Ellipsoids*, Chalmers University, Göteborg, Sweden.

J. Wasen et al. (1996c) *Catalogue of Stereological Characteristics of Selected Solid Bodies, Volume 4: Cubospheres*, Chalmers University, Göteborg, Sweden.

D. Wehbi et al. (1992) Perturbation dimension for describing rough surfaces, *International Journal Machine Tools and Manufacture* 32:211–216.

S. M. Weiss, I. Kapouleas (1990) An empirical comparison of pattern recognition, neural nets, and machine learning classification methods, in *Readings in Machine Learning*, J. W Shavik, T. G. Dietterich (eds.), pp. 177–183, Morgan Kaufmann, San Mateo, CA.

S. M. Weiss, C. A. Kulikowski (1991) *Computer Systems That Learn: Classification and Prediction Methods from Statistics, Neural Nets, Machine Learning, and Expert Systems*, Morgan Kaufmann, San Francisco.

W. T. Welford (1977) Optical estimation of statistics of surface roughness from light scattering measurements, *Optical and Quantum Electronics* 9:269–287.

K. A. Whaler, D. Gubbins (1981) Spherical harmonic analysis of the geomagnetic field: An example of a linear inverse problem, *Geophysical Journal of the Royal Astronomical Society* 65(3):645–693.

D. J. Whitehouse (1994) *Handbook of Surface Metrology*, Institute of Physics Publishing, Bristol, UK.

L. W. Wong, N. Pilpei (1990) The effect of particle shape on the mechanical properties of powders, *International Journal of Pharmaceutics* 59(2):145–154.

X. Xi et al. (2007) Finding motifs in a database of shapes, in W. Jonker, M. Petkovic (eds.) *SIAM Conference on Data Mining, Lecture Notes in Computer Science*, 4721:249–260, Springer, Heidelberg.

S. Xiao et al. (2006) Feature extraction for structured surface based on surface networks and edge detection, *Materials Science in Semiconductor Processing* 9(1–3):210–214.

L. Xing et al. (2006) Overview of image-guided radiation therapy, *Medical Dosimetry* 31(2):91–112.

M.-H. Yang, D. J. Kriegman, N. Ahuja (2002) Detecting faces in images: A survey, *IEEE Transactions on Pattern Analysis and Machine Intelligence* 24(1):34–58.

Z. You, A. K. Jain (1984) Performance evaluation of shape matching via chord length distribution, *Computer Vision, Graphics and Image Processing* 28(2):185–198.

P. Yu et al. (2007) Cortical surface shape analysis based on spherical wavelets, *IEEE Transactions on Medical Imaging* 26(4):582–597.

L. A. Zadeh (1965) Fuzzy sets, *Information and Control* 8:338–353.

C. T. Zahn, R. Z. Roskies (1972) Fourier descriptors for plane closed curves, *IEEE Transactions on Computers* 21(3):269–281.

D. Zhang, G. Lu (2004) Review of shape representation and description techniques, *Pattern Recognition* 37:1–19.

J. Zhang, Y. Yan, M. Lades (1997) Face recognition: Eigenface, elastic matching, and neural nets, *Proceedings of the IEEE* 85(9):1423–1435.

M. W. Zhang, B. J. Guo, Z. M. Peng (2002) Genetic effects on grain shape traits of indica black pericarp rice and their genetic correlations with main mineral element contents in grains, *Journal of Genetics and Genomics* 29(8):688–695.

F. Zhao et al. (2006) Image matching by normalized cross-correlation, *IEEE International Conference on Acoustics, Speech and Signal Processing* II:729–732.

G. Y. Zhou, M. C. Leu, D. Blackmore (1993) Fractal geometry model for wear prediction, *Wear* 170:1–14.

Y. Zhou, A. Kaufman, A. W. Toga (1998) 3D skeleton and centerline generation based on an approximate minimum distance field, *Visual Computer* 14(7):303–314.

A. Ziegler et al. (2010) Opportunities and challenges for digital morphology, *Biology Direct* 5:45.

A. Ziegler et al. (2011) Effectively incorporating selected multimedia content into medical publications, *BMC Medicine* 9:17.

H.-J. Zimmermann (1987) *Fuzzy Sets, Decision Making and Expert Systems*, Kluwer Academic Publ., Boston, MA.

Index

9 781138 072190